Cancers in the Urban Environment

Patterns of Malignant Disease in Los Angeles County and its Neighborhoods

Co-Editors

Wendy Cozen, D.O., M.P.H.   **Myles Cockburn, Ph.D.**
Department of Preventive Medicine
Keck School of Medicine
University of Southern California
Los Angeles, California

Programmer/Analysts

Frances Wang, M.S.   **Yaping Wang, M.D., M.S.**
Department of Preventive Medicine
Keck School of Medicine
University of Southern California
Los Angeles, California

Amy Laurent, M.P.H.
California Cancer Registry
Sacramento, California

With Contributions from

Ronald Ross, M.D.   **Dennis Deapen, Dr. P.H.**
Leslie Bernstein, Ph.D.   **Lihua Liu, Ph.D.**
Department of Preventive Medicine
Keck School of Medicine
University of Southern California
Los Angeles, California

John P. Wilson, Ph.D.   **Christine Lam, M.S.**
GIS Research Laboratory
Department of Geography
University of Southern California
Los Angeles, California

Hans Storm, M.D.   **Gerda Engholm, M.S.**
Niels Christensen
The Danish Cancer Society
Department of Cancer Prevention and Documentation
Copenhagen, Denmark

And With the Support of

The California Cancer Registry
The American Cancer Society, California Division
The National Institute of Environmental Health Sciences
The Norris Comprehensive Cancer Center
Southern California Environmental Health Sciences Center
The Los Angeles County Department of Health Services

CANCERS
in the
URBAN
ENVIRONMENT

*Patterns of Malignant Disease in Los Angeles County
and its Neighborhoods*

THOMAS M. MACK, M.D., M.P.H.

Department of Preventive Medicine
Keck School of Medicine
Norris Comprehensive Cancer Center
University of Southern California
Los Angeles, California

ELSEVIER
ACADEMIC
PRESS

AMSTERDAM • BOSTON • HEIDELBERG • LONDON
NEW YORK • OXFORD • PARIS • SAN DIEGO
SAN FRANCISCO • SINGAPORE • SYDNEY • TOKYO

Elsevier Academic Press
525 B Street, Suite 1900, San Diego, California 92101-4495, USA
84 Theobald's Road, London WC1X 8RR, UK

This book is printed on acid-free paper.

Library of Congress Cataloging-in-Publication Data
Mack, Thomas M.
  Cancers in the urban environment : patterns of malignant disease in Los Angeles County and its neighborhoods / Thomas M. Mack ; with the assistance of Wendy Cozen, . . . [et al.].
    p. cm.
  Includes bibliographical references.
  ISBN 0-12-464351-5 (hardcover : alk. paper)
1. Cancer—California—Los Angeles County.   I. Cozen, Wendy,   II. Title.
  RC277.C3M33   2004
  615.5′999′0979493—dc22   2004002306

British Library Cataloguing in Publication Data
A catalogue record for this book is available from the British Library

ISBN: 0-12-464351-5

For all information on all Academic Press publications
visit our Web site at www.academicpress.com

Printed in China
04 05 06 07 08 09 9 8 7 6 5 4 3 2 1

Contents

*Bolded entries combine several subgroups in a given organ or all those at a given age.

Foreword

Scientific interest in the geographical distribution of disease can be traced back at least 2000 years to Hippocrates and his admonition in "On Airs, Waters and Places" that "Whoever wishes to investigate medicine properly should . . . when one comes into a city to which he is a stranger . . . consider its situation, how it lies . . . and the mode in which the inhabitants live, and what are their pursuits." One fascination of and need for a cancer atlas, therefore, is to explore the idea that variations in cancer rates may reflect the influence of geography and of the behaviours of inhabitants of different areas. This may be instructive in order to test ideas about causation of cancer or, from a public health perspective, to identify geographical areas where action is particularly required to reduce the rate of cancer or to provide facilities to treat it. A second important reason to need such an atlas, in a more modern context, is to address the concerns of inhabitants of particular areas that a type of cancer is in excess in their locality, and that this relates to the presence of some factor that they perceive to be pernicious. Such anxieties seem to occur increasingly often, and it is then valuable to be able to turn to an atlas to discover, at the least, whether or not it is true that the cancer of interest is particularly common in that area, and to what extent there are other areas with similar or greater rates. In brief, to put local concerns in perspective.

There are several general reasons, therefore, why a cancer atlas such as this is valuable. There are others, however, particular to this book. First, if geographical variations are to be valid and interpretable, a long enough period of data needs to be available to give stability of numbers and hence to avoid apparent variations being due mainly to chance. A dataset is also needed of sufficient quality, uniformly across the geographical area, that variations between places are not simply due to differences in completeness or in error rates. It is no small part a consequence of Professor Mack's past work as Director of the Los Angeles Cancer Registry that this atlas fulfills these requirements amply, and that Los Angeles can boast high quality cancer registration across such a large population, now of roughly 10 million people, over a period of 27 years.

A second reason, however, for particular interest in cancer geography in Los Angeles, is one which is the more apparent to an observer from across the Atlantic. It is a constant marvel to non-Americans that the United States manages to be at once such a mélange of different cultures and yet at the same time so American. I suspect that there are few places, if any, of which this is more true than Los Angeles. There is therefore a great fascination in discovering to what extent cancer rates in Los Angeles vary according to the ethnicity and origins of the inhabitants of different parts of the city and their degree of integration into U.S. behaviours. Although Professor Mack has modestly described the book simply as an atlas, it is in fact much more than this, and includes data directly about the relation of cancer risks in Los Angeles to several variables other than geography, including ethnicity and socioeconomic status, as well as showing time trends.

It thus provides a guide to many facets of cancer risks in Los Angeles and a mine of information both for residents, epidemiologists, physicians, and others whose work relates to cancer. It also provides explanations, where this is possible, for why the observed variations in cancer risk have occurred, and more generally of the methods used to investigate the geography of cancer, and the pitfalls and artefacts that must be considered when interpreting such analyses. For those who are not from an epidemiological background, a flavour can be gained of the balance of different factors in affecting cancer risk, and the large contributions that come from behaviours and lifestyles.

The range of cancers covered in the atlas is exceptionally wide, with a degree of subdivision that is more detailed and more aetiologically and clinically useful, than is usual in analyses of cancer registration data: 86 separate anatomical sites and/or histological entities are analysed, plus certain groupings of these entities, and cancer overall by age group.

In publishing the first cancer map of which I am aware of, more than 125 years ago, Alfred Haviland stated that "I feel convinced that by studying the geographical laws of disease we shall know where to find its exciting as well as its predisposing cause, and how to avoid it".

Both for citizens of Los Angeles with an interest or concern about cancer in their city, and for epidemiologists and others with a professional interest, this atlas will provide extensive food for thought on the great variety and complexity of cancer occurrence.

Anthony Swerdlow
Institute of Cancer Research
London, England

Introduction for Scientists

This book is intended for both scientists and laypersons. Because the potential readers in these two groups have different backgrounds and different interests, separate introductions are provided.

It is hoped that this material will facilitate the generation and preliminary evaluation of causal hypotheses. Traditionally, in the field of cancer epidemiology (as opposed to acute disease epidemiology), serious emphasis has been placed on "person" and "time," with neglect of the third member of the triad, "place." It is true that many international comparisons have documented the geographical diversity of cancer risks, but the formulation of specific hypotheses has been difficult because international statistics must be interpreted in light of differences in methodology as well as concurrent differences in genes, habits, and environmental exposure. No previous systematic attempt has been made to compare rates between diverse urban localities. Such comparisons could be quite informative, as long as identical methods of registration and rigorous analysis have been employed.

Here the reader will find a detailed description of the patterns of 84 categories of malignancy, as each has occurred in Los Angeles County over the period 1972–1998. In all, three-fourths of a million cases, registered by a single cancer registry according to standard anatomical and histological definitions, are described by age, sex, calendar time, race/ethnicity, social class, and specific residential locality. The occurrence of each malignancy is assessed according to the overall geographic variation in occurrence, and those neighborhoods at apparent high risk are identified. Assuming pictures to be better teachers than words, graphics are used to summarize the information. For each malignancy, a brief synopsis of the known causes, a summary of the pattern of local occurrence, and a brief interpretation is provided. At the end of the book there is a summary of all conclusions in particular reference to unexpected findings.

The premise is that geographical patterns of cancer occurrence within the diverse neighborhoods of a large urban area are likely to be more scientifically informative than those previously available, which have been based on larger and more homogeneous state, county-aggregate, or county units. Because large U.S. populations of cities or states are composed of a mixture of persons at both high and low risk, the occurrence of most cancers tends to be roughly comparable from state to state or city to city.[1]

Los Angeles County is an especially valuable study milieu, because that population, even more than the populations of most cities, is a mixture of persons of different origins and lifestyles. People do tend to segregate residentially on the basis of common ethnicity, education, income, occupation, and/or lifestyle. Of course, residents of any neighborhood also share a unique common exposure to the local environment.

The potential benefit of this local approach can be appreciated from the example of female breast cancer, for which occurrence is largely determined by the reproductive characteristics known to differ geographically by social circumstance. None of 37 U.S. state cancer registries in 1999,[1] none of the 41 California counties in 1989–1993,[2] and none of the 8

Los Angeles County large service planning areas in 1972–1998 reported an occurrence of female breast cancer more than 10% higher than the national, state, and county averages respectively. However, of the 1619 Los Angeles County census tracts assessed for this book, 377 showed a level of incidence of female breast cancer not only 10% higher, but at least 50% higher than the county average, and for 35 of these, the difference could not be conventionally explained by chance. Such strong contrasts between small areas can reveal more information about differential causation than can the more subtle variations between larger geographical units.

The later sections of this introduction are largely written for laypersons. The definitions and known causes of malignancy are described, as are registries and basic epidemiological procedures. A serious attempt is made to explain the potential role of chance on geographical patterns. There is a guide to the interpretation of the figures and maps, and a brief set of scientific and technical notes, in which the arbitrary choices that have been made are defended.

Cancers are commonplace. More than a third of us will get one. Yet the development of a cancer is insidious, even mysterious. A diagnosis causes even the most confident and effective person to feel powerless. Other diseases–infections, heart attacks, or strokes–are easy to understand as biological or mechanical phenomena. But there is no rational way to explain why cells should suddenly grow out of control in an otherwise healthy body. Of course when it actually occurs, finding an explanation is less important than finding effective treatment.

When cancers appear among the neighbors, the response is different. There is little willingness to regard a "cluster" of cases as commonplace. Knowing little about cancer (but believing that nearly everything is known to science), local residents become fearful. Presuming that any one of them might be next, they demand an explanation. When it cannot be provided, the fear is transformed into frustration and anger.

Scientists have certainly not discovered all the causes of all the cancers, but much has been learned. Most cancers appear in familiar patterns, although the details are not readily accessible to the public. Familiarity with the known causes and predictors of malignancy, and with the level of our scientific ignorance, might serve to reduce the fear and anger.

This book is an attempt to give a wider perspective on our understanding of cancer occurrence, and by doing so to enhance that familiarity. It describes the patterns of the important malignancies that have appeared in the urban environment of Los Angeles County, California. Those patterns are likely to be replicated by patterns in other populations, especially populations in other large urban areas. Particular attention is paid to a search for imbalances not otherwise explained, that might have resulted from a polluted environment. Investigation of pollution as a cause must receive priority, since any resultant intervention could protect whole populations, not just individuals.

The subsequent sections of this introduction explain how a malignant cancer is defined, and summarize what science has learned about cancer causation. The available resources and methods used to assess local risks are described, as are the pitfalls commonly encountered. Special attention is given to the confusing role of chance. Finally there is a guide to the interpretation of the figures and maps to follow, and a brief set of scientific and technical notes for those who require more detail.

Readers should come to recognize the unique nature of each malignancy as evidenced by a unique pattern of occurrence. They will appreciate the technical challenges needed to assess local risk and will realize that differences in risk between localities are usually caused by factors other than pollution. When concerns arise from the number of cases occurring in a neighborhood, residents can find the wherewithal to answer questions such as the following:

- Are the cases in a locality similar enough that they might share a common causation?
- What are the known causes of those particular malignancies?
- Could bad luck explain the apparent excess?
- Would available knowledge about the malignancies and the neighborhood have permitted the prediction of that many cases?

- Are these cancers known to sometimes result from a pollutant?

The detail provided will of course be of special interest to the residents of Los Angeles County. Residents can learn which specific cancers are elevated in their own locality, and can identify other neighborhoods at similarly high risk. They can examine each pattern in the context of local geography, and more easily find an explanation.

Notice that the plural form of cancer is deliberately chosen. It has long been recognized that cancer is really a collective noun and not a single disease, yet in the laypress and among laypersons generally, there is an assumption of uniform causation, as if cancer were a single malady. The word "cancer" itself may be partially to blame. Before the seventeenth century, no distinction was made between those painful lumps and sores that healed, and those that went on to grow, invade, and spread. With the intent to separate out the truly malignant diseases from the others, and to help people predict an outcome, the word "cancer", was resurrected from Galen's Latin and was assigned to the malignant group. This new word was thus introduced into English and into the universal medical vocabulary. Possibly because a single word has always been used, the public has seemed to assume that cancer is a single disease. It might have served posterity better if a less classical and more prosaic plural term, such as "invaders" had come into use.

There are actually hundreds of different malignancies, each defined by a different cell and organ of origin, and each with a unique impact on the life of a person. Each cell has a unique small environment, a unique set of influences on its life cycle and rate of growth, and a unique set of genetic instructions. The more we learn about the different malignancies, the less reason there is to presume that any two of them share the same causes. Even when different malignancies are caused by the same exposure, the mechanisms are likely to differ. For example, smoking is known to cause lung cancer (presumably by means of airborne smoke particles), bladder cancer (presumably by tobacco toxins modified and excreted in urine), and pancreas cancer (presumably by blood-borne chemicals from heated tobacco). Each of these three malignancies is known to occur in nonsmokers and each is known to have other specific causes. In general, the cancers are a group as diverse in origin as are infections.

Of course, there are still common biologic features. Every malignancy evolves from a fundamental genetic change in the programming of the cell, usually causing the daughters of the original cell to abandon normal function and to multiply. This cell replication is temporarily undetected and undetectable in the period following the initial change. Usually, a malignancy only becomes evident after years or even decades, after invading other tissues, and the factors responsible for the initial change are hardly ever still in evidence. Moreover, the growth and development of a cancer commonly requires subsequent encouragement, such as by exposure to another chemical, by the presence of a factor that produces an abnormally rapid cell growth, or by an ineffective repair mechanism. The effects of such influences may also take time to develop, and the environmental causes of an adult's cancer may have been locked into place long before the malignancy becomes evident, even as early as childhood. In fact, sometimes growth is so slow that symptoms are never produced, and the presence of the malignancy is only discovered at autopsy.

To be called a malignancy, whatever the cell of origin, requires that growth be invasive and not confined to the tissue of origin. The cells in some organs can produce noninvasive benign tumors, such as adenomas, but they are likely to be produced under different conditions than malignancies of the same tissue, and only in certain specific cases do they sometimes proceed to malignancy.

Any kind of cell can lose control of normal growth, and the name of a malignancy corresponds to the kind of cell. Carcinomas are malignancies of the cells that form the surface of structures, either the hormone factory cells that line a gland (adenocarcinoma) or the protective surface cells, such as those of the skin or mucous membranes (squamous carcinoma). Sarcomas are malignancies that derive from any of a wide variety of structural cells, ranging from the cells of muscle or fat to the cells that make bone or connective tissue. Leukemias are the blood-borne malignancies of the various blood cells, and lymphomas are solid tumors deriving from cells of the same family.

Environmental and Other Causes of Cancer

No more than two centuries ago, an unusual disease occurrence was popularly presumed to originate in a sinister but vague "miasma" in the air. One disease, malaria, even literally means "bad air." The view that disease comes from an unclean milieu was reinforced by the geographical spread of other infectious diseases, such as the plague and cholera, and the public understandably has tended to generalize this presumption to all disease. More recently, Rachel Carson's *Silent Spring* called attention to the omnipresence of man-made chemicals, and the world quickly became aware of the potential for noninfectious disease produced by pollution.

Some common diseases are obviously produced by pollutants, and some others by the body's exaggerated response to contamination. Hepatitis may be carried by polluted water. Eating paint chips that contain lead can poison children. Asthma represents an exaggerated physiological response to poor air quality. These conditions appear more or less soon after exposure, and the agent acting as the disease trigger is evident. It is easy to assume that this mechanism of disease causation is universal.

Even considering the long and complicated genesis of a cancer, to look for an environmental cause is only reasonable. We know that polluted air can sometimes cause cancer because chemical exposure in a poorly regulated workplace has been shown to do so. Moreover, many widespread man-made chemicals have been shown to cause cancer in animals, and our understanding of biology leads us to presume them capable of doing so in people.

Certain pollutants are known to have caused local clusters of a specific malignancy. Arsenic in water (at higher levels than found anywhere in the United States) has produced bladder cancer in Taiwan and in South America. Use of asbestos mineral as the opaque ingredient in whitewash has produced cases of pleural cancer in certain villages in Southern Europe and New Caledonia. The spill of dioxins at Seveso, Italy probably resulted in cases of sarcoma and lymphoma. Although not strictly chemical pollution, environmental exposure to ionizing radiation, such as that following the nuclear disaster at Chernobyl in the Ukraine, can cause several forms of cancer, such as specific forms of leukemia, lymphoma, and cancers of the thyroid, lung, and breast.

It is true that chemicals known to be capable of causing human cancer are found in the air of U.S. cities, including Los Angeles. These chemicals include arsenic, asbestos, and various dioxins. Benzene, which can cause certain kinds of leukemia, is present, as is hexavalent chromium, which can cause lung cancer, and vinyl chloride, which can cause angiosarcoma of the liver. Even so, no local increase in cancer due to pollution has yet been clearly identified in the United States. Even such highly publicized sites of pollution as the Love Canal, Three Mile Island and those popularized in the movies *Erin Brockovich* and *A Civil Action* did not produce clear evidence of a cancer excess, although each of these examples of irresponsible industrial contamination represented a clear potential danger to local residents and may have produced other medical problems.

This comes as a surprise, especially to laymen. We are repeatedly and accurately told that chemicals can cause cancers, that the American environment is polluted with chemicals, and that cancers commonly occur among Americans. Failure to confirm the expectation that American cancers come from pollution does not encourage confidence in the experts.

This paradox is partly the result of language. The word "environment" comes from a word meaning encirclement, and thus properly refers to the quality of air, water, and soil. The fact is that the air we breathe, the water we drink, and the soils we live on constitute only a part of the contacts our bodies make with the outside world. Other components include the things we touch and breathe in the workplace, the food we eat, the alcohol we drink, the cigarettes we smoke, the hormones or other pills we take, and the infections we contract from the persons to whom we are close. We know from many studies that certain forms of malignancy may be produced by each of these factors. Unfortunately, no word other than environment is available to conveniently refer to all the forms of contact as a group, and we are forced to use that word in two ways, both to indicate just the physical milieu, and alternatively to describe the entire set of our external influences, excluding only the genes from our parents (which of course can also be responsible for a malignancy). When a scientific publication correctly states that the environment is responsible for the majority of cancers, many wrongly interpret that statement to refer solely to pollution by air and water.

There are good scientific reasons why foci of cancer have rarely been linked to environmental carcinogens. The dose levels of truly nasty carcinogenic chemicals are inevitably toxic to the body in other ways, and in developed countries, most are closely regulated for that reason. Even so, carcinogenic chemicals can be identified in our air, but they are usually present in minute quantities, and we recognize their presence only because modern assay techniques have become so sensitive. The quantities are far below the levels that have been known to produce a detectable increase in human cancer. By going back to the very circumstance in which a chemical was first demonstrated to cause cancer (usually an animal experiment or a study of highly exposed workers), it is possible to estimate just how many cases might result from an ordinary low-level environmental exposure. Almost always, the number of cases likely to be caused is very small in relation to the number that would appear in the absence of the exposure.

This is especially true when the carcinogenic chemical is emitted by accident from a single industrial site or dump. Picture a target, made up of concentric circles around a central emission site. The concentration of the escaped chemical is reduced by dilution proportional to the square (or even the cube) of the distance from the site of emission. In the very center, where the concentration is highest, the area is tiny and, consequently, the number of exposed people is small. At the extremity, where the number of exposed persons is largest, the concentration of the chemical is lowest.

For example, a recent investigation of a facility in Northern California[3] sought to estimate the risk posed to people living adjacent to the point where the most potent release of a carcinogen in California had occurred (in this case airborne hexavalent chromium). Overall, about 20,000 persons were presumed exposed because they had lived within 2000 m of the emission site. It was estimated that the exposure over a 10-year period would have increased their risk by a factor (relative risk) of 1.0005, that is, it would have caused one additional case for every 2000 cases of lung cancer that might have occurred anyway. It would take about 200 years for 2000 baseline cases to ordinarily occur among the members of a population of 20,000. Among the roughly 1200 persons living closer to the emission site, within 500 m, one extra case would have occurred for every 166 ordinary cases (relative risk = 1.006). It would take about 277 years for this number of cases to normally occur. The persons living within 125 m of the source of emissions, numbering 60, could expect that one case would appear for every 9 ordinary cases (relative risk 1.11), but to accumulate 9 cases among only 60 persons would ordinarily take 300 years. Thus there was no question that residents were subjected to an unwanted carcinogen, but the likelihood that any actual person would contract cancer as a result was very small.

Although such low doses may produce the occasional cancer, our tools usually do not allow us to detect the really small differences in occurrence that might result. Only if levels of environmental pollution were to expose populations for long periods, and only if the exposed population was sizable, would an increase in case number become evident. Nonetheless, cancers and other diseases caused by environmental pollution are especially important to identify or rule out. They are likely to be more easily preventable, because they can be eliminated from an entire population by means of a central action, much more efficiently than if each individual's behavior had to be separately modified. A legitimate scientific goal of the work required to produce this book is the search for any evidence of malignancies caused by pollution.

Most studies of cancer causation have been initiated in order to explain an obvious difference in the pattern of occurrence, and that has enabled us to learn a great deal about the everyday causes of common cancers, (although we predictably know less about the rarer forms). For that reason we have become very sensitive to contrasts in risk according to age, sex, time, race/ethnicity, and social class, and familiar with the reasons likely to explain them. In some cases factors have been identified that can predict the pattern of occurrence of a malignancy with reasonable accuracy, even though the actual cause and its timing are obscure. Within any large city, and for any given malignancy, some neighborhoods will be found to be at higher risk than others, and we expect that such differences will sometimes occur on the basis of known variations in the population. People in some neighborhoods smoke more, drink more, take more medicines, or have jobs at more dangerous workplaces, than those in other neighborhoods.

In the following figures, each particular pattern of cancer is reviewed in light of available knowledge in a search for unexplained cases. Only by doing so can one hope to (following Hamlet) "take arms against a sea of troubles, and by opposing, end them."

The Los Angeles County Cancer Surveillance Program is the official registry for all cases of cancer occurring in residents of the county. It was founded in 1971 as a resource designed to facilitate study of the causes of cancer. It is supported by the California State Department of Health Services, as part of the California Cancer Registry, and by the U.S. National Cancer Institute, as a component of its SEER (surveillance, epidemiology, and end results) system of nationally representative cancer registries.

Cancers are defined by registry convention as cell growths that can do harm by invasion into adjacent tissues or spread to other parts of the body. All California hospitals and pathologists are required by law to report every cancer diagnosis made in a person living in California. Completeness and accuracy of the case counts are routinely assessed by the California Department of Health Services, the National Institutes of Health, the Centers for Disease Control and, because the numbers and rates are published periodically in an international publication, the World Health Organization. In Los Angeles County, reports are estimated to be complete to a level in excess of 99% for all diagnoses since 1971. This volume describes the pattern of cancers that occurred from 1972 to 1998. At the time of preparation, the cases occurring between 1998 and the year of publication had not yet been fully counted and assimilated into the registry. However, the more recent cases of any particular malignancy would be few in number and would be unlikely to alter any of the local patterns.

In addition to details about the malignancy itself, the registry gathers information about the affected person available in the hospital chart that might help to clarify the pattern of cancer occurrence. These items include age, sex, ethnicity, occupation, birthplace, and residential address as of the day of diagnosis. The address enables the registry, using census information, to roughly characterize the neighborhood level of social class, or socioeconomic status (SES). Because no direct evidence about the income and education of each reported case is available, it must be presumed that the personal level of "social class," that is, level of education and income, is similar to that of the typical person living in that neighborhood. The registry and each of these practices is described more fully in the section *Scientific and Technical Notes*.

All the information gathered about affected persons is kept in locked file cabinets and inaccessible computers. No personal identifiers such as name or social security number are ever released except to scientists after a formal process of authorization. Thus all the facts are kept highly confidential, with access restricted to closely supervised staff members, and are always reduced for dissemination into statistical frequencies of the sort presented in this book.

Each figure in this book employs one of two standard measures of disease occurrence. The most important index produced by a disease registry is the "incidence rate," which describes the level of *per capita* occurrence of a given cancer in a population. It is based on the number of cases occurring within a given period, as a proportion of the average number of persons estimated by the census to be present during each year over the period in which the cancers have occurred. Thus ten persons present during one given year count for the same as one person present during each of ten years. For this reason, person-years comprise the actual units at risk. Because there is a need to accurately estimate the number of residents in order to calculate this per capita incidence rate, the units used to define neighborhoods are the units (census tracts) defined and described by the U.S. Census Bureau.

Cancers vary greatly in occurrence depending on age, and some census tracts have a preponderance of young people and some a preponderance of older people. There is, therefore, a need to make sure that comparisons are made fairly. The rate for each age group is separately calculated, and then combined as though each census tract, as well as the county as a whole, contained the same proportion of people in each age group (for a discussion of other problems of unfair comparison see the section *Comparisons Between "Apples" and "Oranges"*).

Census tracts are defined on the basis of several considerations, which include not only population size, but logistical considerations designed to facilitate counting. Some are therefore extremely small, and others are devoid of any ordinary residential population. Beginning with all 1717 census tracts in Los Angeles County used by the census in 1990, a total of 98 census tracts were eliminated. Excluded were census tracts with an absent or very small population (too small to accurately measure an incidence rate), and census tracts in which the population counts varied markedly from decade to decade and even from year to year (making it impossible to accurately estimate the number of persons at risk). The census tracts excluded on this basis largely represented hospitals, prisons, schools, airports, harbor facilities, parks, and commercial sites.

Thus all estimates in this book are based on information about the residents of 1619 residential census tracts for which populations are available and accurate. These census tracts contain, on average, about 5000 persons at any given time, and range in size for any given year from as few as 500 to as many as 10,000 persons. Each person's residence at the time of cancer diagnosis was coded to one of these 1619 census tracts, geographically defined as of the 1990 census, even if their place of residence had been assigned to a different census tract by a previous census.

When there is need to compare one incidence rate to another, the conventional method is to employ a "standard" rate as a common denominator. In this case, the most convenient and useful standard is the average rate for all residents of the county as a whole. It is not as perfect as a purist would like, because the people in each individual census tract of interest are also included in the total and therefore in the standard population, but every individual census tract is so small in comparison with the total that comparisons are still very accurate. The great advantage is that all census tracts can be compared using the same standard.

Such a comparison is usually summarized by a ratio, calculated by dividing the rate in the census tract of interest by the standard rate. This ratio is usually called a "relative risk.". Both incidence rate and relative risk are universally accepted measures of disease occurrence. The relative risk may be calculated directly, or alternatively, it may be estimated by a measure called the "standard incidence ratio." This is not a ratio of rates but of case numbers; the number observed to occur divided by the number expected, that is, the estimated number of cases that would have occurred had the population consisted of persons just like those in the standard county as a whole. It is often prudent to estimate the relative risk for a census tract twice, first using the entire county population as the standard, and secondly using only the subpopulation of persons living in neighborhoods (census tracts) known to be at the same level of income and education as the census tract of interest. This allows consideration of the possibility that any observed difference might simply be due to a difference in income or education between the census tract of interest and the entire county. This means the latter result is "adjusted" for social class (see the section *Scientific and Technical Notes*).

The information about each cancer provided in this book is purely descriptive. That means that each pattern of cancer is based only on knowledge of the occurrence of the cancer and of the population of Los Angeles County. No special knowledge of the causes of a cancer in any particular person is incorporated, nor for that matter is any information about the exposure of any person. Only the speculative discussion that accompanies the description of each malignancy takes such outside knowledge into account.

A total of 84 different categories of cancer are described, including 72 uniquely different malignancies and 12 combinations according to age, organ system, or histology that are likely to be of interest (see the section Scientific and Technical Notes). The choices are based on frequency of occurrence and on the available information bearing on the specifics of causation. Each cancer has been examined separately by sex, because sometimes the causes of a cancer in one sex are somewhat different from the causes of the same cancer in the other sex.

Criteria were selected in order to identify census tracts at high risk. To be identified as high risk each such tract must be shown to have a relative risk of at least 1.5, that is, a rate at least 50% higher than that expected, either on the basis of the county as a whole or on the basis of all census tracts of comparable income and education. Moreover, this relative risk estimate must be based upon at least three cases and must exceed the conventional (95%) upper limit of statistical confidence around the expected number (the next section, *Errors Due to Chance*, will explain this in more detail). If chance alone were responsible for an observed difference between a census tract of interest and the county as a whole, the number of cases in a census tract designated at high risk would fail to exceed the number expected in 95 of every 100 census tracts of the same size and composition. The choice of these criteria is discussed in more detail in the section *Scientific and Technical Notes*.

The same sequence of figures is provided for each combination of sex and type of cancer. The rates in Los Angeles County are compared to rates in other communities, and within the county population the differences according to age, sex, race/ethnicity, calendar time, and social class are illustrated. The range of relative risk estimates for all the census tracts in the county is shown and the number of cases in each census tract is compared to the number that is likely to have been produced solely by chance. Those census tracts fulfilling the high-risk criteria are identified, and among them the time trend is assessed, and a comparison is made between the risk to males and the risk to females. These comparisons are also identified on maps. Each figure depicting high-risk census tracts is first prepared using a comparison with the county as a whole. The comparison is then repeated after adjusting for social class in order to evaluate the latter as a factor underlying the high risk. A more detailed discussion of the interpretation of each figure is provided in the section *Guide to the Figures*.

It is upsetting to learn that the local rate of a cancer is higher than average, and it is really important to recognize that chance alone might be responsible. Here we are misled by the word "average." Again going back several centuries, that word referred historically to the insurance charge assessed to the London residents who received goods by ship. The charges were evenly assessed for fairness, but the money was pooled into payments to cover individual losses. Shippers understood that these insurance charges were made uniform, at the average level, in order to spread the cost, but recognized that their losses, and their reimbursements, might range unpredictably from low to high. For better or worse, our democratic heritage has led us to presume that misfortune also should be evenly distributed, and we have come to expect that under normal circumstances every person, every family, and indeed every community should suffer the same level of average unhappiness. The truth is that plain old ordinary bad luck really does exist. The troubles of an individual, of a family, and even of a community are usually either above or below the average level due to (as Hamlet also said) "the slings and arrows of outrageous fortune."

People certainly acknowledge the influence of good and bad luck in the comparisons they make every day in their nonscientific lives. They do so when shopping, when investing, when driving, and most obviously when playing games of chance.

A child may ask why a new box of cereal with raisins does not seem to have as many raisins as usual. Although the price of raisins and the degree to which they influence sales may determine the *average* number in a shipment, no manufacturer can examine the content of every box, and chance is likely to determine whether the number of raisins in a specific box is higher or lower than average. No child would notice a box with an excess, but we all pay attention if the number seems too low, and it is left to the consumer to decide whether to buy another box or write a letter of complaint.

It is not always obvious whether to blame a hardship on bad luck or deliberate unfairness. If a grocer guarantees that no more than 10% of his peaches will be bad but a consumer finds that one of two peaches pulled from a basket is rotten, more serious concerns are raised. If that customer pulled out eight peaches, of which two were rotten, he might still feel that the grocer was cheating him because 2/8, or 25%, is a much higher figure than 10%. However, a statistician, assuming the accuracy of the 10% rate, would say that eight peaches chosen at random from any full basket would probably be as likely as not to contain at least two rotten peaches by bad luck alone.

There are as many gamblers as shoppers. Card players know that in five-card poker, a hand with four aces ought to appear only once in a long while. From the composition of a deck of cards, we can estimate how often four aces should be dealt (once in about every 54,000 five-card hands). Although that is a very low frequency, every once in a while a coincidence will occur. If an opponent gets four aces twice in the same game, there are two possible explanations: chance or cheating. Some players might reject chance as an explanation and call the police. Others might wait until the third or fourth such stroke of "luck." The more extreme the run of bad luck, the less credible is chance as an explanation.

So it is with the cancers diagnosed in the residents of a neighborhood. An unexpectedly large number of cases appearing among the residents of a small community within a brief period might reflect a common local carcinogenic exposure, but might alternatively result from chance. No matter how unusual an event, chance can never be completely excluded. Even an event as extreme as four successive hands of four aces might happen by chance once in a zillion games of poker. The important question is not whether an excess number *could* happen by chance but *how often* it is likely to happen.

Thus the likelihood that a finding could occur by chance must be considered together with a measure of how often it would occur. An event occurring normally only once in a century is more compelling than one occurring annually. In the case of cancer, the appearance of five times the normal number of cases in a community provides more compelling evidence than does the appearance of twice the number. However, such comparisons are tricky. If only 1 case is expected, those comparisons would be with either 5 or only 2 observed cases, a far less compelling difference than if 100 cases are expected, and the analogous comparison is with either 500 or 200 cases.

As with an opponent's poker hands, it becomes prudent at some arbitrary point to reject chance and assume the alternative, even in the absence of absolute certainty. A generally accepted point at which the line is drawn by scientists is the point where chance would produce the observed extreme occurrence less often than 5 times out of every 100 tries (2.5 times for an extremely high occurrence, and 2.5 times for an extremely low one). By convention, such an observation is designated "statistically significant." Otherwise, the result is ignored under the assumption that chance

could explain the result and more meaningful explanations, while possible, are not worth acting upon. (Sometimes the significance criterion is drawn more conservatively at the level of one time per hundred, or more liberally at ten times per hundred.)

When considering community statistics, most experts would pay little attention to an increase any less than 50% (a factor or relative risk of 1.5), not because lower increases are presumed unimportant, but because they are so hard to distinguish from artifact given the "noise" in the system.

Instead of the 2.5% value used to demarcate the statistical significance of an extremely high rate, scientists often find it useful to use the same math theory to report a measure of *confidence* in the proposition that the observed value is in fact due to chance. Such "confidence limits" provide the most extreme values that still would allow chance to be considered a reasonable explanation. A set of confidence limits can be calculated to go around each numerical estimate of increased or decreased risk. An 80% increase in risk can be described as a relative risk of 1.8, and could be accompanied by 95% confidence limits of (a low of) 1.05 and (a high of) 3.42. Such limits tell us that one can have confidence to a level of 95%, and that the actual increase lies somewhere between a factor of 1.05 (a 5% increase) and 3.42 (a 242% increase).

Whether measured by the 5% criterion or by the equivalent confidence limits, the greater the experience, that is, the more observations, the less likely that chance is responsible for a given difference. If the 80% example above was based on many more observations, the 95% confidence limits might contract to (a low of) 1.71 and (a high of) 1.86. (More about both of these statistical criteria can be found in the section *Scientific and Technical Notes*.)

The methods of seeking to distinguish chance from a real difference can be illustrated clearly with dice. A throw of two dice, on average, results in some combination totaling seven dots. About 5% of the time the dice will show the extreme values of either two sixes (the most dots) or two ones (the fewest dots). Because some persons regard two ones ("snake-eyes") to be unlucky, let us consider that roll to be a bad outcome, temporarily standing in for an excess of cancer cases. It turns out that one can expect snake-eyes to result by chance in about 2.8% (once out of every 36 throws) of all honest throws of two dice. Figure A shows the result of 1000 throws of two such dice, with the green bar representing the 28 snake-eye results that have occurred by chance. On a single throw, the appearance of snake-eyes might represent either the one chance appearance out of every 36 or, like the four aces referred to previously, something more ominous than chance. From a single throw, one has no idea which it is.

Now consider what might happen if the same dice were loaded unfairly, with the face showing a six (the face opposite the one with a single dot) heavier than the other five faces and thus the one most likely to land face down. If that were true, more snake-eyes would appear than chance would predict. Figure B shows the result of 1000 throws, this time with loaded dice. Instead of the expected 28 snake-eyes, the tendency for dice to fall with the heavy side down produced 20 extra snake-eye results, making 48 in all, a number 1.7 times that expected. One can see in the figure that the range of results is asymmetrical, even though it would not be evident from the result of any single throw.

Now imagine if the same trial of 1000 throws of the dice were repeated 100 times, each time with a different set of dice, in an attempt to identify those dice that might be unfairly loaded. One could record the number of snake-eyes that appeared in each trial. Figure C shows how these results might appear. Because the average number of snake-eye results from 1000 throws is 28, each of the 100 results is shown in comparison to the average number expected each time (shown by the green cross-line). However, not all trials would produce the average number of snake-eyes. We anticipate that most results will appear just above, just below, or squarely on the expected number (represented by the green bar in Figure A). We can identify the point (shown by the light gray cross-line) at which the observed excess number of snake-eyes results would exceed a 50% excess (relative risk of 1.5), and the point at which it exceeds the upper 95% confidence limit (the dark gray cross-line). There is no longer 95% confidence that chance has produced the deviation from expected. Those dots that represent trials in which the relative increase exceeds both the relative risk of 1.5 and the 95% upper confidence limit are colored red. Because these dots represent statistically significant findings, prudence demands that they be considered to represent loaded dice, rather than chance effects, even though the red dot just above the 1.5 line might well still be the result of chance. The uppermost red dot, however, reflects a result (essentially that shown in Figure B) that is more than 1.7 times expected and well beyond the 95% upper limit of confidence. This indicates that the dice used in that throw should be presumed loaded.

Figure D shows how the number of cases of a cancer appearing in a census tract corresponds to the results of dice throws. Whereas the expected number of snake-eyes on every pair of honest dice is fixed at 2.8%, because all dice have the same pattern of dots, the size of Los Angeles County census tracts, and therefore the number of cases to be expected, is highly variable from census tract to census tract. Thus Figure D shows not a single vertical line of dots, but a complete array of dots, with each vertical line of dots appearing around a central point on the green line representing the expected for that size census tract, and the entire green "expected" line representing the range in size of all census tracts. At each horizontal point, the dots arranged vertically represent the range in number of

Figure A: Results of 1000 throws of two dice.

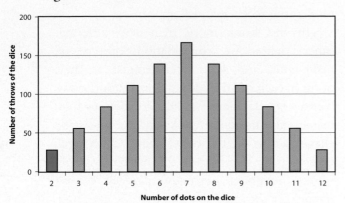

Figure B: Results of 1000 throws of two loaded dice.

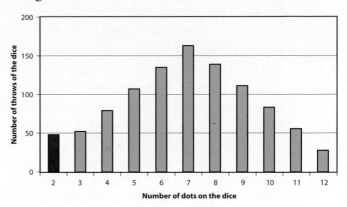

Figure C: Number of two's (snake-eyes) observed in 100 trials using different dice, each trial of 1000 throws.

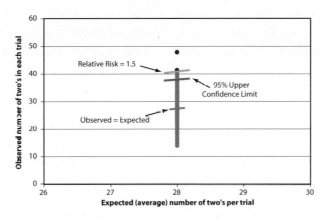

Figure D: Census tracts at high risk of colon cancer according to the number of observed and expected cases.

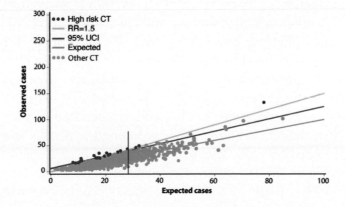

observed cases of colon cancer in Los Angeles County from 1972 to 1998. Just like the range in observed dots seen in Figure C, dots that reflect at least a 50% increase over the expected number (above the light gray line), as well as an excess over the 95% upper confidence limit around the expected number (the dark gray line), are colored red, indicating a statistically defined presumption of high risk to the residents. These dots correspond to the red snake-eyes dots in Figure C. Although some of these seemingly high-risk census tracts still could be positioned by chance, we prudently screen them out from the others for more detailed scrutiny, recognizing that they are more likely than the others to represent something real.

Note in Figure D that the smaller the neighborhood (i.e., the smaller the number expected cases), the greater the role of chance in determining numbers that seem especially high or low. As the expected number becomes larger to the right of the figure, the spread of observed numbers around that expected number diminishes. Thus as indicated above, the number of cases that is expected greatly influences the likelihood that a given observed number can be explained by chance. For this reason, differences in disease frequency are more difficult to verify when populations are small than when counties, states, or countries are compared.

For each of the malignancies described in this book, Figure 7 corresponds directly to Figure D. The only major difference is that the green dots, representing census tracts not presumed to be at high risk, are not shown.

Of course the policy of prudently assuming the worst about extremely high rates expected no more often than 2.5% by chance implies that if a comparison *was* actually repeated 100 times (e.g., if 10 different cancers were tested simultaneously in each of 10 different neighborhoods), 5 statistically significant differences would in fact appear *just by chance*. Thus not all "significant" results are really significant, and the more comparisons that are made, the more likely that extreme results due to chance will appear.

In this book more than 160 different cancer definitions (combinations of type of malignancy and sex) are examined as they occur in each of 1619 different neighborhoods (census tracts). Thus we would expect to find many extreme differences on the basis of chance alone. At the conventional 2.5% level of statistical significance for extremely high results, assuming (wrongly) that all census tracts were the same size and that the expected incidence of a given cancer in each neighborhood is exactly the same as the average occurrence in the whole county, the observed number of cases would prove "significantly" high over 6000 times. In fact, many fewer significant high levels actually appear, because many census tracts are quite small, many malignancies are rare, and steps have been taken to screen out those uninterpretable minor increases (of less than 50%) and those that are based on very few cases.

Although every census tract fulfilling the high-risk criteria is independently identified, factors other than chance are suspected only when the pattern of high-risk census tracts gives additional evidence. Such evidence may consist of adjacent census tracts that aggregate into high-risk combinations, census tracts at high risk that tend to appear in only one region, census tracts that show extremely high risk, or census tracts in which the high-risk criteria are satisfied for the members of both sexes.

People are often concerned about and report "clusters" of cancer occurring in groups even smaller than an entire census tract, such as among the people in a single city block or a single office. Because of the very small number of cases to be expected in a small group of people, even a few cases may seem to be a large excess in comparison with the number expected. When an ordinary single statistical comparison between the rate in such a small group and the standard county rate is calculated, the relative risk may be very high and well above the upper confidence limit. For example, 3 cases of breast cancer occurring within a 2-year period in the same Los Angeles office among 25 women averaging 50 years of age represents about a 600-fold increase over the expected number. Such an increase would be expected to occur by chance no more often than once in nearly 4000 offices.

Here, however, the statistical comparison is deceptive. About 4.5 million women live in Los Angeles County, and about half of them work. Assuming that 10% of all women work in such large offices, that means that there are about 18,000 offices and on average, chance would produce 3 or more cases in at least 5 offices in any given 2-year period. Over a 10-year period, one would expect that number in 25 different offices. Thus when multiple cases are *already known* to have occurred in one small sample of people among many, a simple statistical test will be misleading. In effect the question is only posed because an extreme occurrence was *preferentially* selected for comparison, and the statistically significant result is therefore a foregone conclusion.

This is an example of the phenomenon described as "Texas sharpshooting," a term that refers to the joke about the Texas marksman who stepped up after shooting and drew the target around the place where the bullets entered the barn. If the reader still has doubts about the meaning of a statistical test in this circumstance, consider the most extreme situation. If the 80-year-old wife of a statistician were to develop breast cancer, the annual rate of occurrence in her house, since she is the only one woman who lives there, would be 400 times the expected rate, but the same rate would prevail in about 3000 other houses in Los Angeles County in the same year.

Finally, when it is suggested that chance is responsible for the geographic pattern of a cancer, it *never* means that chance is responsible for the *occurrence* of the cancer. When the pattern seems to be consistent with chance, it just means that no evidence of an alternative explanation is available. For an individual person, a cancer can be due to any number of causes that do not show up as a pattern in the population. Even cancers that seem to have been scattered randomly in the population are undoubtedly due to specific biological causes. For example, if a cancer were caused by a food that everybody eats, there would be no evidence of an unusual geographical pattern, even though a specific cause was actually responsible.

Rates must be compared without a "stacked deck," that is, without any built-in inequity that would produce an inappropriate or unfair comparison (between "apple" and "orange" neighborhoods) and predetermine the result. Such a result often can be predicted in advance, either by knowing that different means were used to identify the subjects, or that different means were used to measure the exposure. Carrying on the everyday analogy, it can be described in the context of a shopper trying to identify the best peaches for the dollar. A bias, or inappropriate comparison, might occur if one dishonest vendor among many vendors hid all his rotten fruit at the bottom of the basket, or, in the opposite direction, if an overly picky customer assumed that each imperfectly shaped peach from one vendor was rotten.

In the case of the comparisons in this book, the collection and classification of cancers by hospitals and then by the cancer registry are done without reference to characteristics of the person, including the place of residence. One kind of unfair comparison can be introduced by errors in the census. Because we need to know the number of people in a neighborhood who are at risk for a cancer, we must use the census tract as the source of that number, even though errors in the census sometimes occur. The mistaken identification of a high-risk neighborhood ("false positive") is more likely than missing one ("false negative"), because the census is more likely to miss people than overcount them. If a neighborhood actually includes more residents than were reported by the census, or more residents of advanced age (older persons being at higher risk of most cancers), more cases will be found than would have been expected, producing an overestimate of the level of risk. Errors are especially likely when rapid changes in population have occurred, because the number of expected cases cannot be accurately estimated. It is because there are likely to be many small differences between the census figures and the actual population, as well as because of the many small differences likely to exist by chance, that when census tract population estimates are involved any observed excess of a cancer smaller than a 50% increase is generally considered unreliable.

A second kind of unfair comparison can result from local variation in the efficiency of the diagnostic process. Pathologists generally use hard and fast criteria to decide what is and what is not a cancer. An individual pathologist generally applies the same criteria to all specimens reviewed, irrespective of the age, sex, ethnicity, social class, or geographical location of the patient's residence. However, some criteria are somewhat subjective and may change slightly from pathologist to pathologist, and therefore from hospital to hospital, neighborhood to neighborhood, and from time to time, thus resulting in a net geographical imbalance.

More often, there can be local variation in the selection of persons to be sent for biopsy in the first place. For most cancers, which progress inexorably, the earlier identification of small lumps in persons from one neighborhood, and the later identification of the same kind of lumps after they have grown larger in persons from a neighborhood with less timely medical care, will just be reflected in an earlier average stage at diagnosis, but not in a difference in the incidence of occurrence. However, if a detectable malignancy is known to sometimes grow very slowly or not at all, and commonly never progresses to a symptomatic stage, even over a lifetime, an apparent geographic gradient may result when particularly efficient or aggressive diagnosticians preferentially serve one locality or region.

Cases missed by a cancer registry such as that in Los Angeles County are very few. The registry accepts as an eligible cancer any cell growth that threatens harm by evidence of invasion into adjacent tissues or remote parts of the body. While a negligable number of such malignancies are missed, the registry does not count non-invasive tumors.

One minor source of error resists any simple solution and prevents any claim of perfect accuracy. If cancers took a single day to develop, virtually no exposed cases would be missed and local risk estimates would be perfect. But if every person exposed to an important carcinogen were to move away before diagnosis, local risk could never be measured, because no means of accurately counting the annual number of exposed persons would be available.

In fact, the reality is imperfect, but not as bad as that. Cancers may take decades to appear, and although some people at risk of cancer change their place of residence, most stay put. If a map demonstrates that a locality is clearly at increased risk of a malignancy, it usually indicates that enough persons exposed to the local causes of that malignancy are still living there. This may only be because the average length of residence is rather long, or it may be that despite the turnover in the population, a set of similarly exposed persons have tended to congregate in that place according to a common personal preference. Personal characteristics (i.e., ethnicity, social class) do tend to remain constant features of a neighborhood, even if the actual persons change with time. Thus a cluster of cancer cases caused by a transient pollutant is probably easier to miss after a few years

of population turnover than one that can be predicted on the basis of the permanent characteristics of the place or the local residents.

In fact, residential moves are concentrated according to age, especially common in the period between the age when young persons leave high school and the age attained at the time that their own kids are starting school. Because cancers are more common among older adults, and because such persons move less frequently, differences in the level of neighborhood risk can be seen. Of course, malignancies do appear in persons well after a move away from the place of exposure, but the probability of such an error is about equally common in every location, and the end result is one of slight underestimation of local risk. Those cancers caused by factors in childhood but diagnosed in middle or late adulthood, and which are unrelated to the reasons for choosing a place to live, are the ones most likely to be missed.

Unexpected cases of a particular origin can sometimes go unrecognized for other reasons. In any census tract, an extra case or two caused by a local factor simply may not push the relative risk high enough to meet the arbitrary high-risk criteria. Moreover, we depend upon evidence of local aggregation to identify neighborhoods at particularly high risk. Other patterns of local exposure might exist. For example, if residence within a certain distance of a high-voltage power line, an underground seepage, or a freeway were to result in increased risk, the resultant cases might be divided with only a case or two in each of many census tracts, forming a pattern that reflected the exposure but would be invisible on the map.

Distinctive geographical patterns of cancer excess are identified in this book for about half of the individual malignancies, including some, such as pleural mesothelioma, which are commonly diagnosed many decades after exposure. It is likely that other, less striking patterns exist, but were not detectable by the relatively crude methods available.

These unavoidable losses are regrettable, but there are two important caveats. One is that if a distinctive geographical pattern is shown, and is unlikely to be due to chance or improper comparison, then it is very likely to be linked in some way to the causes of the cancer. In other words, while residential changes between the causal exposure and the appearance of disease could result in missing a truly high-risk neighborhood, they could not falsely identify one.

The second caveat is that if the map shows no local increase in risk, then any increase is probably too small to be detected by more crude means, such as a neighborhood survey.

Such counts inevitably distort estimates of personal risk because they are usually based in part on cases diagnosed while living elsewhere, and on cases exposed elsewhere but diagnosed soon after moving in. In contrast, the registry is comprehensive, carefully identifying the residential address and the social class of cases *at the time of diagnosis*, and making an accurate count of the number of people in the neighborhood who would have been counted in that neighborhood if they had been diagnosed. Diagnoses may occur after moving away from the neighborhood of exposure, but they cannot be included to make an assessment more accurate, because all the out-migrants cannot be systematically located, even if the appropriate interval since exposure was known. Moreover, the only appropriate standard of comparison is based on the total, because every other neighborhood count is based on cancers occurring not only among long-term residents but also among recent homebuyers.

Examination of the risk to a particular census tract or tracts from a specific malignancy, using this book or otherwise, can result in one of only three possible conclusions. First, there may be evidence of an increased risk, but that increase would be predicted on the basis of available information. Secondly, there may be no evidence of an increased risk, neighborhood concerns to the contrary notwithstanding. Finally, there may be an unexplained consistent increase in risk, with or without concern about a specific local source of carcinogens.

In the first case, there is likely to be no group action called for because the reasons for the high risk are familiar, and in most cases they can only be addressed on a personal basis. In the second case, frustration on the part of residents is likely, if their own observations had led them to expect confirmation of high risk. It is as though a noise was heard downstairs in the night and the policeman who responded to a call found nothing unusual. There could still be a burglar somewhere, and analogously there could still be a source of local carcinogens, but in each case the absence of supporting evidence discourages the allocation of investigational resources.

The third situation, however, may justify further investigation. Virtually all residents of Los Angeles County live in a census tract that is at high risk for one or more malignancies. Sometimes the reason for a geographic pattern is not explained by chance, bias, or any known confounding factor; but the magnitude of the threat is not great, and the need to investigate further can be questioned. On the other hand, no one wants to miss an opportunity to recognize and eliminate an environmental cause of cancer. There is no clear rule to guide the allocation of investigational resources, and priorities are inevitably formulated not only by scientific plausibility and the magnitude of the potential impact, but by public and political pressure.

The obvious first step is to list all the possible factors that could be responsible, including the special characteristics of residents, the unique characteristics of the locale, the local sources of carcinogen, and, especially, the exposures that are already known to produce that particular malignancy. Because science proceeds by ruling out possible explanations and not by ruling them in, each possibility should be considered in light of all available information and prioritized in order of plausibility.

When assessing a potentially beneficial but potentially dangerous medicine, the random assignment of persons into a treated and an otherwise identical untreated group is the preferred method of investigation. Needless to say, dangerous exposures cannot be assigned that way. Studies of cancer causation proceed by observing the natural human experience. This means that some flaws in the process are inevitable, and repetition is usually necessary before results are generally accepted. Actually devising and carrying out such a study in the context of a locally observed risk can be exasperating for both scientists and residents. It is exasperating for scientists, because a good study of individuals is inevitably difficult, expensive, and time-consuming. It is exasperating for both, because the results of a local study are likely not to be definitive.

In explanation of the high risk to a neighborhood, two separate questions are of interest, one general and one specific. The general question asks whether a given exposure is capable of causing a given malignancy; the specific one asks whether the local exposure is responsible for the observed high local risk.

Obtaining an answer to the general question is usually feasible, if sufficient resources are available. It requires a measure of the strength of any association between exposure and malignancy, and the gathering of evidence to rule out chance, bias, or confounding as possible explanations for that association. It need not be performed in the specific locality. In fact there are two reasons why a local study of the general association is unlikely to be useful. First, the number of subjects required to exclude chance is unlikely to be available locally. Secondly, having generated the hypothesis to be studied in a given locality, use of the cases in that locality will bias the result of any subsequent study conducted there, since a link between the suspected exposure and the disease is already known to be present. It will be necessary to identify a different setting or settings in which an identical exposure occurs in which persons with and without the exposure can be named, and in which an investigation of individuals can be conducted. Only after confirmation of the general relationship can the local role of that exposure be seriously considered.

One of two common study methods may be used. The most obvious ("cohort" study) proceeds by identifying a group of persons with the hypothesized exposure, as well as a group of persons comparable in every pertinent respect except for the exposure. Personal information about this and all other pertinent exposures is gathered from all subjects. One then must maintain contact with the members of both groups over a period long enough to produce enough cases of the malignancy to show, or to dismiss, a pertinent general difference in risk between the groups.

The negative aspects of this approach are the high expense required to assemble the necessary number of individuals, the inevitable levels of noncomparability between the exposed and the unexposed, and the time it takes to collect enough cases to reach firm conclusions.

The alternative ("case-control" study) is more commonly used, for example, in the process of investigating an outbreak of infectious disease. Proceeding backward instead of forward, one identifies cases of the malignancy and comparable healthy persons. Information about all the pertinent exposures is then collected in retrospect from each subject. This approach is cheaper and faster, but the information gathered from memory may be less accurate, or that from the cases may be more complete or biased than that gathered from the healthy control persons.

Two problems common to both methods can be anticipated. No matter how hard the investigator tries, and no matter which design is chosen, the number of cases *exposed* to an unusual factor may be very small, because only a small proportion of the population is likely to have been thus exposed. Secondly, it is frequently difficult to find healthy persons willing to participate in a "normal" (i.e., unexposed or unaffected) comparison group. Today, healthy persons are often reluctant to provide investigators, however well-intentioned, with personal information, and those that do participate may not be very comparable to the exposed or affected subjects.

Still other problems may occur. Measures of exposure obtained prospectively may become outdated and obsolete before the appearance of cases. On the other hand, measures obtained long after the fact may be inaccurate because efforts to reduce exposure have been successful in the interim. Exposures that result from personal choice, such as cigarette usage, are more easily measured than exposures that are strictly environmental, such as air pollution, which can only be approximated using the residential address. The reasons for personal choices are often obscure, and may be linked to other determinants of the cancer. For these and other reasons, scientists rarely are convinced of a newly discovered environmental cause of a cancer without the results of several consistent independent studies.

After having established (or having obtained from the literature) a general link between the hypothesized exposure and the specific disease outcome, assessing the role of the exposure in the local circumstance still remains. Because the actual number of cases in a neighborhood is usually small, it is rarely possible to make a definitive statement. Nonetheless it is certainly possible to assess the comparative prevalence of the exposure locally, estimate the relative frequency of cases based on the best available measure of the link between exposure and disease, and evaluate whether that frequency is consistent with the excess cases that have been observed. It is also possible to estimate what proportion of local cases can be attributed to the exposure, and compare that proportion with the expected proportion prevailing in other settings.

For more detail about analytic observational studies, see an introductory textbook of epidemiology, such as that by Rothman.[6]

The following section describes the information provided about each malignancy. For each sex that is affected, two pages of figures are provided to illustrate the pattern. Each set of two pages contains nine different figures. If census tracts at high risk are identified, two additional pages are provided to describe them with maps and male–female comparisons.

For Each Gender

Figure 1: Age-adjusted incidence rate by place. This provides a comparison of the overall incidence rate of that malignancy in Los Angeles County to the incidence rate prevailing among the residents of three other U. S. communities and one foreign country. San Francisco and Utah were chosen because they have, respectively, the highest and lowest rates of cancer (all kinds combined) in the United States. The SEER group combines the information from ten registries, ranging from Connecticut to Hawaii, and provides the best available information about the United States as a whole. The rate from Denmark, a country with cancer incidence data of high quality, provides an international perspective. In order to avoid differences based on age, the rate from each of the five places is prepared as if the age distribution of each population were identical, namely to that of the United States in 1970. This method of adjusting to a common age distribution is maintained in the next three figures as well.

To compare the incidence to men in Los Angeles to that in women, use the male and female versions of this figure. When doing this, pay careful attention to the two scales, because rates in one sex are often very different from those in the other sex.

Figure 2: Age-adjusted incidence rate from 1972 through 1998. This provides the trend in incidence rate over five periods from 1972 through 1998, adjusted to the same 1970 age distribution. In each of these figures the trend is compared to the trend over the same period for all cancers combined. Because the rates for all cancers combined are much larger than the rates for any single cancer, the line depicting the trend for all cancers combined has been made to fit the scale for the single cancer by converting each age-specific rate into a convenient fraction of the total. Because malignancies occur with highly variable rates, that fraction differs from malignancy to malignancy.

Figure 3: Age-adjusted incidence rate by age and race/ethnicity. This provides two kinds of information. For each of the four major race-ethnic categories in the Los Angeles population (African-American, European American "whites", Asian-American, and Latino), the trend in incidence according to age is shown, and with the same figure the race/ethnicity groups can be compared to each other. The Asian-American category includes all East and Southeast Asian ethnicities (Chinese, Japanese, Korean, Filipino, Vietnamese, and other Southeast Asians) in a single category. The smaller groups of South Asians (Indians, Pakistanis, etc.) and of Pacific Islanders are excluded. Most cancers become more frequent as men and women age, usually beginning in the late 40s. Major exceptions to this rule are specifically mentioned in the description of each local pattern. As with Figure 1, to compare the pattern by age among men to that among women requires side-by-side examination of the male and female versions of Figure 3, paying attention to the two scales.

Figure 4: Age-adjusted incidence rate by social class. This shows the rate for that malignancy in each of the five categories of census tract aggregated according to social class, from high to low. Social class categories are based on the ranking of census tracts according to an index formed using the average income of households and the average education of the adult residents reported to live in each census tract (see the section *Scientific and Technical Notes*).

Figure 5: Distribution of the relative risk values for all census tracts. This illustrates the extent of variation in local risk within the county. The horizontal axis shows the relative risk measure (factor by which incidence differs from that for all census tracts of the same social class) for each of the 1619 census tracts, enumerated on the vertical axis. The colorless bars on the left end of each figure (census tracts with a relative risk lower than 1.0) represent neighborhoods with a level of risk that is average or below average. Those with full color (red) on the right side of each figure (with a relative risk of 1.5 or higher) are those with an apparent elevation in risk of at least 50%. The census tracts represented in the middle of the figure by bars with a shade that is in between are those with a relative risk between 1.0 and 1.5, that is, with level of risk higher than average, but still below the designated threshold of 1.5. Notice how many of the 1619 census tracts seem to be at either very low risk or very high risk, especially for the majority of cancers that are relatively rare. The relative risks are ratios that ignore the actual number of cases that each estimate is based upon, and most of the census tracts seemingly at very high or low risk only appear to be that way because of the large chance variation

around the small average number of expected cases (often a fraction of 1.0).

Because this figure is meant to show the degree to which chance causes so many census tracts to show an extreme relative risk, Figure 5 is most informative when comparing one malignancy to other malignancies. For this reason, the figure is not usually discussed in the description of the local pattern of each malignancy.

Figure 6: Census tracts by the number of cases per tract. This describes the number of census tracts according to the number of cases that have occurred in comparison with the number that would have been expected by chance. Based on the overall county rate and the size of each census tract, an estimate was made of how many census tracts would be expected to produce a given number of cases (i.e., 0, 1, 2, 3, etc.) by chance alone, and this number is reflected in the left-most of each contiguous set of three bars. The same process was then repeated with adjustment for social class, basing each estimate this time not on the all-county rate, but on the rate for the combined census tracts of the same social class (middle bar). These two gray bars representing the unadjusted and adjusted expected number of tracts by case number are to be compared with the green bar on the right of each set, which represents the number of census tracts actually observed to have that many cases. When the green bar is taller than the accompanying gray bars, there are more census tracts than expected with that number of cases. Because the left–right position of the three bar sets reflects the number of expected cases and therefore the general size of the census tract population, the sets of bars on the far left generally represent smaller census tracts, and those on the far right represent larger ones. Because the values given for all bars of the same color (light gray, dark gray, and green) always add up across the whole figure to equal the same total number of cases, every time a bar is higher in one location, a bar or bars of the same color must be equivalently lower elsewhere in the figure. In general, systematic (i.e., nonrandom) increases in geographical risk will show up as green bars that are higher than the gray bars at both ends of the figure. When chance is the only factor determining the geographical distribution of cases, the matched green and gray bars will all be of roughly the same height. Specific census tracts cannot be identified in this figure. When describing this result in the context of individual malignancies, it will usually be described as showing a strong, medium, or weak effect of nonrandom distribution.

Figure 7a and b: Census tracts at high risk according to the number of cases (a) unadjusted and (b) adjusted for social class. This enumerates those census tracts found to be at high risk according to the specified criteria. Each red dot in Figure 7a represents a census tract that fulfills the two principal criteria of a high-risk tract: (1) a greater than 50% increase in risk (i.e., a relative risk greater than 1.5), and (2) measured risk higher than the upper 95% confidence limit, showing confidence that the number of excess cases is higher than the number consistent with chance alone. The dots are plotted according to the observed and the expected number of cases, and lines are provided that represent the two criteria, as well as one indicating the expected number of cases. The distance between each dot and the nearest of the two criterion lines crudely indicates just how extreme the high risk in that census tract is. Figure 7b describes the high-risk census tracts with red dots in the same way, but after having adjusted for social class, the difference between the arrangement of dots in the two figures roughly indicates how much of the apparent high risk is due to social class alone.

Figure 8: Risk over the period for high-risk census tracts relative to all census tracts. This demonstrates the change over time in the relative risk to the populations of the combined high-risk census tracts. The dashed red line shows the overall change, and the solid red line represents the change after adjustment for social class. The constant black line along the bottom, constant at a level of 1.0, represents the overall trend in the county (or in the aggregated census tracts of equivalent social class) as a baseline (essentially the incidence trend from Figure 2, flattened out into a straight line).

This figure requires special care to interpret, because it is based on a ratio. When the red lines are straight and parallel to the straight black baseline, the trend over time in the census tracts at high risk is the same as the trend over time (Figure 2) in the county as a whole (even though the actual *level* of risk may be very different). When a line in Figure 8 increases or decreases, it indicates that the trend in the high-risk census tracts departs in that direction from the overall trend. For example, if the trend in Figure 2 is decreasing over time, but the trend in Figure 8 also drops, it means that the incidence in the residents of high-risk census tracts is decreasing even more dramatically than it is in the county as a whole. If Figure 2 shows a decrease but Figure 8 shows an increase, incidence in the high-risk census tracts is either not decreasing as much as it is in the county as a whole, or is actually increasing. Inflections in a Figure 8 line indicate that relative to the overall trend, the time trend in the high-risk census tracts changed in degree or direction over the course of the period.

The shape of this trend is subject to great chance variation when the number of high-risk census tracts is small, so it is not described as part of the local pattern unless there are more than a few high-risk census tracts.

For Both Genders

Figure 9: Map of census tracts at high risk (without adjustment for social class). This shows all census tracts that are high risk (with at least a 50% increase, i.e., a relative risk of at least 1.5), at a level above the upper 95% confidence limit. Those census tracts at high risk for both men and women (in red) are distinguished from those that are at high risk only for men (blue) and those that are at high risk only for women (yellow). Because it is certain that some of the high-risk census tracts have received that designation by chance alone, the goal is to identify an overall geographical pattern that is unlikely to represent chance. The features of importance when interpreting the map are whether the high-risk census tracts are near or adjacent to one another, whether they are concentrated in one area of the county, and the number of red census tracts that are at high risk for both males and females. No specific statistical criteria are employed (see the section Scientific and Technical Notes).

Interpretation of the pattern of high-risk census tracts depends on knowledge of the geography and demography of Los Angeles County. Figures depicting the limits of statistical power, and the geographical distribution of annual case frequency, race/ethnicity, and social class are provided in the section *Scientific and Technical Notes*.

Looking casually at maps of this large county may lead to a common mistake. The size (area) of a census tract does *not* reflect the number of persons in it, nor does it reflect the number of cases upon which the high-risk judgment is based. It simply represents the physical area in which people live. Some parts of Los Angeles County, particularly in the north, are thinly populated, and each census tract covers a large area. In fact, for a few of those northern census tracts, and for tracts adjacent to the harbor in San Pedro, the census population figures do not accurately count the number of persons actually at risk of being diagnosed with a cancer. Because risk in these tracts cannot therefore be accurately assessed, these tracts are always depicted as not known to be at high risk. (For a more complete discussion of the method of mapping, see the section *Scientific and Technical Notes*.)

Figure 10: Male–female correlation between the relative risks for high-risk census tracts (without adjustment). This provides an assessment of residential risk in each high-risk census tract by comparing the relative risk for males on one axis to that for females on the other (based on census tracts with at least one sex showing a relative risk greater than 1.5), whether or not the other two high-risk criteria are met. The black dot indicates the position of a theoretical census tract in which both sexes are at the average risk of all persons in Los Angeles County. If these two relative risks were to be correlated by census tract, as they would be if local exposure variations occurred in parallel for men and women, the points would form a line extending from low (lower left) to high (upper right). Neither chance, nor exposures that are likely to be specific for one sex, such as those from hormones or the workplace, would produce such an effect. Because the census tracts with high risk in both men and women can be located on the maps by color and are described in that context, Figures 10 and 12 are not mentioned in the description of the local pattern.

Figures 11 and 12: These are identical to Figures 9 and 10 after adjustment for social class. Comparisons between the maps in Figures 9 and 11 or between the patterns in Figures 10 and 12 provide additional measures of the degree to which the high-risk designation is dependent upon social class.

A brief discussion precedes each set of figures. It provides a background summary of known or probable causes, a description of the pattern presented by the figures, and an interpretation of this pattern in light of the background, pointing out those findings that are not easily explained.

Background

The internationally assigned code for each malignancy is provided, as is the number of cases available for study. If not obvious, the reasons why that particular cancer was selected for presentation are given. When there is general agreement among experts about the known causation of a malignancy, this information is summarized. The factors previously known to predict or alter incidence, and any other exposures that are strongly suspected to be causal, are noted. References are not provided, and the reader who wishes a more detailed discussion of causation is referred to a current textbook of cancer epidemiology (such as that by Schottenfeld and Fraumeni[4] or that by Adami, Hunter, and Trichopoulos[5]).

Local Pattern

The observed pattern of occurrence within Los Angeles County is then summarized. The incidence is compared to incidence in other populations, and the pattern of occurrence by sex, age, race/ethnicity, and social class is described. The degree to which geographic variation in risk seems to be due to factors other than chance is roughly indicated. The characteristics of those census tracts meeting the high-risk criteria are discussed, including in respect to degree of risk, gender, social class, and time trend. Finally, the geography of those high-risk census tracts is described in relation to the geographical and demographic characteristics of the population of Los Angeles County. It may be useful to compare each map with the three maps of annual case frequency, social class, and race/ethnicity provided in the section *Scientific and Technical Notes*.

Even though our methods are designed to single out only those high-risk census tracts *least* likely to be due to chance, the number of comparisons is so large that many of the census tracts selected will still be included by virtue of chance. There is no hard and fast rule available to tell us whether this is true for any given tract or set of census tracts. A judgment must be made on the basis of whether an individual census tract shows an extreme excess of cases, whether the high-risk census tracts are confluent or clearly near to each other, whether both sexes seem to be affected in parallel, and whether the pattern of high-risk tracts corresponds to meaningful demographic or geographical boundaries.

Having decided that chance is unlikely to be the explanation for a concentration of cases, bias or other forms of inappropriate comparison must be considered. One plausible possibility is a disparity between the population of the census tract derived from census counts and the population claiming residence in the tract at the time of diagnosis. Another possibility would be the presence of local providers who more comprehensively screen patients for small asymptomatic malignancies that are likely to progress slowly if at all (such as certain cases of chronic lymphocytic leukemia or of carcinoma of the thyroid or prostate).

Thumbnail Interpretation

It is important to emphasize that judgments about the reasons for each nonrandom geographical distribution, as well as the reasons for variations according to by calendar time, race/ethnicity, age, sex, and social class, are based on the characteristics of groups, not individuals. It is *never* possible to attribute causes to the malignancy of an individual person.

One always hopes that a map can provide a pattern that identifies a previously unrecognized local source of carcinogens so that the source can be eliminated and future cancers prevented. In actuality, the situation is rarely that simple. About one-third of Americans (and the same proportion of residents of other relatively wealthy countries) are likely to get a cancer at some time in their lives. We therefore expect all cancers as a group to occur at roughly comparable levels among lawyers and garbage workers, African-Americans and German-Americans, and the residents of particularly polluted and exceptionally clean neighborhoods. It is only in relation to specific malignancies that variation may be great enough to be informative.

Having decided that a pattern of occurrence, geographical or otherwise, is unlikely to be due to chance or artifact, we consider whether it would have been predicted by the known characteristics of affected persons. When a known cause, such as cigarette smoking, is likely to be influential, the pattern of occurrence is compared to the known pattern of long-term cigarette smokers. Only if that pattern does not explain the cancer pattern are unknown factors considered. In some cases we have made only moderate progress in identifying specific causes, but we have learned to recognize universal predictors of disease (or risk factors), and the observed pattern is compared to the pattern expected on that basis.

For example, a lower level of social class is a strong predictor of cervical cancer. We know that most cases are caused by one of several forms of papillomavirus, transmitted from person to person, and that social class, as a risk factor, is a surrogate for the higher likelihood of past exposure. In addition, poorer women have less opportunity for cervical screening, and therefore later detection of premalignant lesions occurs. We certainly cannot judge whether an individual woman is likely to have been either exposed or protected, but knowing that cervical cancer is more common among less educated and less affluent women permits us to predict the pattern of occurrence in the population. It also allows us to prioritize persons for screening programs. If we wish to test, for example, whether or not smoking also causes cervical cancer, we must not mistake the link between cervical cancer and social class for the link between cervical cancer and smoking, which also may be related to social class.

Such noncausal risk factors must be taken very seriously when a malignancy is evaluated in relation to an environmental carcinogen, because a factor such as social class might incorrectly either suggest or mask an important link to the environment. When a neighborhood with a specific environmental exposure is shown to be at high risk of a particular cancer, the finding may only have appeared because people in that neighborhood are especially vulnerable on the basis of ethnicity, social class, or lifestyle. For example, if it is known that African-Americans smoke more than other residents, a higher rate of lung cancer in a neighborhood with a high proportion of African-Americans cannot provide evidence that the nearby smokestacks are responsible. On the other hand, if such a high rate occurs in persons of an ethnicity known to smoke less than average, suspicion must be directed to other explanations.

The factors exemplified here by ethnicity are referred to as "confounding" factors, and their influence may be partially reduced by adjustment for social class. If neither chance, nor bias, nor an obvious confounding factor seem to provide a plausible reason for an observed local excess of cancer, we are left to speculate on the reason, consider the alternatives that might explain it, and search for additional facts. The construction of this book has already prompted several such explorations. The magnitude of the risk and the pattern of high-risk neighborhoods can and should be used to guide a more detailed scientific study of individual persons. The considerations involved in such a search are discussed in the next section. To read more about the difficulties in evaluating the magnitude of environmental risks, see the book entitled *How Much Risk*" listed in the bibliography at the end of this book.

To find out which census tract contains a specific address, and to locate that census tract on a map, resources outside this book will be required. The 1990 census tract definitions have been used here, because it was relatively simple to place persons with addresses from the 1970s and 1980s into the appropriate 1990 census tract. The Census Bureau no longer provides a service on the Internet to help allocate addresses to 1990 census tracts, although older paper publications available in many libraries may be of assistance. For the time being, one can enter an address into a search Web site maintained by the Federal Financial Institutions Examination Council (FFIEC) to get both the number of the census tract corresponding to a given address and a map showing the location and outline of the whole census tract. Be sure to enter the proper year (1990) in order to identify the census tracts as they are described here. If a guide to the 1990 census tracts is no longer available, it is reasonably safe to use maps of the 2000 census tracts, because most of them are the same. The current FFIEC Web site address is http://www.ffiec.gov/geocode/Geocode SearchMapping.htm.

The goals of this volume require that the information provided be accurate and helpful, but that at the same time the complexity of the maps and the analysis be kept to a minimum. A number of technical decisions warrant explanation. These include the methods used by the registry, the definition of cancer entities, the basis for social class classification, the use of statistical procedures, the choice of high-risk criteria, the method of producing Poisson-generated expected numbers with which to compare observations in Figure 6, the choice of a mapping strategy, and the absence of formal cluster analysis.

Methods of the Los Angeles County Cancer Surveillance Program

Los Angeles County is the largest county in the United States, with a population in 2004 if more than 9 million persons. It is one of the most ethnically and socioeconomically diverse urban areas in the world, and is extremely heterogeneous in both respects at the neighborhood level.[7] It serves as the economic center for a population twice that size in Southern California, and a wide range of industries, occupations, cultural practices, and levels of education are represented. Medical care for cancer patients is uniformly available, if not always timely, and standards for medical records and pathology services are high. The Cancer Surveillance Program employs standard registry methodology.[8] It achieves more than 90% complete registration of all cases within about 12 months of diagnosis and more than 99% within 18 months. All cancers are classified by both anatomic site and histological cell of origin. Place of residence, not place of diagnosis, is the address of record, and meticulous care is taken to exclude duplicate reports from analytic files (second primary cancers in the same person are counted separately). For this volume, all cases from 1972 through 1998 have been reassigned to census tracts based on the 1990 tract boundaries. The accompanying map shows the distribution of census tracts according to the annual number of all cancer diagnoses combined, a reflection of the size of each census tract population.

Annual mid-year population estimates according to those same 1990 census tract definitions were prepared using information from each census (1970, 1980, 1990, 2000), with attention to the more detailed estimates made annually by the California Department of Finance Demographic Research Unit. Race is classified from hospital records, and Latinos, that is, persons of Spanish-speaking Latin-American cultural heritage, are distinguished by comparing the names of all cases with names on the Spanish-surname list maintained by the U. S. Census Bureau. Although this comparison is done irrespective of race, the number of Latino African-Americans in Los Angeles County is small, and Spanish-surnamed Asians (Filipinos) and Pacific Islanders (Guamanians) are separately classified. Census undercounts of minority populations do not exceed 5%. The registry therefore provides data well suited for small-area studies.[9]

Incidence rates by sex, race, and calendar time for the county as a whole and for individual census tracts were prepared by first

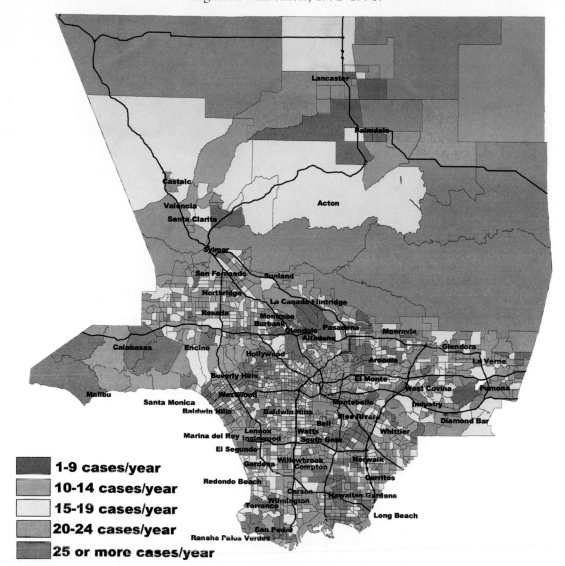

Figure E: Census tracts of Los Angeles County by the average annual frequency of residents diagnosed with cancer, 1972–1998.

1-9 cases/year
10-14 cases/year
15-19 cases/year
20-24 cases/year
25 or more cases/year

calculating age-specific rates. These were combined by direct adjustment to the age distribution of the 1970 U. S. population. To directly calculate the relative risk for each census tract, the resultant adjusted rate for the census tract was divided by the analogous rate for the county as a whole. Estimates of the number of cases to be expected over the period in each census tract were obtained by applying all-race rates for the entire county to the population of each census tract by age, calculating

the expected number of cases, and combining them to produce an indirectly age-adjusted total. That is directly compared to the observed number in the form of a standardized incidence ratio (SIR), roughly equivalent to a relative risk.

Registries of the National Cancer Institute SEER system (including the registries in the San Francisco Bay Area and Utah)[10] employ essentially identical methodology and the same system of classification (more can be learned at its Web site address http://www. seer.cancer.gov). The Danish Cancer Registry uses an equivalent method of data collection and employs the same method of classification, with a few minor exceptions, as noted in the discussion of certain malignancies.

Definition of Cancer Entities

Historically, rates of cancer usually have been provided for each of the malignancies on a list of only 20 to 30 entities. Most of these standard rubrics are based on anatomic site of occurrence, and the cases represented included all malignancies found to originate in that organ, regardless of the histological cell of origin. As time has advanced, the hematological (blood cell) malignancies and certain sarcomas have been broken out to form separate entities that cut across anatomic location. The Los Angeles registry has complied with the general conventions for joint publication, but for other purposes, malignant neoplasms have been categorized in more detail, using as a basis the cell and organ of origin and the available information about known or presumed differences in causation. That classification was the origin of the version used in this book. If reliable studies exist in which histological entities show differences in occurrence between subpopulations, differences in survival, or links to different genetic or environmental agents, the malignancies have been described separately.

The malignancies described here have all been defined using the histological cell of origin as well as the anatomic site of primary occurrence, and are all framed using the *International Classification of Diseases for Oncology, Second Edition*.[11] Of the 84 rubrics described, 12 are not members of the mutually exclusive set of malignancies formed by the remaining 72, but are aggregates of those, combining certain anatomic or age subgroups. The 72 mutually exclusive entities do not account for all malignancies, because even in this large experience, a few tumors are sufficiently unique and sufficiently uncommon to be neither combined with others nor represented separately.

The classification of non-Hodgkin lymphomas (malignancies of the T and B lymphocytes) is always particularly troublesome. Pathologists have changed the approach to this classification over the period covered by this book, and some completely new subgroups are now recognized. To combine the best of new knowledge with historical consistency, we have used a classification that combines elements of the older Working Formulation with the newer revised European and American lymphoma (REAL) and WHO-revised classification.

Definitions of Social Class (SES)

Social class as used here is a convenient but crude categorization, based on the characteristics of a group (in this case the population of each census tract) rather than the characteristics of the individual. For each U. S. decennial census (1970, 1980, and 1990; the 2000 census results were not available at the time of analysis), each of the 1619 census tracts (according to 1990 census definitions) was ranked according to two standard census variables: "average years of schooling" (for adults) and "median household income" (for households). These ranks were then summed, and the sums ranked.[8] These final rankings were then divided into five equal parts (quintiles), each arbitrarily comprising one social class. Class 1 includes census tracts with the highest education–income rankings, and Class 5 represents census tracts with the lowest education–income rankings. The accompanying map shows the geographical distribution of these five classes.

The incidence of a malignancy in each census tract has routinely been compared to the incidence in the combined census tracts of the same social class, as well as to the overall county incidence. Admittedly, such an adjustment is never perfect. Social class is only a

Figure F: Census tracts of Los Angeles County by the social class of residents diagnosed with cancer, 1972–1998.

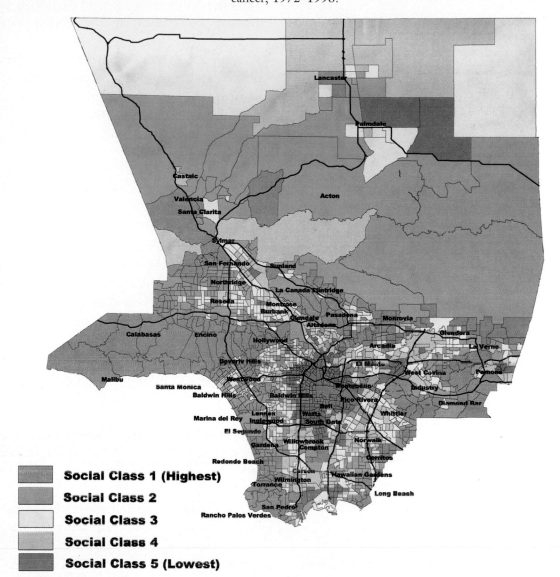

Social Class 1 (Highest)
Social Class 2
Social Class 3
Social Class 4
Social Class 5 (Lowest)

crude, if convenient, way of broadly classifying people when education, or income, is presumed to be an important predictor of risk. Of course it is never really education or income that is of causal concern, but some more direct factor linked to education or income: for example, a habit more common among the poor, or a wealthy person's ability to buy potentially dangerous products. Some errors are inevitable, because the adjustment can only be based on the characteristics of an entire census tract, and those might differ from the characteristics of individual cases. Moreover, close neighbors may differ in exposure, and some census tracts include a range of heterogeneous neighborhoods. Although minor shifts are frequent, adjustment for social class does not usually produce major alterations in high-risk pattern, suggesting that social class is only occasionally important as a predictor of risk.

Race and Ethnicity

Most census tracts include persons of different race/ethnicity, and in any case both logistical and statistical considerations prevent the separate adjustment for race/ethnicity that is done for social class. However, sometimes race/ethnicity does play an important role in the pattern of a neoplasm. This may sometimes occur because of genetic heritage, but more often it seems to happen on the basis of differential cultural practice. Even perfect adjustment for social class does not adjust for race/ethnicity, and an observed difference between localities can sometimes be partly explained on that basis. Although the ethnic distribution of the Los Angeles population is too complex a subject for this book, there is a need to know roughly how the various groups are geographically distributed. We have therefore also produced a map of census tracts from which a single race/ethnicity has comprised a majority of the cancer cases occurring over the entire period. Most census tracts are allocated to one of four broad ethnic groups: Latinos, African-Americans, Asian-Americans, and European-Americans (whites). Those census tracts lacking a majority of cases from any single group are separately designated.

Statistics

Many different comparisons can be made from Figures 1–4 by sex for each of the 84 rubrics. One would therefore expect, all things equal, that for each comparison, a few of the 84 versions would show statistical significance at the conventional level of 95% by chance alone. Therefore no notations of statistical significance are provided. Statistical comparisons between age-specific rates, genders, race/ethnic groups, regions, social classes, and points in time are available elsewhere.[12] Moreover, each figure would become much more complex and confusing if statistical elements were added. Almost all the rates are based on such large numbers that if the eye can detect a substantial difference, it is likely not to be explained by chance at the conventional level.

Figure 5 provides a simple range of values, with no explicit comparison, as do Figures 7 and 8, and neither the comparisons of skew in Figure 6, nor the relative risk correlations in Figures 10 and 12, permit a simple test of significance.

The maps in Figures 9 and 11 are based on standard incidence ratio estimates, so that patterns of the various cancers can be compared to one another. Each expected frequency is based on rates from a complete set of census tracts (either the entire county, or all those census tracts with a specific social class designation). These standards are always inclusive of the tract being described, so that the standard incidence ratio is technically only an estimate of the risk relative to other census tracts. However, because each census tract is only one of 1619, or (when adjusted for social class) one of about 324, and because no very large census tract is shown to be at an extreme level of risk for any individual malignancy, the estimates of relative risk are rather accurate.

No attempt has been made to formally employ a statistical distinction between chance and alternative explanations for a geographical pattern, and the interpretations provided are speculative. However, the average tract has five separate borders. If this were an invariable rule, and if as many as 20 high-risk census tracts were to be found, the probability that a single pair of them would be adjacent on one side strictly by chance is roughly 0.24, and the probability that three such combinations occur is 0.01. The probability of one combination increases to about 0.36 if there are 30 high-risk census tracts and 0.48 if there are 40. For 30, the probability of three is still below 0.05, and for 40 it drops below 0.05 to 0.03 for

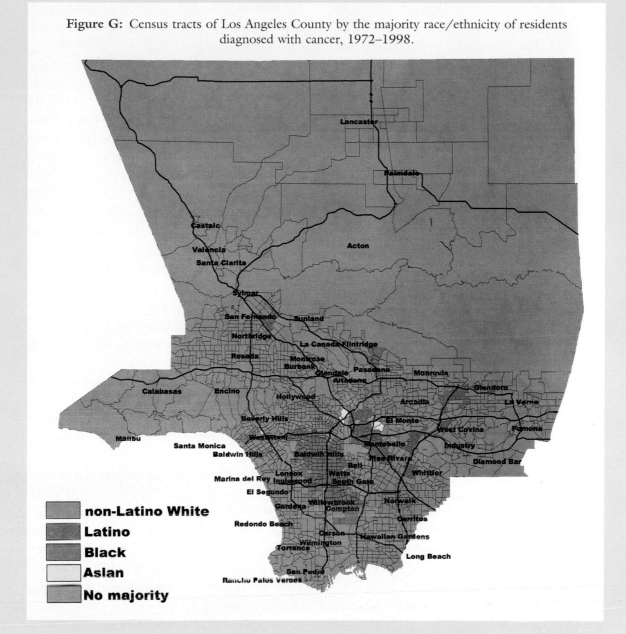

Figure G: Census tracts of Los Angeles County by the majority race/ethnicity of residents diagnosed with cancer, 1972–1998.

5 such combinations. When only a few contiguous combinations appear among a large number of high-risk census tracts, chance may offer a reasonable explanation, but when there are many, especially if geographically concentrated in one subregion, chance is an unlikely explanation. Hard and fast rules cannot be easily set, because census tracts are highly variable in size and shape as well as population size. As is described below in the context of in-

dividual malignancies, most of these apparently nonrandom patterns are consistent with known predictors of disease occurrence. See the section *Cluster Investigation* below for more discussion.

Choice of High-Risk Criteria

Statistical criteria have only been employed to designate census tracts at high risk. To do so, arbitrary threshold choices of relative risk (1.5) and alpha upper confidence limit criterion (95%) have been made, and at least 3 cases over the entire period of 27 years are required. Other thresholds could have been chosen. A relative risk of 1.25 would identify many more census tracts at a much lower average risk, and one of 2.0 would identify many fewer census tracts at a much higher average risk. Similarly, if a more restrictive statistical criterion, say the upper 99% confidence limit, were chosen, a much smaller proportion of comparisons would be "significant" by chance alone, and some large differences based on the experience of small census tracts would be missed. On the other hand, if a more inclusive criterion, such as the upper 90% confidence limit, had been chosen, a substantial proportion of census tracts appearing by chance would be mixed in with those truly at high risk.

To more fully understand the significance of these criteria, the following tables describe the actual results that would have occurred for selected cancers known to differ in overall frequency and in degree of nonrandom occurrence. The first two tables both describe common female cancers. Census tracts differ dramatically in risk of breast cancer, but not in risk of colon cancer. The third and fourth tables describe uncommon cancers affecting males, with Kaposi sarcoma, but not acute lymphoblastic leukemia (ALL) known to vary greatly in local risk. Each cell shows the number of census tracts that would actually be designated as at high risk using the criteria indicated, as well as the average number of expected cases responsible.

If all 1619 census tracts in the county were the same size, and if each cancer varied in frequency only by chance (i.e., if census tracts were selected by the criteria only on the basis of chance, and none were truly at high risk), the use of a relative risk of 1.01 (i.e., the smallest risk higher than 1.0) would serve to identify 164 census tracts using a 90% confidence interval, and 82 and 16 census tracts using the 95 and 99% criteria, respectively. This ratio of

10 to 5 to 1 would be roughly maintained, albeit with smaller numbers, as the relative risk cut-off is increased.

In practice, any reasonable relative risk criterion that might be chosen would be higher than 1.01, and in any case these ratios are altered because the census tracts vary in size, ethnicity, and social class. Thus the ratios in the left-hand column appear closer to 4 to −2 to 1 than 10 to −5 to 1 for the more or less randomly distributed colon cancer and for lymphoblastic leukemia. These vertical gradients tend to disappear as a higher relative risk is selected and the proportion of census tracts identified by chance alone diminishes. The gradients are also reduced as more census tracts exceed the threshold on the basis of truly high risk, such as, in the tables for breast cancer and Kaposi sarcoma. This reduction occurs regardless of the relative risk criterion, because fewer of these census tracts are ever selected solely by chance.

The number of census tracts seemingly at high risk of either colon or breast cancer also varies greatly, as expected, across the horizontal gradient of relative risk, although it follows no simple mathematical gradient. This variation is nearly absent in the remaining two tables, both representing malignancies that are quite rare, because only a few unexpected cases, whether occurring by chance or otherwise, are required to satisfy even the highest level of relative risk.

Empirically, to choose a relative risk criterion lower than 1.5 would be to designate as many as 10% of the census tracts at high risk for common malignancies, and to choose one higher than 1.5 for a common malignancy would result in the selection of very few census tracts.

For an uncommon malignancy without any dramatic geographical variation, use of the 99% upper confidence threshold would restrict the census tracts fulfilling the criterion to a very small number, no matter what level of relative risk was employed. For a rare cancer with

High Risk Census Tracts Identified
(Average Number of Expected Cases per Tract)
Using Various Relative Risk and Upper Confidence Limit Criteria

FEMALE COLON CANCER

	>RR = 1.25	>RR = 1.50	>RR = 1.75	>RR = 2.00
>90% UCL	93 (24.6)	48 (16.0)	18 (11.0)	4 (6.2)
>95% UCL	58 (25.7)	38 (17.4)	16 (11.9)	3 (7.6)
>99% UCL	24 (27.3)	20 (20.4)	10 (13.8)	2 (9.4)

FEMALE BREAST CANCER

	>RR = 1.25	>RR = 1.50	>RR = 1.75	>RR = 2.00
>90% UCL	172 (59.0)	37 (58.1)	10 (20.1)	6 (13.2)
>95% UCL	164 (78.1)	35 (61.3)	7 (25.5)	3 (21.1)
>99% UCL	120 (85.8)	33 (64.2)	6 (28.6)	3 (21.1)

MALE ACUTE LYMPHOBLASTIC LEUKEMIA

	>RR = 1.25	>RR = 1.50	>RR = 1.75	>RR = 2.00
>90% UCL	50 (1.4)	49 (1.4)	49 (1.4)	49 (1.4)
>95% UCL	27 (1.3)	27 (1.3)	27 (1.3)	27 (1.3)
>99% UCL	16 (1.3)	16 (1.3)	16 (1.3)	16 (1.3)

MALE KAPOSI SARCOMA

	>RR = 1.25	>RR = 1.50	>RR = 1.75	>RR = 2.00
>90% UCL	152 (4.9)	152 (4.9)	150 (4.8)	140 (4.8)
>95% UCL	140 (4.9)	140 (4.9)	140 (4.9)	135 (4.8)
>99% UCL	122 (4.9)	122 (4.9)	122 (4.9)	122 (4.9)

a clear geographic pattern, such as Kaposi sarcoma, it really would not make any difference which set of criteria was chosen, because the number of census tracts meeting the criteria does not vary. Therefore either a 90% or a 95% upper confidence limit would provide informative maps if coupled with a relative risk criterion of 1.5. We have therefore chosen to use a 95% upper confidence threshold, not only because it is the most common conventional standard, but because the interpretation is likely to be easier if based on the pattern of census tracts less likely to include those included only by chance.

It is true that the two criteria selected are imperfect for an unavoidable reason—because census tracts vary substantially in size, large tracts at a given level of relative risk are more likely to pass any statistical criterion than small tracts. On the other hand, small tracts generate a very small expected number, and at a given level of statistical confidence the observed relative risks are more likely to exceed the criterion.

In all likelihood, most if not all of the meaningful geographical patterns described below would be evident no matter which of these criteria were adopted.

The accompanying figure illustrates the power from 27 years of cases in Los Angeles County available to detect minimally high-risk census tracts, according to the size of the census tract and the level of incidence of the malignancy. Increases of the order of 50–100% can be detected, even in small census tracts, for any but the most uncommon malignancies.

Preparation of Figure 6

It has long been recognized that one crude measure of the environmental causation of a disease is the degree to which occurrence varies from place to place, especially if that variation is difficult to explain on the basis of either chance or genetic susceptibility. The efforts of Doll and Peto[13] to use international and migration-based comparisons to make such assessments were marred by the inevitable differences in the methodology underlying the collection of data. The examination of non-random differences within the area served by a single large cancer registry can serve a similar purpose, but without that liability. To assess whether neighborhoods tend to produce the

number of cases that chance would dictate, that number has been estimated using a Poisson distribution, the statistical method of estimating the effect of chance when the average number of cases in a census tract is small. In Figure 6 we estimate the number of census tracts expected to produce each given number of cases for purposes of comparison with the observed number.

The average size of census tracts in Los Angeles County could not serve as the basis for the Poisson calculation, because of the large variation in population size. For this reason a composite Poisson distribution was created. For each of the 1619 census tracts, the expected age-specific rate of cases per person-year (from the county averages) was applied to the array of age-specific frequencies from that tract to get the average number of expected cases per tract for census tracts of that size. Then a Poisson distribution for 1619 census tracts of that size and age distribution was created. After repeating that process 1619 times, using the population of each tract to generate such a Poisson matrix, each such matrix was divided by 1619 and all 1619 results were added together to form a single composite expected distribution. This process was then repeated for each census tract using social-class-specific estimates of age-specific incidence instead of the overall county age-specific rates, summing the results to obtain a single expected distribution, this time adjusted for social class.

Mapping Strategy

Maps of health events can be generally divided into spot maps and area maps. With a "spot" placed on a map at the point of an individual event, spot maps are visually appealing and easy to understand, but they cannot be used to examine the geographical occurrence of a rare disease because the pattern of spots might reflect not just the concentration of individual events but the concentration of

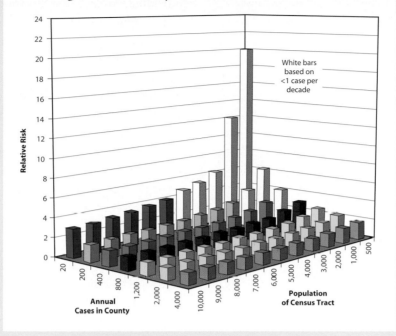

Figure H: Lowest census tract relative risk detectable in 27 years (requiring that 95% upper confidence level be exceeded) according to annual county cases and the size of the census tract.

the persons at risk of an event. Health event maps have more commonly employed area, or choropleth, maps, which are capable of showing differences between rates using different colors or shadings for the different areas, that is, different defined populations, according to the various levels of risk. Most such maps have been produced for the purpose of displaying patterns thought to correspond to specific patterns of exposure, or for the purpose of identifying spatial patterns of co-occurrence, commonly "autocorrelations" or associations between the rates of adjacent areas.

Although they are commonly used, attractive, and generally informative, subtle problems interfere with the detailed interpretation of these area maps. A major problem is that different political units have populations of different size, resulting in different levels of statistical confidence in the mapped level of risk.

The most obvious way of reducing the heterogeneity of statistical confidence is to increase the number of events covered by each mapped area. This strategy also obviates errors that are produced because methods of disease event registration often improve or otherwise change over relatively short intervals or within short distances, and not necessarily in parallel from place to place or interval to interval. Such inconsistency, together with concerns about the distortions attributable to chance, have driven mapmakers to use large geographical units, such as provinces or counties, and/or to bracket extended periods of time. In North America, with a high level of internal migration and thus of population homogeneity, the use of larger geographical units usually means that the units are demographically, culturally, and even environmentally rather similar to each other, thus minimizing the meaningful geographical variation.

Most of those responsible for making cancer maps have had a mandate to describe at least one entire country. As a consequence they have chosen to map measures of mortality rather than of incidence, both because they are uniformly available, and because the different areas mapped are more likely to have been served by comparable methods of vital statistics registration. Mortality rates are heavily influenced by the care with which vital statistics are gathered, as well as by the quality of care, and are not optimal for the study of causation.

As the basis for categorizing areas, maps may employ disease frequency, rates of mortality or incidence, or levels of relative risk, the latter usually based on standard incidence or mortality ratios. Maps of health events sometimes employ incidence rates as the basis for mapping so that the consumer understands the variation in the absolute magnitude of risk. It is difficult for readers to then compare and contrast the patterns of occurrence of different conditions, because the resultant different scales produce nonparallel distortions of pattern. Patterns of color may be used to enforce uniformity,[14] but readers must be alert to very subtle messages.

The use of larger areas, the focus on mortality, or the choice of a particular measure to map does not completely solve the problem of variable statistical confidence. Maps of significance alone may be useful for hypothesis testing, but are of little use to laymen. The issue of significance may be addressed by the use of bivariate maps (with both level of risk and a statistical parameter jointly mapped),[15] by the exclusion of small populations, by alteration of boundaries in order to render areas demographically comparable,[16] or by adjustment (smoothing) of the observed categorizations.[17] The latter is often done using empirical Bayes' methods, either adjusting to a global or to a local standard.[18] None of these solutions is completely adequate, and none solves an additional serious problem, namely that the visual interpretation of complex choropleth maps is hindered greatly not only by the confusion caused by the multiplicity of classifications and the complicated sequence of cut

points, but even by the choice and sequence of the multiple colors or shadings.[19]

The present task is to describe the risk of individual small neighborhoods, measured over a long period by a uniform method of incidence registration. The choice is to map relative risk (standard risk ratio) so that comparisons between the nonrandom occurrences of different diseases can provide information about parallel etiology. This map is a relatively simple hybrid between a spot map and an area map. Spatial/demographic units are mapped according to risk level, but it is a spot map in the sense that only the locations of clearly adverse conditions, that is, only high-risk census tracts, are portrayed. The observations have not been modified (i.e., by smoothing) nor have observations been excluded on the basis of population size, although some exclusions were based on demographic inaccuracy. In short, the maps were designed to be easily interpreted to focus the attention of both residents and decision makers.

These methods minimize, but do not completely avoid, the distortion produced by statistical variability. Even though exclusions are based strictly on uncertainty (i.e., lack of statistical confidence), the smaller the population of a census tract, then the fewer the cases, and the less likely the tract is to be included on a map, on statistical grounds. The larger the population, then the larger the number of cases, and the easier it is to exclude chance as the explanation for an increase. We have attempted to minimize confusion by ignoring subtle variations in risk. All the census tracts that are colored are selected on the basis of a single criterion of risk level, and a single criterion of statistical uncertainty.

Cluster Investigation

Statistical maneuvers of variable complexity have long been used to search for general evidence of spatial or time-space clustering, and to identify specific examples of such clusters. Underlying these efforts are explicit or implicit etiological hypotheses of clustered causation.[20] Such hypotheses may postulate transmission of a carcinogenic organism between geographically contiguous dwellings, emission of carcinogens from a local man-made point source, or common neighborhood exposure to natural geological carcinogens or to polluted materials (air, water, food, other material) that are preferentially distributed to the residents of a restricted locality. There are precedents for time-space clusters of cancer diagnoses, but those that have been clearly verified are limited to a few specific exposures, such as ionizing radiation, either released accidentally or as a weapon of war; asbestos, or a similar mineral used as housing material or decoration; or arsenic contamination of local water supplies.

Interestingly, much of the motivation for the statistical search for clusters has been based on concern about the person-to-person transmission of infectious agents. For studies of cancer, this is misdirected, because when such transmission has occurred, it occurred long before and in no constant relation to the time of diagnosis. Transmission of a carcinogen usually has occurred between individuals not likely to live in geographic proximity at the time of diagnosis, and not even at a time when the recipient, now the patient, was residing at the current address. We now know that the known and suspected infectious causes of cancer are either transmitted at a very remote time and place by mechanisms other than casual contact (hepatitis B virus, human papillomavirus, HIV), or are ubiquitous agents that produce tumor growth only under specific host conditions (Epstein-Barr virus).[21] Unlike ordinary infectious diseases, the latent period between exposure and the diagnosis of most cancers is measured in decades, producing great variation in the latent period and effectively eliminating any distinction between spatial and time-space clustering.

Infectious agents aside, the local air- or water-borne spread of emitted carcinogen from an occult point source is usually transient and results in dilution of the agent as it fills a space proportional to the square or even the cube of the distance from the source.[22] Thus only those persons exposed very near the source, who occupy a small area and thus are inevitably few in number, are likely to receive a substantial dose, even from a powerful carcinogenic emission. The occasional resultant case or cases are likely to be very difficult to distinguish from background cases unless the cancer in question is very rare and the risk per unit dose very high. For more discussion of this issue, see the *Scientific American* article in Reference 23.

For these reasons the measures of risk for adjacent census tracts are not likely to correlate on the basis of either temporary point-source emissions or person-to-person transmission. Moreover, they are very likely to correlate for other reasons, because persons of similar race/ethnicity, social class, occupation, and lifestyle often choose to reside in communities that transcend small area boundaries. Small numbers of aggregated census tracts at high risk must therefore be interpreted with caution.

However, an environmental explanation does gain biological credibility if a carcinogen is distributed over a wide area and over the long term. A broad pattern of excess risk is more likely to be meaningful, especially if it cuts across diverse areas in a pattern consistent with prevailing air, water, or commercial distributions. No formal attempt to statistically assess nonrandom geographical distribution has been attempted here (see the above section *Statistics*), but the mapping itself does identify clear evidence of such broad nonrandom occurrence. Although the criteria for high-risk employed here are arbitrary, they have been applied independently for each census tract and a tendency for a large number of census tracts at high risk to geographically aggregate is unlikely to be due to chance. Special attention has been paid to the few instances of broad geographical aggregation that have no obvious explanation. Whatever the advantages over other forms of mapping for this purpose, the search for nonrandom variation by the conservative method of mapping discrete localities on the basis of a simple risk dichotomy has important disadvantages. Useful distinctions between degrees of risk are ignored, and much of the information from the pattern of occurrence fails the dichotomous threshold distinction and is therefore wasted. Formal methods of cluster analysis that take advantage of more empirical information could be used to corroborate or add to the suggestions of nonrandom occurrence presented here. They too are imperfect, lacking optimal means of accounting for the detailed distribution of the population at risk or the number of potential comparisons, but the use of such methods could refine the findings from these maps.

Bibliography

1. U.S. Cancer Statistics Working Group. (2002). United States Cancer Statistics, 1999 Incidence. Atlanta: Department of Health and Human Services, Centers for Disease Control and Prevention, 64–65.
2. Morris, C., and Wright, W. (1996). Breast Cancer in California. Sacramento: California Department of Health Services, California Surveillance Section, 12.
3. Anonymous. (2003). Public Health Assessment, Evaluation of Exposure to Historic Air Releases from the Abex/Remco Hydraulics Facility. Sacramento: California Department of Health Services.
4. Schottenfeld, D., and Fraumeni, Jr., J. (1996). *Cancer Epidemiology and Prevention.* New York: Oxford.
5. Adami, H.-O., Hunter, D., and Trichopoulos, D. (2002). *Textbook of Cancer Epidemiology.* New York: Oxford.
6. Rothman, K. (2002). *Epidemiology.* New York: Oxford.
7. Allen, J., and Turner, E. (1997). *The Ethnic Quilt.* Northridge: California State University.
8. Liu, L., Zhang, J., Deapen, D., Bernstein, L., and Ross, R. (2001). Cancer in Los Angeles County; Incidence and Mortality by Race/Ethnicity. Los Angeles: Los Angeles County Cancer Surveillance Program, University of Southern California.
9. Swerdlow, A. (1992). Cancer incidence data for adults. In: *Geographical and Environmental Epidemiology: Methods for Small Area Studies,* P. Elliot, J. Cuzick, D. English, and R. Stern, Eds., pp. 51–62. Oxford: New York.
10. Program SR. (2003). SEER*Stat software (www.seer.cancer.gov/seerstat). Bethesda: National Cancer Institute.
11. Percy, C., Van Holten, V., and Muir, C. (1990). *International Classification of Diseases for Oncology.* Geneva: World Health Organization.
12. Parkin, D., Whelan, S., Ferlay, J., Raymond, L., and Young, J. (1997). *Cancer Incidence in Five Continents, Volume VII.* Lyon: International Agency for Research on Cancer.
13. Doll, R., and Peto. R. (1981). The causes of cancer. *J. Natl. Cancer Inst.,* 66:1197–1312.
14. Smans, M., Muir, C., and Boyle, P. (1992). *Atlas of Cancer Mortality in the European Economic Community.* Lyon: International Agency for Research on Cancer.
15. Mason, T., McKay, F., Hoover, R., Blot, W., and Fraumeni, J. (1975) *Atlas of Cancer Mortality for US Counties: 1950–69.* Washington D.C.: U. S. Government Printing Office.
16. Elliott, P., Wakefield, J., Best, N., and Briggs, D., eds. (2000). Spatial epidemiology; methods and applications. In: *Spatial Epidemiology: Methods and Applications.* Oxford: New York, 3–14.
17. Smans, M., and Esteve, J. (1992). Practical approaches to disease mapping. In: *Geographical and Environmental Epidemiology: Methods for Small Area Studies,* P. Elliott, J. Cuzick, D. English, and R. Stern, Eds. pp. 141–150. Oxford: New York.
18. Clayton, D., and Bernardinelli, L. (1992). Bayesian methods for mapping disease risk. In: *Geographical and Environmental Epidemiology: Methods for Small Area Studies,* P. Elliott, J. Cuzick, D. English, and R. Stern, Eds., pp. 205–220. Oxford: New York.
19. Walter, S. (1993). Visual and statistical assessment of spatial clustering in mapped data. *Statis. Med.,* 12:1275–1291.
20. Wakefield, J., Kelsall, J., and Morris, S. (2000). Clustering, cluster detection, and spatial variation in risk. In: *Spatial Epidemiology; Methods and Applications,* P. Elliott, J. Wakefield, N. Best, and D. Briggs, Eds., pp. 128–152. Oxford: Oxford, England.
21. Mueller, N., Evans, A., and London, W. (1996). Viruses. In: *Cancer Epidemiology and Prevention,* D. Schottenfeld and J. Fraumeni, Jr., Eds., pp. 502–531. Oxford: New York.
22. Shy, C. (1996). Air pollution. In: *Cancer Epidemiology and Prevention,* D. Schottenfeld and J. Fraumeni, Jr., Eds., pp. 406–417. Oxford: New York.
23. Trichopoulos, D., Li, F., and Hunter, D. (1996). What causes cancer? *Sci. Amer.* 275:80–87.

Oropharyngeal Carcinoma

ICDO-2 Code Anatomic Site: C 0–6, 9–14
ICDO-2 Code Histology: 8000–8586
Age: All
Male Cases: 13348
Female Cases: 6681

Background

Cancers of the mouth, tongue, gums, oral cavity, and pharynx are more common in smokers, drinkers, and especially drinkers who smoke. Chewing tobacco and *paan* (a mixture with tobacco chewed in Asia and the Pacific) clearly increase risk of these cancers as does poor oral hygiene and the consumption of very hot drinks, such as mate, a South American tea. Dietary fruit and vitamin C and E supplements may decrease the risk of these cancers. Lip cancer is more common in outdoor workers and is probably related to sun exposure. Cancer of the nasopharynx (the extension of the throat up behind the nose) is also related to certain dietary practices, such as the childhood consumption of salted fish, as practiced by Cantonese people. Neither lip nor nasopharyngeal cancers account for more than a very small number of all oropharyngeal cancers in Los Angeles.

Local Pattern

Cancers of the oropharynx are nearly three times as common among men as women. They occur at about the same rate in Los Angeles as in other parts of the country, although rates among women in Utah are rather low. Latino and Asian-American women have identically low rates of these cancers, although middle-aged Latino men are subject to higher rates than Asian-American men of the same age. Incidence among African-Americans and whites of both sexes are higher, although the cases among African-American women occur somewhat earlier in life than those among white women. Men, but not women, from lower social class neighborhoods experience higher incidence than others. The incidence of oropharyngeal cancers has decreased over the period, both in Los Angeles County as a whole and among the residents of high-risk census tracts. Figure 6 shows a moderate nonrandom excess of census tracts with unexpectedly few or unexpectedly many cases. Census tracts meeting the high-risk criteria are numerous and widely scattered throughout the county. There are several small groups of contiguous high-risk census tracts in the Santa Clarita Valley, in the San Gabriel Valley, in the South Bay south of Los Angeles Airport, and especially in the southeast corner of the county east of the Long Beach freeway. In the latter area there are several census tracts with high risk to members of both sexes.

Thumbnail Interpretation

In America, oropharyngeal cancers mostly result from cigarette smoking, alcohol consumption, and dietary inadequacy. We would predict that incidence should decrease over time, and that men, and persons of lower social class, especially poorer African-Americans, would be at higher risk. The pattern of occurrence is generally consistent with that prediction. Although squamous carcinoma of the esophagus and carcinoma of the larynx share the same set of known causes, the geographical distribution is not identical. Many possible reasons might explain these differences. Among these are use of chewing tobacco, the relative mix of smoking and drinking, and the specific products consumed, and any as yet unidentified causes.

Figure 1: Age-adjusted incidence rate by place.

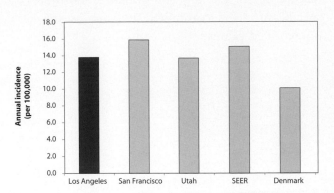

Figure 2: Age-adjusted incidence rate over the period.

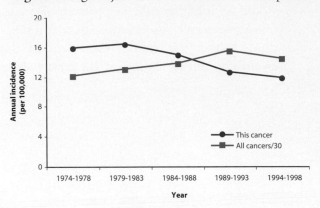

Figure 3: Age-adjusted incidence rate by age and race/ethnicity.

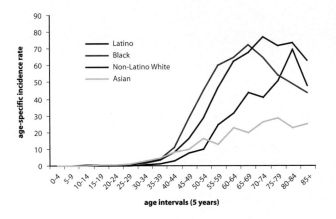

Figure 4: Age-adjusted incidence rate by social class.

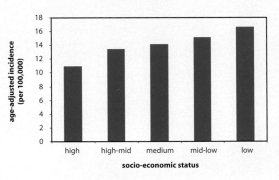

Figure 5: Distribution of the relative risk values for all census tracts.

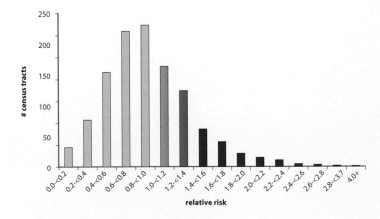

Figure 6: Census tracts by the number of cases per tract.

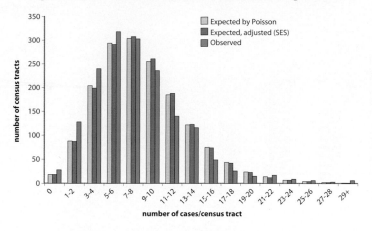

Figure 7a and b: Census tracts at high risk by the number of cases. (a) Unadjusted and (b) adjusted for social class.

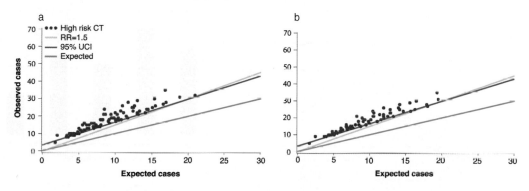

Figure 8: Risk over the period for high-risk census tracts relative to all census tracts.

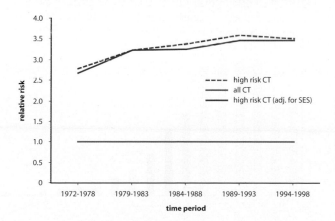

Figure 1: Age-adjusted incidence rate by place.

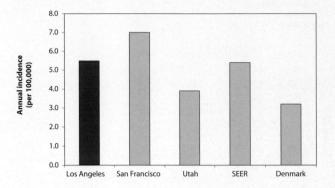

Figure 2: Age-adjusted incidence rate by age and race/ethnicity.

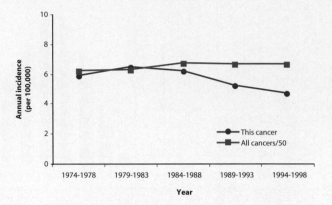

Figure 3: Age-adjusted incidence rate over the period.

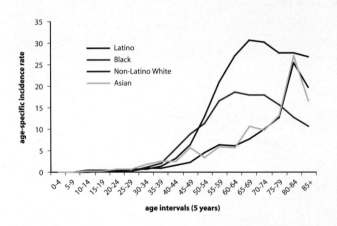

Figure 4: Age-adjusted incidence rate by social class.

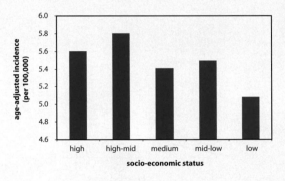

Figure 5: Distribution of the relative risk values for all census tracts.

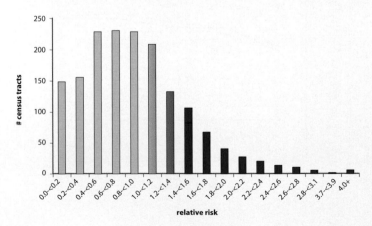

Figure 6: Census tracts by the number of cases per tract.

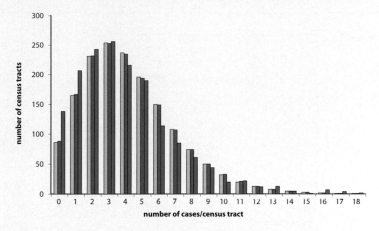

Figure 7a and b: Census tracts at high risk by the number of cases. (a) Unadjusted and (b) adjusted for social class.

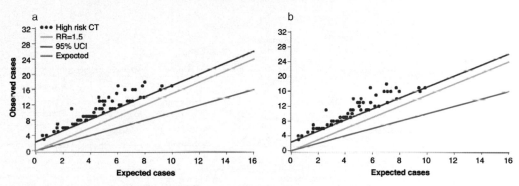

Figure 8: Risk over the period for high-risk census tracts relative to all census tracts.

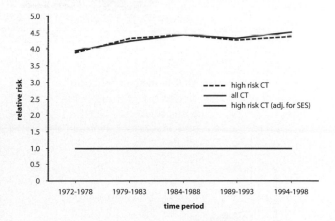

Figure 9: Map of census tracts at high risk.

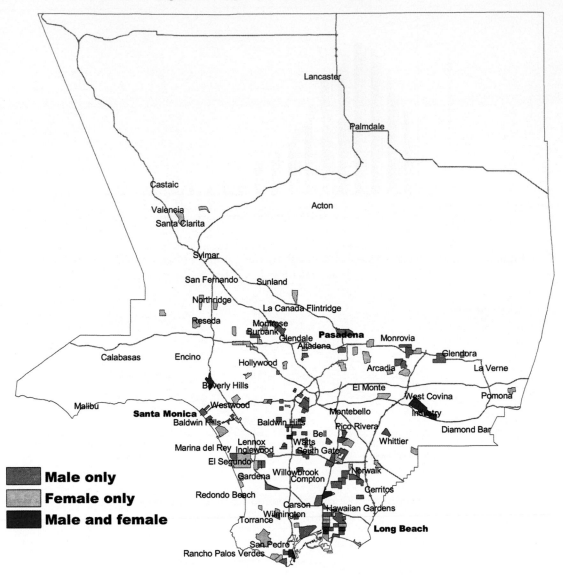

Figure 10: Male-female correlation between the relative risks for high-risk census tracts.

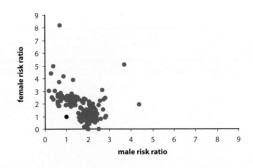

Figure 11: Map of census tracts at high risk, adjusted for social class.

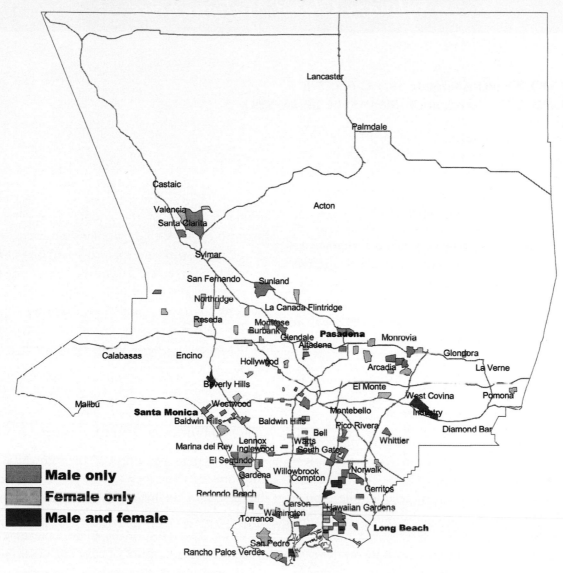

Figure 12: Male-female correlation between the relative risks for high-risk census tracts, adjusted for social class.

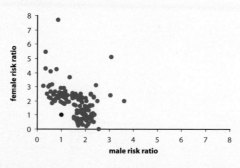

Salivary Gland Malignancies

ICDO-2 Code Anatomic Site: C 6.9, 7–8
ICDO-2 Code Histology: 8000–8586, 8933–8941
Age: All
Male Cases: 1050
Female Cases: 920

Background

The causes of salivary gland malignancies are unknown, although some have suggested that ionizing radiation may play a causal role. There are also suggestions that certain occupational exposures, such as those prevalent in the rubber industry, might contribute to increased risk.

Local Pattern

These malignancies are slightly more common among men than women, and slightly less common in Los Angeles County than in other areas of the country. All racial/ethnic groups are affected to about the same level, but persons of higher social class are more frequently affected. Among the male residents of high-risk census tracts, incidence was higher in the earlier part of the period, although the rates have been stable over the entire period among residents of the county as a whole. Figure 6 shows a slight nonrandom excess of census tracts with unexpectedly few or unexpectedly many cases. High-risk census tracts are scattered throughout the county with few that are contiguous and none showing high risk for both sexes. No census tract stands out on the basis of a particularly high number of excess cases.

Thumbnail Interpretation

Other than the unspecified occupational exposures of some men, no speculation about the reason for the male excess is available. The relatively early onset may reflect the timing of unknown exposures, or the anatomy or developmental characteristics of cells in the salivary glands. No systematic pattern of geographical occurrence is apparent, and therefore no local source of causation can be proposed.

Figure 1: Age-adjusted incidence rate by place.

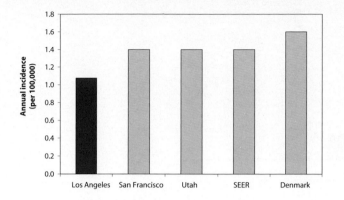

Figure 2: Age-adjusted incidence rate by age and race/ethnicity.

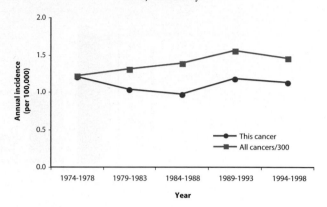

Figure 3: Age-adjusted incidence rate over the period.

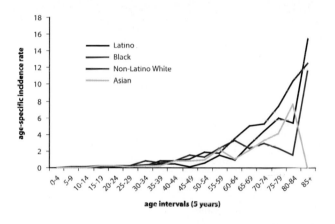

Figure 4: Age-adjusted incidence rate by social class.

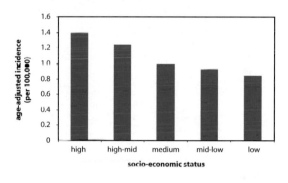

Figure 5: Distribution of the relative risk values for all census tracts.

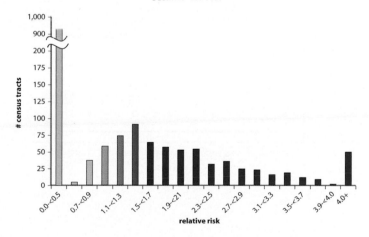

Figure 6: Census tracts by the number of cases per tract.

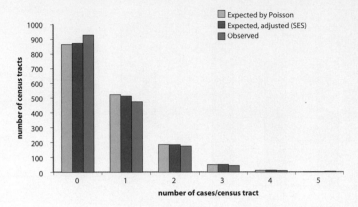

Figure 7a and b: Census tracts at high risk by the number of cases. (a) Unadjusted and (b) adjusted for social class.

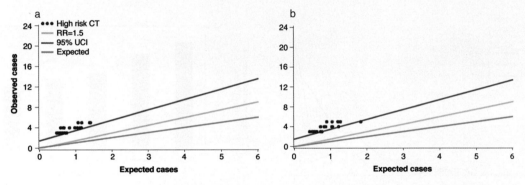

Figure 8: Risk over the period for high-risk census tracts relative to all census tracts.

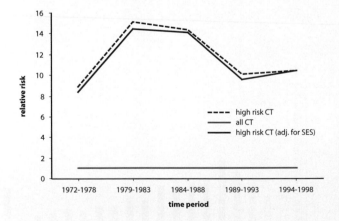

Figure 1: Age-adjusted incidence rate by place.

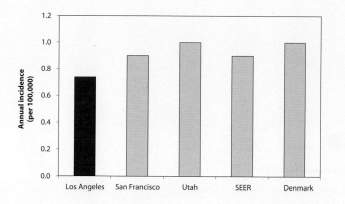

Figure 2: Age-adjusted incidence rate by age and race/ethnicity.

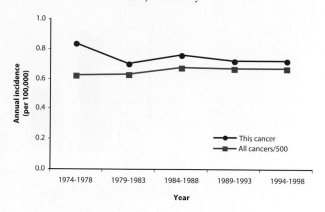

Figure 3: Age-adjusted incidence rate over the period.

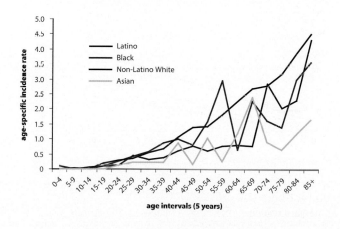

Figure 4: Age-adjusted incidence rate by social class.

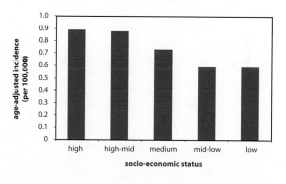

Figure 5: Distribution of the relative risk values for all census tracts.

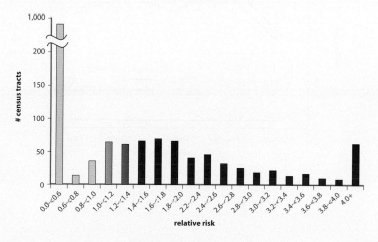

Figure 6: Census tracts by the number of cases per tract.

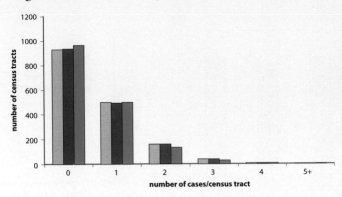

Figure 7a and b: Census tracts at high risk by the number of cases. (a) Unadjusted and (b) adjusted for social class.

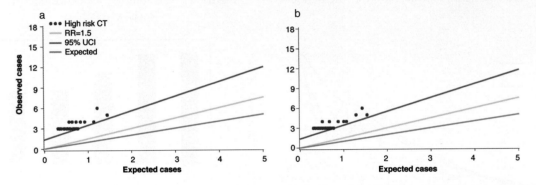

Figure 8: Risk over the period for high-risk census tracts relative to all census tracts.

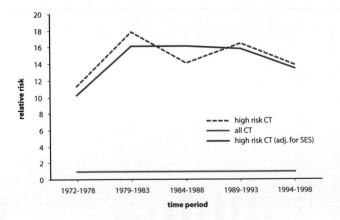

Salivary Gland Malignancies

Figure 9: Map of census tracts at high risk.

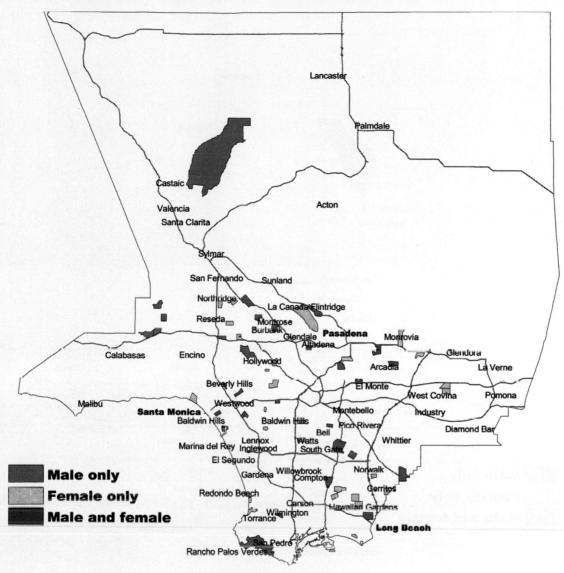

Male only

Female only

Male and female

Figure 10: Male-female correlation between the relative risks for high-risk census tracts.

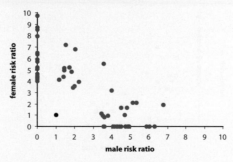

Figure 11: Map of census tracts at high risk, adjusted for social class.

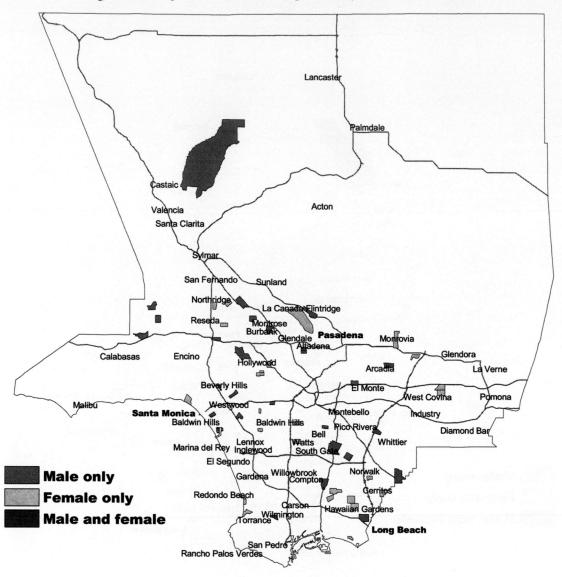

Figure 12: Male-female correlation between the relative risks for high-risk census tracts, adjusted for social class.

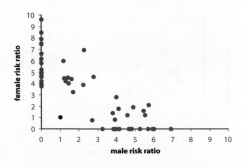

Squamous Carcinoma of the Esophagus/Gastric Cardia

ICDO-2 Code Anatomic Site: C 15, 16.0
ICDO-2 Code Histology: 8000–8076
Age: All
Male Cases: 3663
Female Cases: 2144

Background

Tobacco smoking and alcohol consumption are known to cause this carcinoma, and both may interact with genetic factors to further enhance the risk. Lower consumption of fruits and vegetables may also increase risk. In some populations, the combined effect of these exposures fails to explain the difference in risk between men and women, suggesting that other causes may be important.

Local Pattern

Esophageal cancer is substantially more common in Los Angeles and San Francisco than in other regions, especially Utah, and is much more common among men, African-Americans, and persons of lower social class. Incidence rates rise in early middle age. Over time, incidence rates have been decreasing in the county as a whole, although there has been little change among the residents of high-risk census tracts. There is a moderate nonrandom excess of census tracts with unexpectedly few or unexpectedly many cases (Figure 6). A substantial number of census tracts are shown to be at very high risk. A large aggregate of high-risk census tracts, including several showing excess risk among both men and women, appears in the predominately African-American region of South-Central Los Angeles. Adjustment for social class only partly reduces the size of this aggregate.

Thumbnail Interpretation

Knowing that squamous cancers of the esophagus among Americans are related to habits of cigarette smoking, alcohol consumption, and dietary inadequacy allows us to predict that incidence would have decreased over time, and that a higher frequency would be observed among men, African-Americans (known as a group to include a higher proportion of smokers), and persons of lower social class. The pattern of occurrence generally is consistent with that expectation, as is the geographical distribution.

Figure 1: Age-adjusted incidence rate by place.

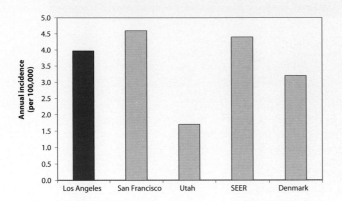

Figure 2: Age-adjusted incidence rate by age and race/ethnicity.

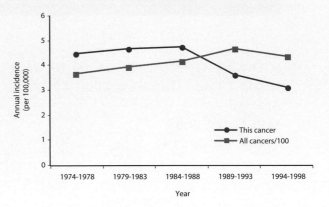

Figure 3: Age-adjusted incidence rate over the period.

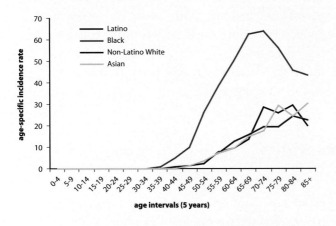

Figure 4: Age-adjusted incidence rate by social class.

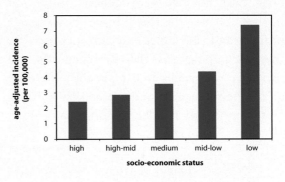

Figure 5: Distribution of the relative risk values for all census tracts.

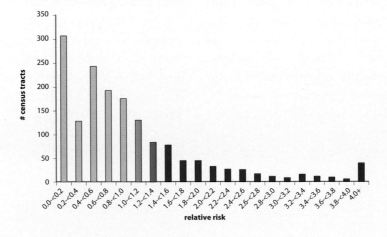

Figure 6: Census tracts by the number of cases per tract.

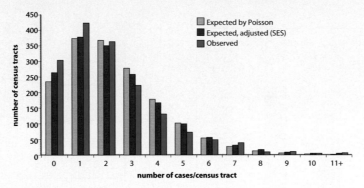

Figure 7a and b: Census tracts at high risk by the number of cases. (a) Unadjusted and (b) adjusted for social class.

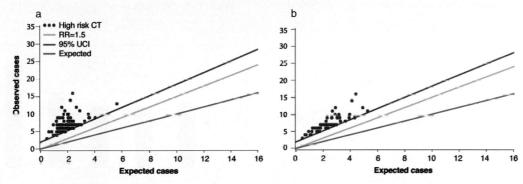

Figure 8: Risk over the period for high-risk census tracts relative to all census tracts.

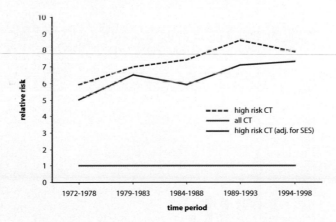

Figure 1: Age-adjusted incidence rate by place.

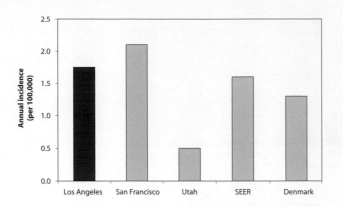

Figure 2: Age-adjusted incidence rate by age and race/ethnicity.

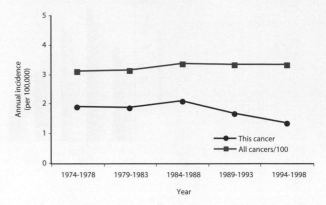

Figure 3: Age-adjusted incidence rate over the period.

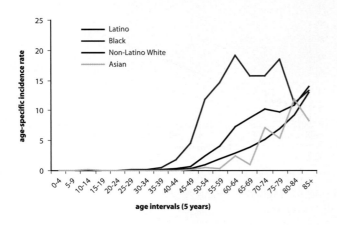

Figure 4: Age-adjusted incidence rate by social class.

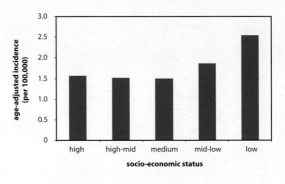

Figure 5: Distribution of the relative risk values for all census tracts.

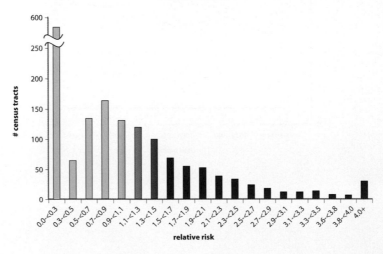

Figure 6: Census tracts by the number of cases per tract.

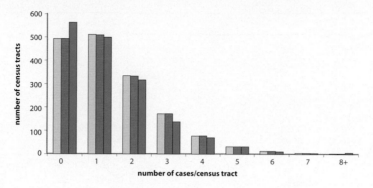

Figure 7a and b: Census tracts at high risk by the number of cases. (a) Unadjusted and (b) adjusted for social class.

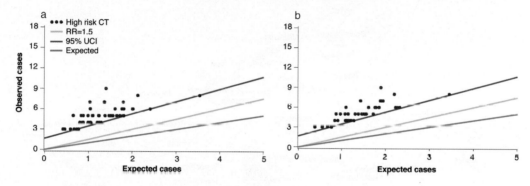

Figure 8: Risk over the period for high-risk census tracts relative to all census tracts.

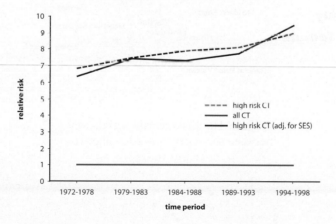

Figure 9: Map of census tracts at high risk.

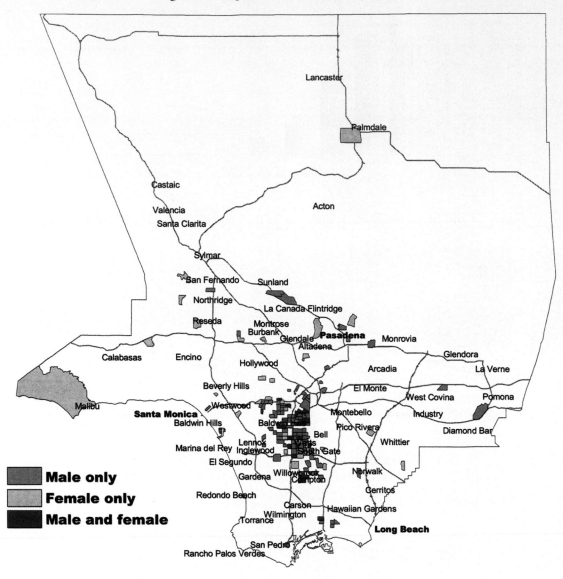

Figure 10: Male-female correlation between the relative risks for high-risk census tracts.

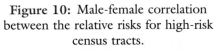
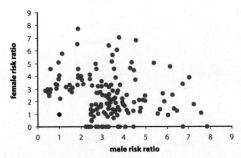

Figure 11: Map of census tracts at high risk, adjusted for social class.

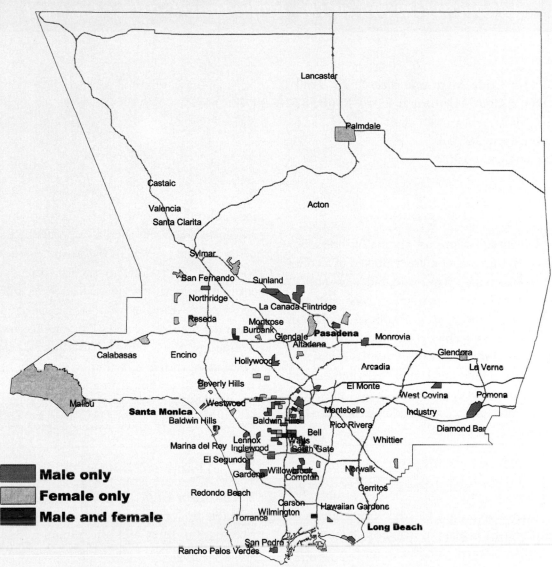

Figure 12: Male-female correlation between the relative risks for high-risk census tracts, adjusted for social class.

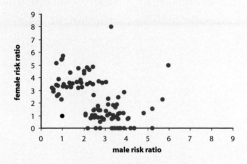

Adenocarcinoma of the Esophagus/Gastric Cardia

ICDO-2 Code Anatomic Site: C 15, 16.0
ICDO-2 Code Histology: 8140–8560, 8570–8573
Age: All
Male Cases: 3851
Female Cases: 1002

Background

Unlike squamous carcinoma of the esophagus, this tumor is not closely linked to alcohol use. Tobacco may play a role, but a less important one than for squamous cancers. Risk may be higher in those who are obese. A medical condition that seems to be important in the causation of these adenocarcinomas is gastroesophageal reflux (acid reflux), wherein acid from the stomach travels up into the lower esophagus, causing changes in the tissue structure.

Local Pattern

Adenocarcinoma of the esophagus and gastric cardia is many times more common in men than women, and occurs in Los Angeles County at about the same rate as in other parts of the country. Incidence among men begins to increase in middle age. Whites and Latinos are at clearly higher risk than the members of other groups. The incidence rate in the county increased over the period, especially among men, although among the residents of high-risk census tracts this increase appears to have been restricted to early in the period. Figure 6 shows only a modest nonrandom excess of census tracts with unexpectedly few or unexpectedly many cases. A number of census tracts passed the high-risk criteria. They are scattered throughout the county with no regional aggregation and few examples of contiguity. The pattern does not change after adjustment for social class.

Thumbnail Interpretation

The higher risk to males and whites is characteristic of this malignancy, as is the increase over time and the diminished risk among those of lower social class, even though the malignancies of adjacent tissues in the esophagus and lower stomach are both strongly linked to lower social class. No obvious nonrandom determinant distinguishes the observed high-risk census tracts from the others.

Adenocarcinoma of the Esophagus/Gastric Cardia: Male

Figure 1: Age-adjusted incidence rate by place.

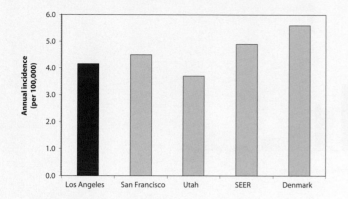

Figure 2: Age-adjusted incidence rate by age and race/ethnicity.

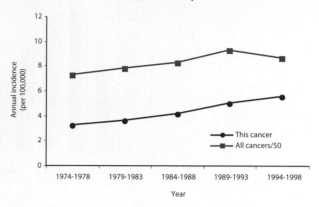

Figure 3: Age-adjusted incidence rate over the period.

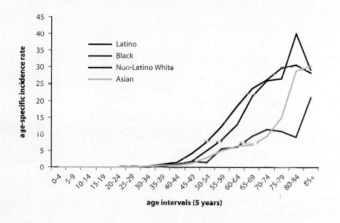

Figure 4: Age-adjusted incidence rate by social class.

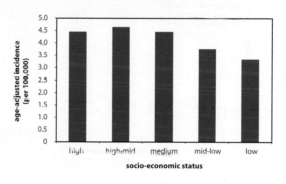

Figure 5: Distribution of the relative risk values for all census tracts.

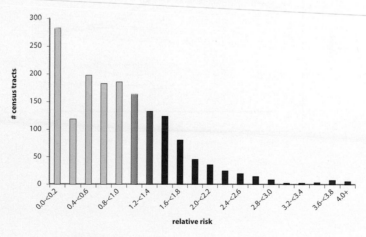

Figure 6: Census tracts by the number of cases per tract.

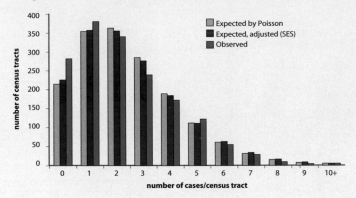

Figure 7a and b: Census tracts at high risk by the number of cases. (a) Unadjusted and (b) adjusted for social class.

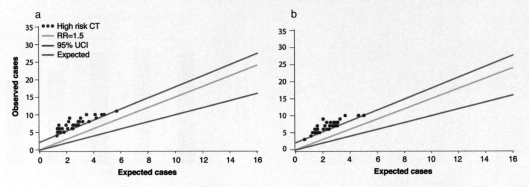

Figure 8: Risk over the period for high-risk census tracts relative to all census tracts.

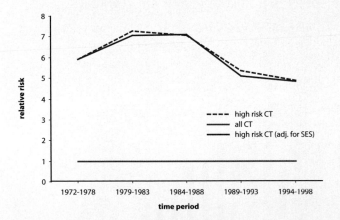

Figure 1: Age-adjusted incidence rate by place.

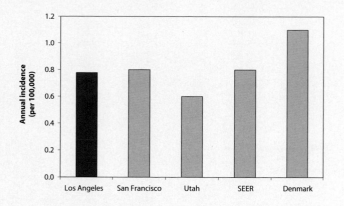

Figure 2: Age-adjusted incidence rate by age and race/ethnicity.

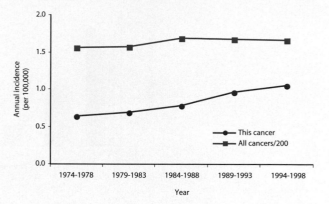

Figure 3: Age-adjusted incidence rate over the period.

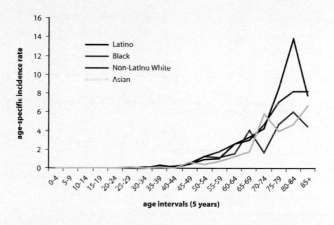

Figure 4: Age-adjusted incidence rate by social class.

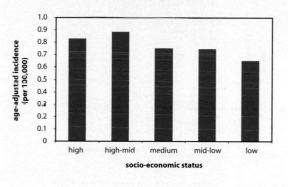

Figure 5: Distribution of the relative risk values for all census tracts.

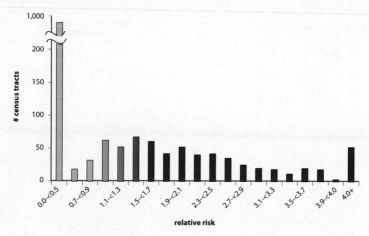

Figure 6: Census tracts by the number of cases per tract.

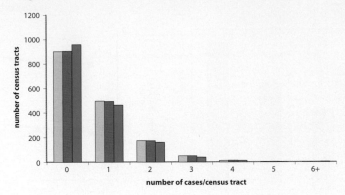

Figure 7a and b: Census tracts at high risk by the number of cases. (a) Unadjusted and (b) adjusted for social class.

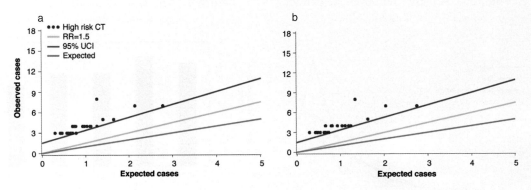

Figure 8: Risk over the period for high-risk census tracts relative to all census tracts.

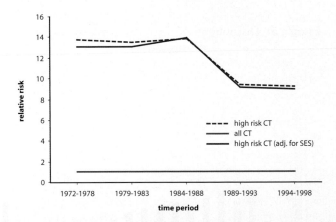

Figure 9: Map of census tracts at high risk.

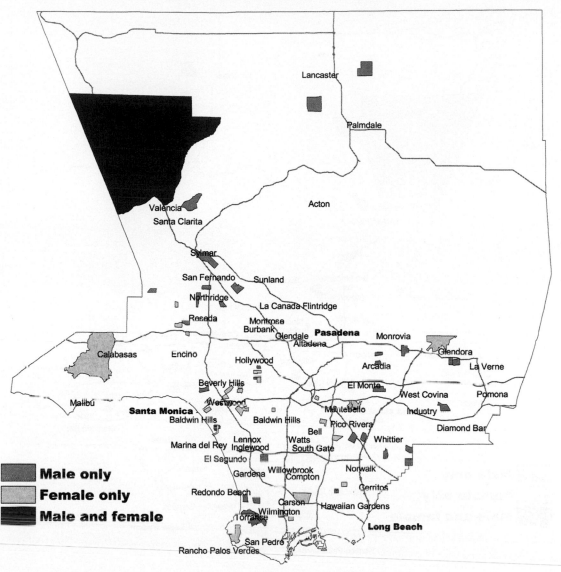

Figure 10: Male-female correlation between the relative risks for high-risk census tracts.

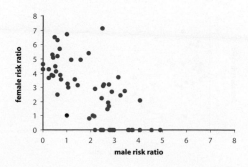

Figure 11: Map of census tracts at high risk, adjusted for social class.

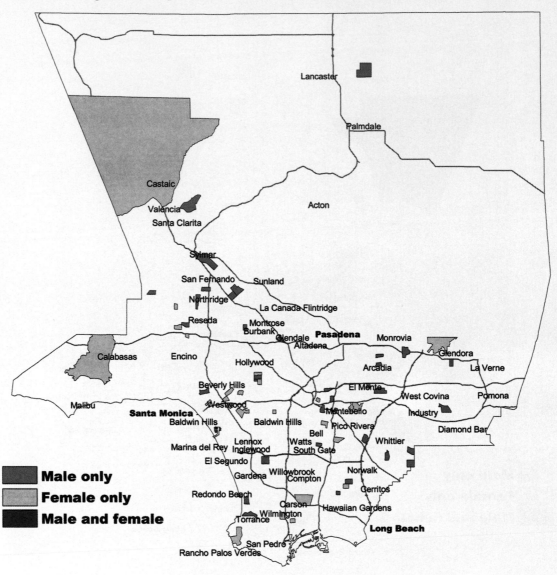

Figure 12: Male-female correlation between the relative risks for high-risk census tracts, adjusted for social class.

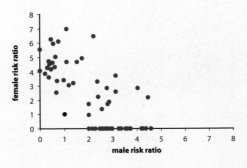

Adenocarcinoma of the (Lower) Stomach

ICDO-2 Code Anatomic Site: C 16.1–16.9
ICDO-2 Code Histology: 8000–8560, 8570–8573
Age: All
Male Cases: 9070
Female Cases: 7106

Background

Stomach cancer has long been thought to be a result of dietary factors, including excessive consumption of salted foods, starches, and foods containing nitrites, and deficiencies of fruits and vegetables. More recently it has become apparent that the single most important cause is a bacteria, *Helicobacter pylori*, which is acquired early in life, probably from family members.

Local Pattern

Gastric adenocarcinoma increases in frequency generally with age and is twice as common among men as women. It occurs with equal frequency in Los Angeles County and most areas of the country, other than Utah, where incidence is lower. Incidence rates among white men and women are lower than among men and women of the other three common racial/ethnic groups. Incidence is somewhat higher among persons of lower social class. The occurrence of this malignancy has gradually decreased, both throughout the county as a whole and among the residents of high-risk census tracts. Figure 6 shows a moderate nonrandom excess of census tracts with unexpectedly few or unexpectedly many cases. High-risk census tracts are densely concentrated in South Central Los Angeles, East Los Angeles, the west part of San Gabriel Valley, Koreatown, and Chinatown. After adjustment for social class, some of these census tracts are excluded, but aggregations of contiguous census tracts are still seen in South Central Los Angeles, East Los Angeles, and the San Gabriel Valley.

Thumbnail Interpretation

Stomach cancer has long been known to occur more commonly among men, persons born in developing countries, and persons of lower social class. The current conventional explanation for this pattern is that the causative organism, *H. pylori*, spreads more easily within families under conditions of reduced hygiene. The observed pattern of occurrence is consistent with that explanation, because the census tracts at high-risk tend to be those with large numbers of immigrant Latinos and Asian-Americans. The reason for the higher risk among men is unknown.

Figure 1: Age-adjusted incidence rate by place.

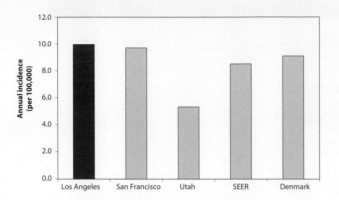

Figure 2: Age-adjusted incidence rate by age and race/ethnicity.

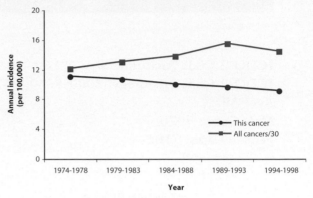

Figure 3: Age-adjusted incidence rate over the period.

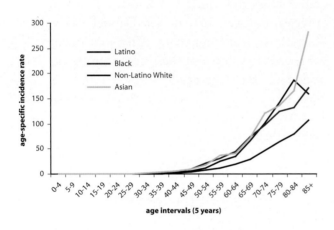

Figure 4: Age-adjusted incidence rate by social class.

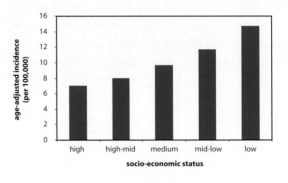

Figure 5: Distribution of the relative risk values for all census tracts.

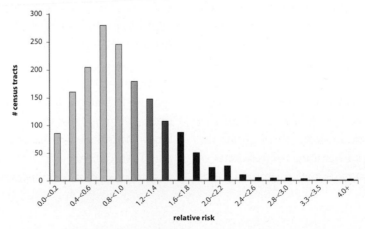

Figure 6: Census tracts by the number of cases per tract.

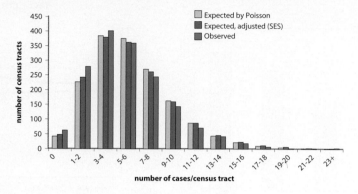

Figure 7a and b: Census tracts at high risk by the number of cases. (a) Unadjusted and (b) adjusted for social class.

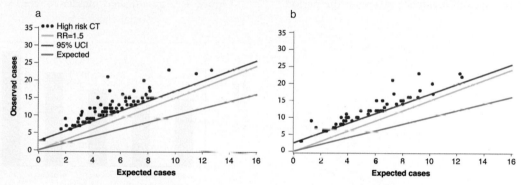

Figure 8: Risk over the period for high-risk census tracts relative to all census tracts.

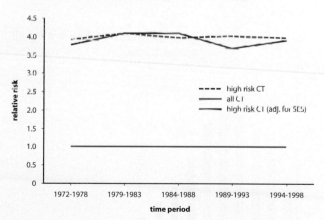

Figure 1: Age-adjusted incidence rate by place.

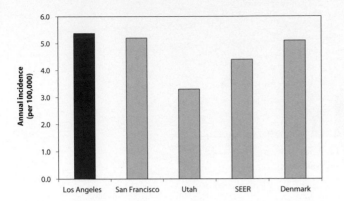

Figure 2: Age-adjusted incidence rate by age and race/ethnicity.

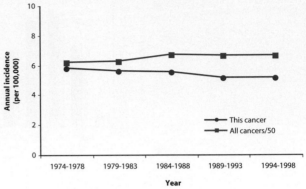

Figure 3: Age-adjusted incidence rate over the period.

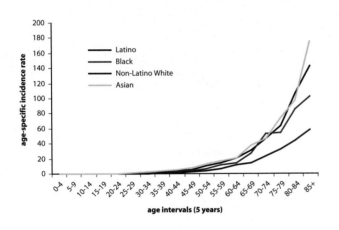

Figure 4: Age-adjusted incidence rate by social class.

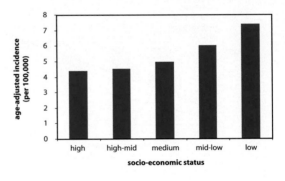

Figure 5: Distribution of the relative risk values for all census tracts.

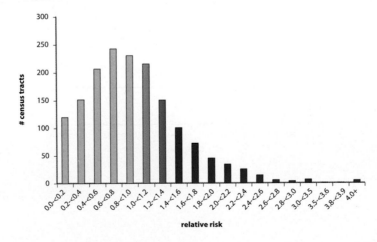

Figure 6: Census tracts by the number of cases per tract.

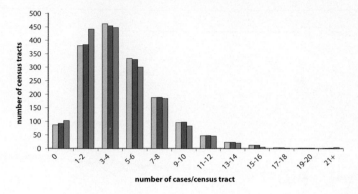

Figure 7a and b: Census tracts at high risk by the number of cases. (a) Unadjusted and (b) adjusted for social class.

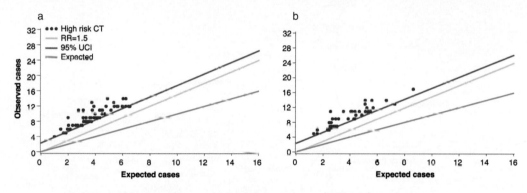

Figure 8: Risk over the period for high-risk census tracts relative to all census tracts.

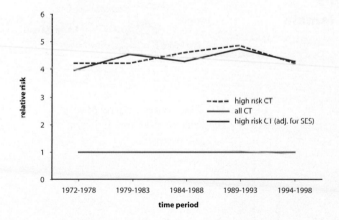

Figure 9: Map of census tracts at high risk.

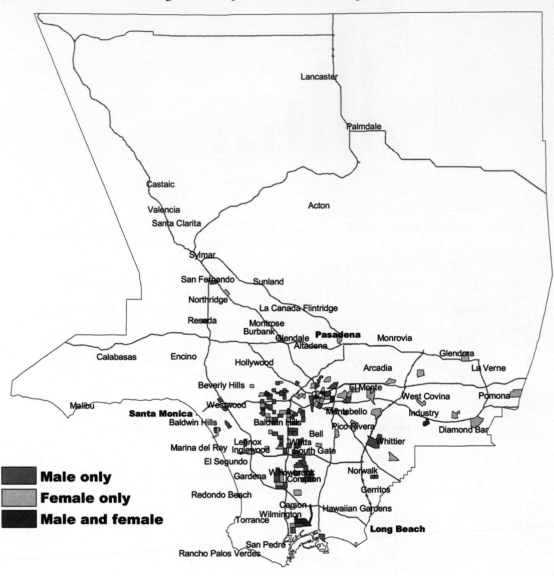

Figure 10: Male-female correlation between the relative risks for high-risk census tracts.

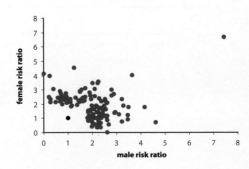

Figure 11: Map of census tracts at high risk, adjusted for social class.

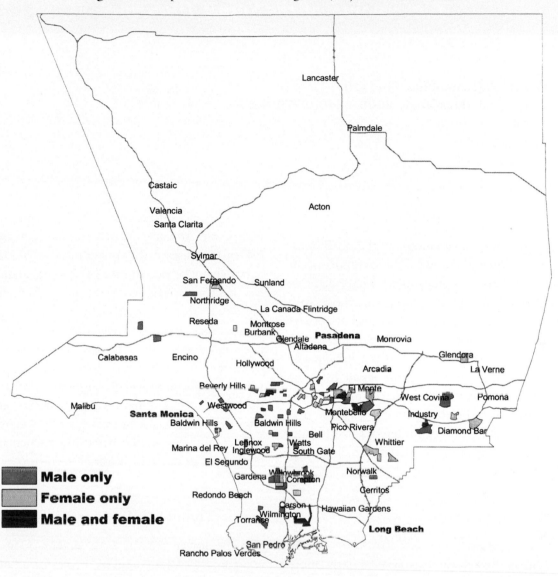

Figure 12: Male-female correlation between the relative risks for high risk census tracts, adjusted for social class.

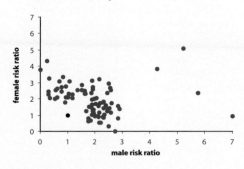

Carcinoma of the Small Intestine

ICDO-2 Code Anatomic Site: C 17
ICDO-2 Code Histology: 8000–8560, 8570–8573
Age: All
Male Cases: 570
Female Cases: 507

Background

The causes of carcinoma of the small intestine are largely unknown. These tumors occur more commonly in persons who have had carcinoma of the large intestine, in those with certain rare genetic conditions, and in persons with Crohn's disease, an autoimmune condition of the small intestine that is at least partly under genetic control. Most small intestine carcinomas occur in the region of the small intestine near the ducts that empty bile and pancreatic fluid into the lumen of the intestine, leading to the suspicion that these fluids play a causal role.

Local Pattern

Small bowel carcinoma is slightly more common in men than women. In contrast to carcinomas of the large bowel, those of the small bowel have been diagnosed less frequently in Los Angeles County than in other parts of the country, especially in comparison with Utah. African-Americans are at increased risk of this disease. No relation to social class is evident. The occurrence of this malignancy has been relatively constant in the county. Figure 6 shows no nonrandom excess of census tracts with unexpectedly few or unexpectedly many cases. Relatively few census tracts meet the high-risk criteria, and they are scattered throughout the county with no contiguous tract combinations, both before and after adjustment for social class. No census tract stands out on the basis of a particularly high number of excess cases.

Thumbnail Interpretation

The lower risk in Los Angeles County compared to Utah suggests a role for a habit more common among the members of Latter-day Saint congregations. However, the specific reason for that difference, or for any feature of the local pattern, is unknown.

Figure 1: Age-adjusted incidence rate by place.

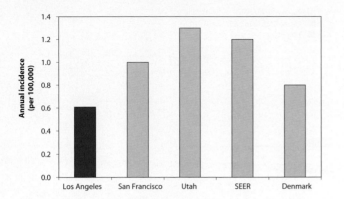

Figure 2: Age-adjusted incidence rate over the period.

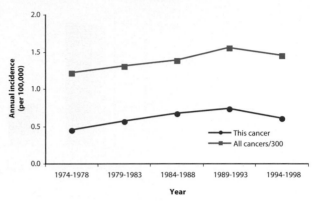

Figure 3: Age-adjusted incidence rate by age and race/ethnicity.

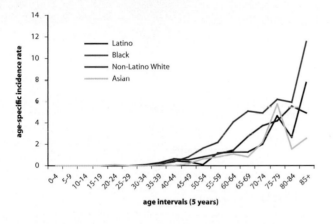

Figure 4: Age-adjusted incidence rate by social class.

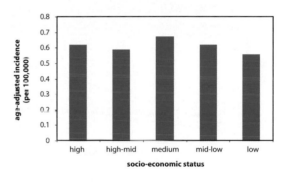

Figure 5: Distribution of the relative risk values for all census tracts.

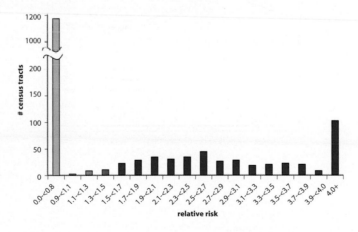

Figure 6: Census tracts by the number of cases per tract.

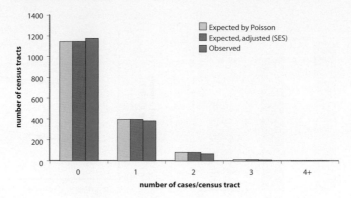

Figure 7a and b: Census tracts at high risk by the number of cases. (a) Unadjusted and (b) adjusted for social class.

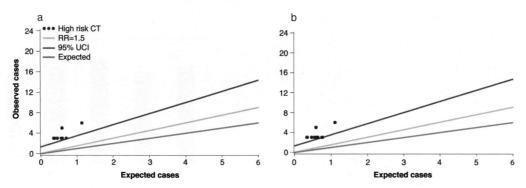

Figure 8: Risk over the period for high-risk census tracts relative to all census tracts.

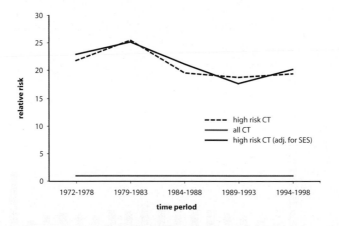

Figure 1: Age-adjusted incidence rate by place.

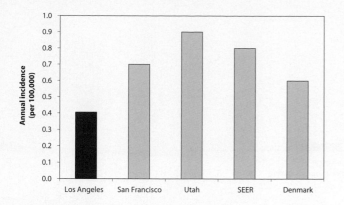

Figure 2: Age-adjusted incidence rate over the period.

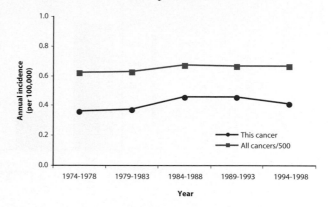

Figure 3: Age-adjusted incidence rate by age and race/ethnicity.

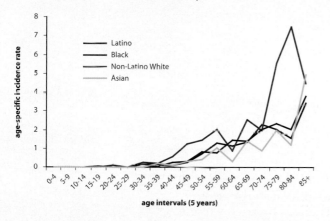

Figure 4: Age-adjusted incidence rate by social class.

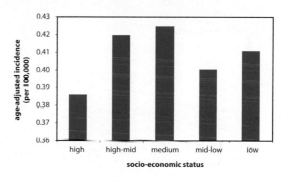

Figure 5: Distribution of the relative risk values for all census tracts.

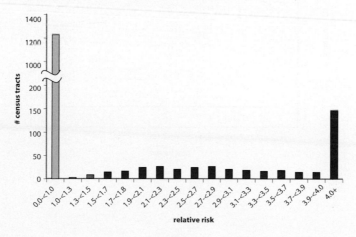

Figure 6: Census tracts by the number of cases per tract.

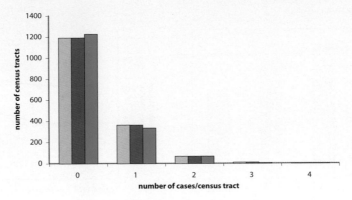

Figure 7a and b: Census tracts at high risk by the number of cases. (a) Unadjusted and (b) adjusted for social class.

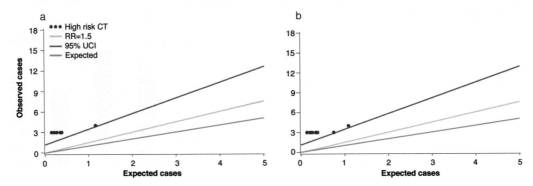

Figure 8: Risk over the period for high-risk census tracts relative to all census tracts.

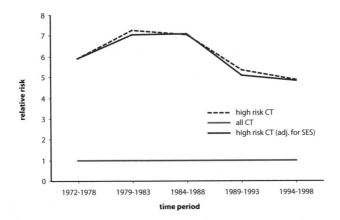

Figure 9: Map of census tracts at high risk.

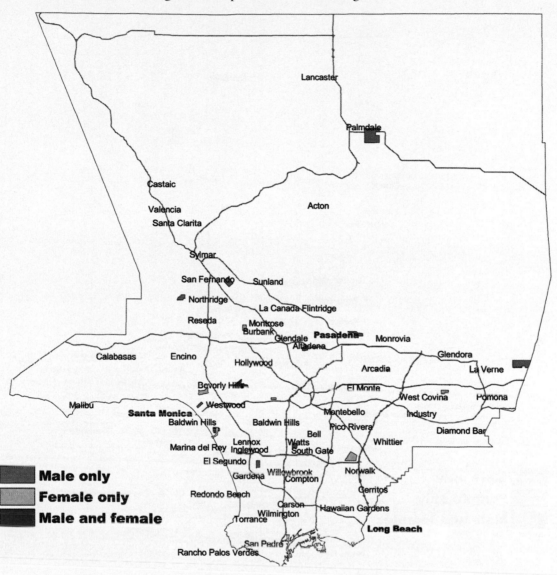

Figure 10: Male-female correlation between the relative risks for high-risk census tracts.

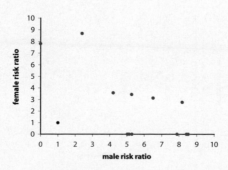

Figure 11: Map of census tracts at high risk, adjusted for social class.

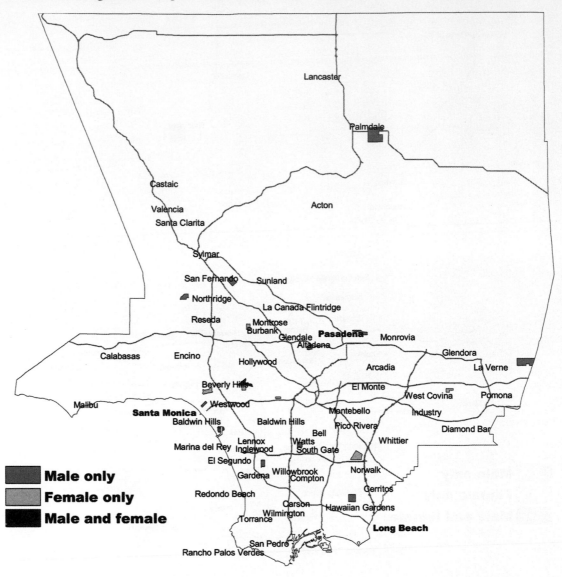

Figure 12: Male-female correlation between the relative risks for high-risk census tracts, adjusted for social class.

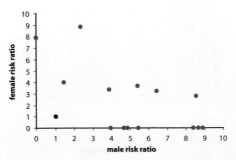

Colon Carcinoma, Total (Both Upper and Sigmoid)*

ICDO-2 Code Anatomic Site: C 18
ICDO-2 Code Histology: 8000–8560, 8570–8573
Age: All
Male Cases: 31885
Female Cases: 35680

Background

A small proportion of colon cancers are clearly genetic in origin, either based on membership in one of the rare families who develop the colonic polyps that develop into carcinomas, or based on the somewhat more common genetically determined "nonpolyposis" colon cancer (probably still accounting for less than 5% of all cases). Diet has long been suspected to be an important element in causation, based on international comparisons and the recognition that adult immigrants from developing countries tend to be at lower risk. However, while careful investigators have linked more animal fat and fewer dietary vegetables to colon carcinoma in some studies, the relationships have not been fully verified. Some inconsistent dietary evidence also suggests that consumption of calcium or dietary fiber offers protection, and that consumption of well-done or barbecued red meat is associated with higher risk. Lower risk has been consistantly linked to physical exercise, regular use of aspirin, and use of hormones at menopause have consistently appeared to lower risk. Ionizing radiation and alcohol consumption have both been found associated with subsequent colon carcinoma, but play a very small role. Because polyps and adenomas can be detected by screening before they develop into carcinomas, some persons may go on to develop colon cancers only because of inadequate access to preventive medical services.

Local Pattern

Incidence of colon cancer is somewhat higher among men than among women, and is about as common in Los Angeles County as in other places. There is little difference by social class and only a minor decrease in incidence over time, both generally and among the residents of high-risk census tracts. Risk among whites and African Americans is identical, and is higher than that among Latinos and Asian-American residents, both of which groups include a large proportion of immigrants. Figure 6 shows only a slight non-random excess of census tracts with unexpectedly few or unexpectedly many cases. A small number of census tracts meet the formal high-risk criteria, none by any substantial margin. The census tracts fulfilling the high-risk criteria are scattered over the county. There are few contiguous high-risk census tracts, and a small number show high risk for both males and

*Bolded entries in Table of Contents combine several subgroups in a given organ or all those at a given age.

females. No census tract stands out on the basis of a particularly high relative risk and number of excess cases. After adjustment for social class, a modest concentration of high-risk census tracts appears among the neighborhoods of South Central Los Angeles.

Thumbnail Interpretation

This malignancy is more evenly distributed among the population of Los Angeles County than is any other common tumor. Nonetheless, the evident lower risk to Latinos and Asian-Americans and the number of census tracts with unexpectedly few cases may indicate that census tracts with large numbers of low-risk immigrants have been relatively spared. The reason for the cluster of high-risk neighborhoods in South Central Los Angeles is unknown, although it is consistent with a local maldistribution of dietary factors hypothesized to be either causal or protective for colon cancer.

Colon Carcinoma, Total (Both Upper and Sigmoid): Male

Figure 1: Age-adjusted incidence rate by place.

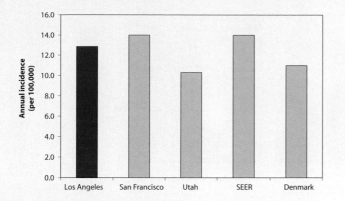

Figure 2: Age-adjusted incidence rate over the period.

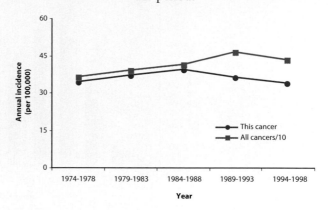

Figure 3: Age-adjusted incidence rate by age and race/ethnicity.

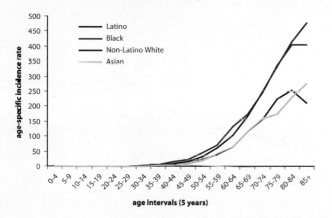

Figure 4: Age-adjusted incidence rate by social class.

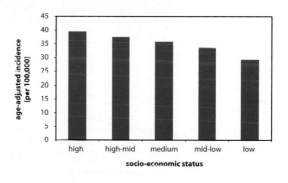

Figure 5: Distribution of the relative risk values for all census tracts.

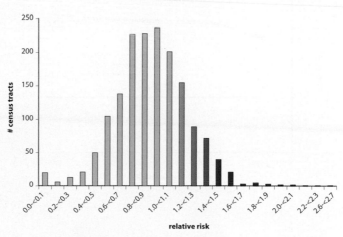

Figure 6: Census tracts by the number of cases per tract.

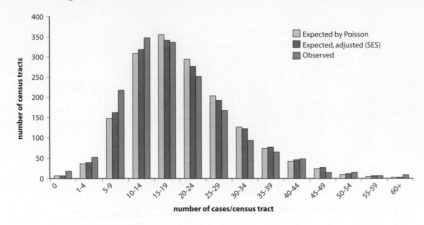

Figure 7a and b: Census tracts at high risk by the number of cases. (a) Unadjusted and (b) adjusted for social class.

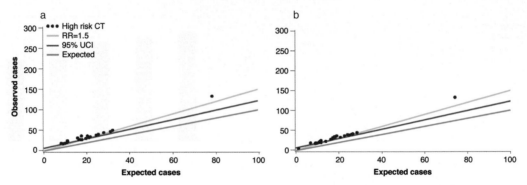

Figure 8: Risk over the period for high-risk census tracts relative to all census tracts.

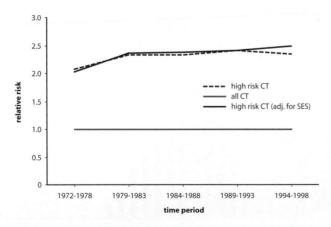

Colon Carcinoma, Total (Both Upper and Sigmoid): Female

Figure 1: Age-adjusted incidence rate by place.

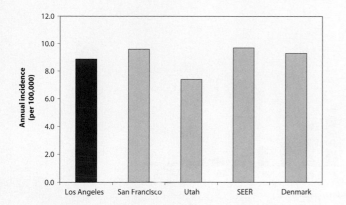

Figure 2: Age-adjusted incidence rate over the period.

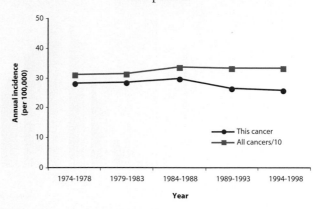

Figure 3: Age-adjusted incidence rate by age and race/ethnicity.

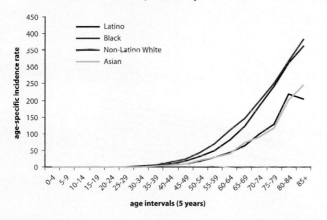

Figure 4: Age-adjusted incidence rate by social class.

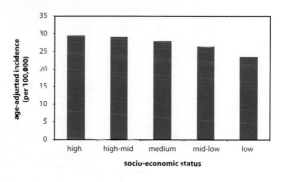

Figure 5: Distribution of the relative risk values for all census tracts.

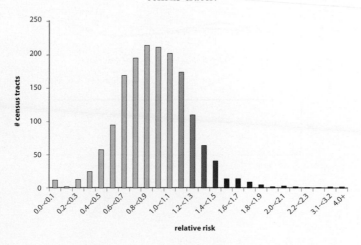

Colon Carcinoma, Total (Both Upper and Sigmoid): Female

Figure 6: Census tracts by the number of cases per tract.

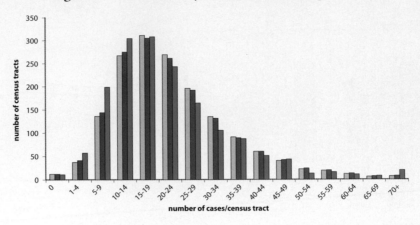

Figure 7a and b: Census tracts at high risk by the number of cases. (a) Unadjusted and (b) adjusted for social class.

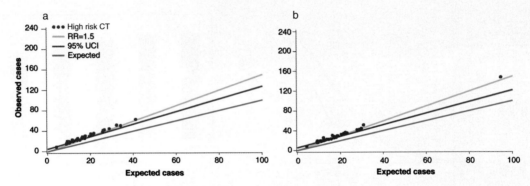

Figure 8: Risk over the period for high-risk census tracts relative to all census tracts.

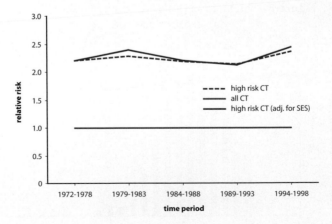

Figure 9: Map of census tracts at high risk.

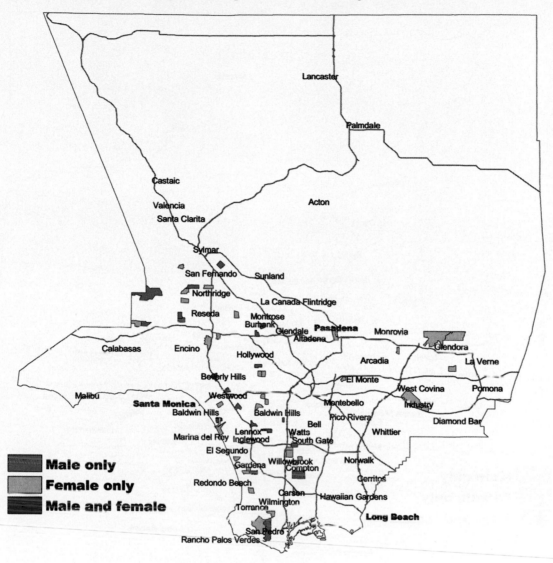

Figure 10: Male-female correlation between the relative risks for high-risk census tracts.

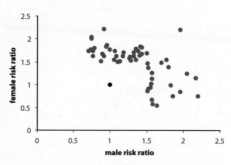

Colon Carcinoma, Total (Both Upper and Sigmoid)

Figure 11: Map of census tracts at high risk, adjusted for social class.

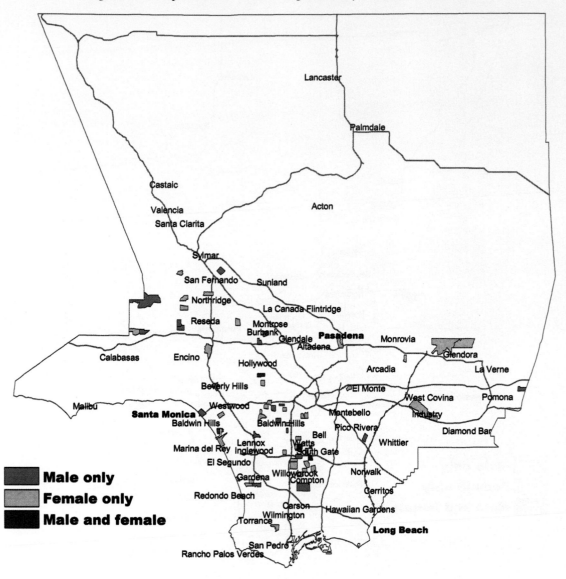

Figure 12: Male-female correlation between the relative risks for high-risk census tracts, adjusted for social class.

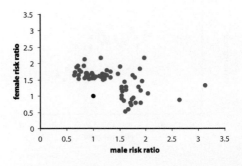

Carcinoma of the Upper Colon

ICDO-2 Code Anatomic Site: C 18.0–18.6
ICDO-2 Code Histology: 8000–8560, 8570–8573
Age: All
Male Cases: 18418
Female Cases: 22081

Background

See colon carcinoma total. Each of the statements made there are also pertinent to carcinoma of the upper colon. The colon was separated into two parts because the function of the organ changes over the course of the colon, and because there are some differences between the causes of carcinoma of the rectum, the latter joined to the lower end of the colon, and the colon generally.

Local Pattern

Cancers of the upper colon increase in frequency with age and are slightly more common among men than women. They occur in Los Angeles County and other regions with roughly equal frequency. These malignancies are more common in whites and African-Americans than in the other groups. They have occurred with a constant incidence over the period, both in the county as a whole and among the residents of high-risk census tracts. They are slightly more common among men of higher social class. Figure 6 shows a moderate nonrandom excess of census tracts with unexpectedly few or unexpectedly many cases. High-risk census tracts are scattered throughout the county, with small contiguous clusters and occasional census tracts showing high risk for both sexes. Before and after adjustment for social class, the only concentration of these contiguous census tracts is found in South Central Los Angeles.

Thumbnail Interpretation

Lower risk to Latinos and Asian-Americans and the evidence of census tracts with unexpectedly few cases may indicate that census tracts with large numbers of low-risk immigrants have been spared. The reason for the cluster of high-risk neighborhoods in South Central Los Angeles is unknown, although it is consistent with a local maldistribution of dietary factors hypothesized to be either causal or protective for colon cancer.

Figure 1: Age-adjusted incidence rate by place.

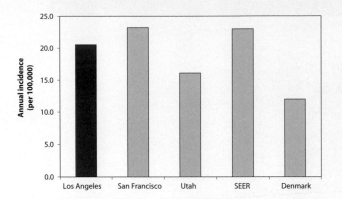

Figure 2: Age-adjusted incidence rate over the period.

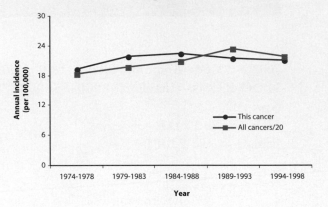

Figure 3: Age-adjusted incidence rate by age and race/ethnicity.

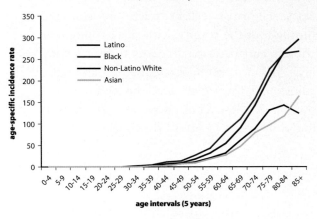

Figure 4: Age-adjusted incidence rate by social class.

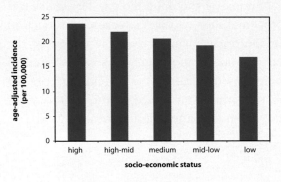

Figure 5: Distribution of the relative risk values for all census tracts.

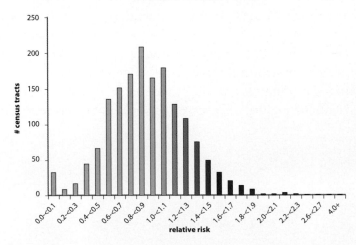

Figure 6: Census tracts by the number of cases per tract.

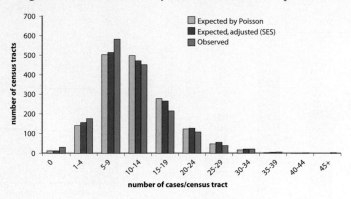

Figure 7a and b: Census tracts at high risk by the number of cases. (a) Unadjusted and (b) adjusted for social class.

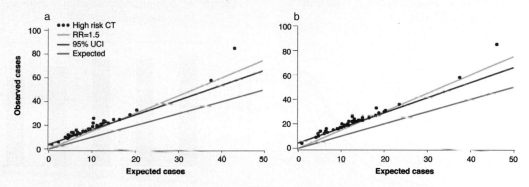

Figure 8: Risk over the period for high-risk census tracts relative to all census tracts.

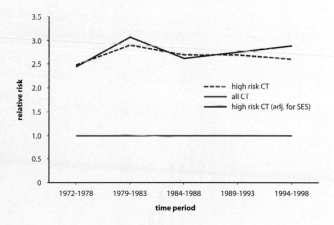

Carcinoma of the Upper Colon: Female

Figure 1: Age-adjusted incidence rate by place.

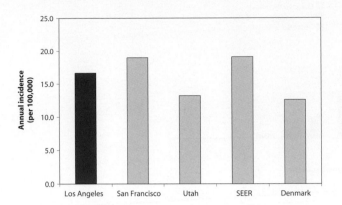

Figure 2: Age-adjusted incidence rate over the period.

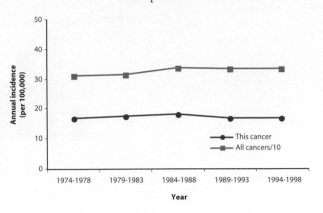

Figure 3: Age-adjusted incidence rate by age and race/ethnicity.

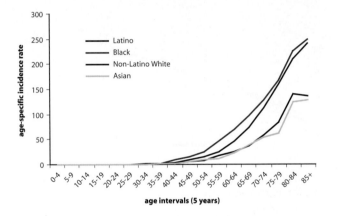

Figure 4: Age-adjusted incidence rate by social class.

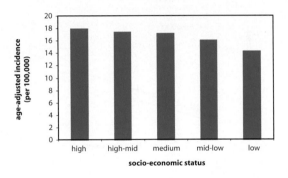

Figure 5: Distribution of the relative risk values for all census tracts.

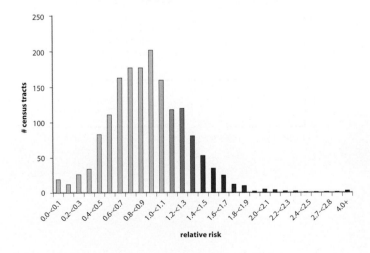

Figure 6: Census tracts by the number of cases per tract.

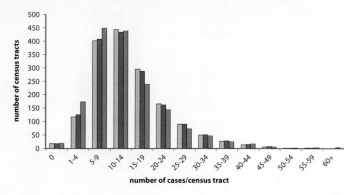

Figure 7a and b: Census tracts at high risk by the number of cases. (a) Unadjusted and (b) adjusted for social class.

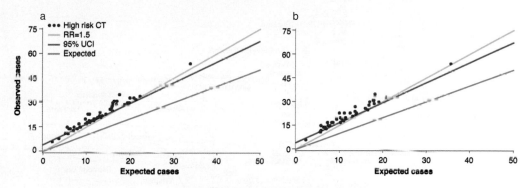

Figure 8: Risk over the period for high-risk census tracts relative to all census tracts.

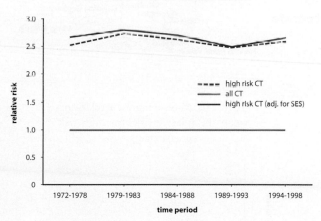

Figure 9: Map of census tracts at high risk.

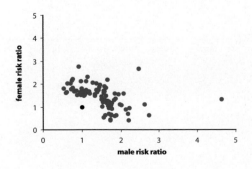

Figure 10: Male-female correlation between the relative risks for high-risk census tracts.

Figure 11: Map of census tracts at high risk, adjusted for social class.

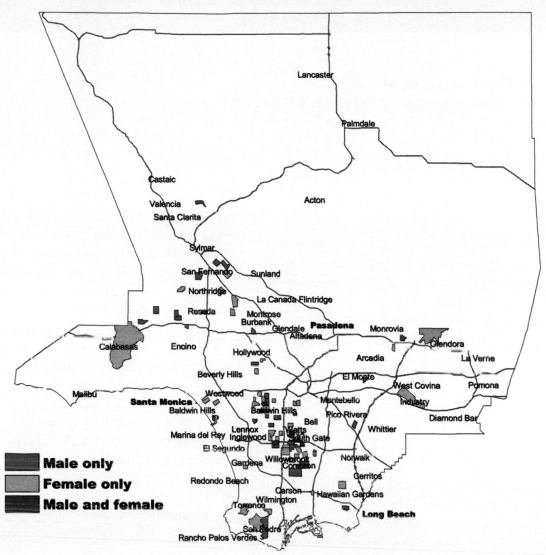

Figure 12: Male-female correlation between the relative risks for high-risk census tracts, adjusted for social class.

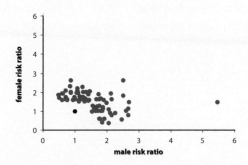

Carcinoma of the Sigmoid Colon

ICDO-2 Code Anatomic Site: C 18.7
ICDO-2 Code Histology: 8000–8560, 8570–8573
Age: All
Male Cases: 11610
Female Cases: 11205

Background

See colon carcinoma total. Each of the statements made there are also pertinent to carcinoma of the sigmoid colon. We have described the two parts of the colon separately because they differ slightly in function, and because of evidence of differential causation between the colon and the rectum, anatomically and physiologically contiguous to the sigmoid colon.

Local Pattern

Cancers of the sigmoid colon are more common than those of the upper colon, increase in frequency with advancing age, and appear slightly more often in men than in women. They occur in Los Angeles County and other regions with roughly equal frequency. Sigmoid cancers, in contrast to those of the upper colon, are only slightly more common among whites and African-Americans than among others, and slightly less common among those of the lowest social class. They have occurred with constant incidence over the period, both in the county as a whole and among the residents of high-risk census tracts. Figure 6 shows a moderate nonrandom excess of census tracts with unexpectedly few or unexpectedly many cases. High-risk census tracts are scattered throughout the county. There are few contiguous clusters of them and few census tracts at high risk for the members of both sexes. No census tract stands out on the basis of a particularly high relative risk and number of excess cases. There is no concentration in South Central Los Angeles.

Thumbnail Interpretation

The pattern of occurrence of this malignancy is intermediate between that of the upper colon and that of carcinoma of the rectum. This is evident in the relative occurrence by sex, social class, racial/ethnic group, and by the absence of any systematic geographical distribution.

Carcinoma of the Sigmoid Colon: Male

Figure 1: Age-adjusted incidence rate by place.

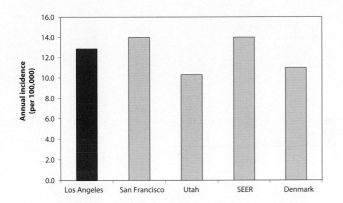

Figure 2: Age-adjusted incidence rate over the period.

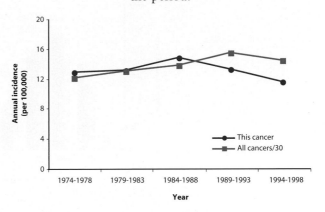

Figure 3: Age-adjusted incidence rate by age and race/ethnicity.

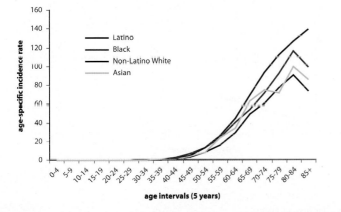

Figure 4: Age-adjusted incidence rate by social class.

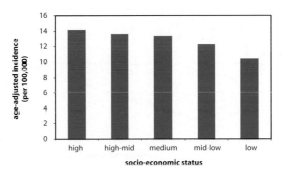

Figure 5: Distribution of the relative risk values for all census tracts.

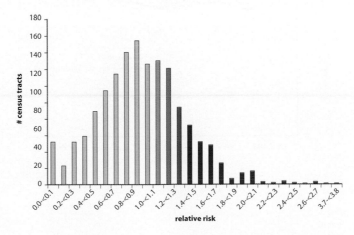

Figure 6: Census tracts by the number of cases per tract.

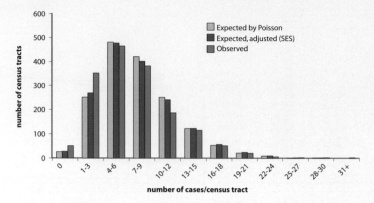

Figure 7a and b: Census tracts at high risk by the number of cases. (a) Unadjusted and (b) adjusted for social class.

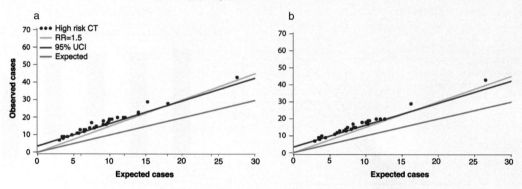

Figure 8: Risk over the period for high-risk census tracts relative to all census tracts.

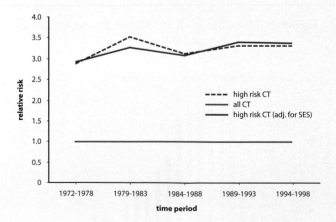

Carcinoma of the Sigmoid Colon: Female

Figure 1: Age-adjusted incidence rate by place.

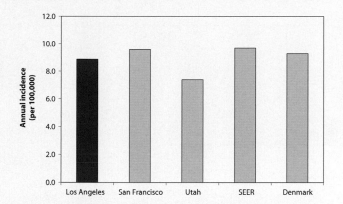

Figure 2: Age-adjusted incidence rate over the period.

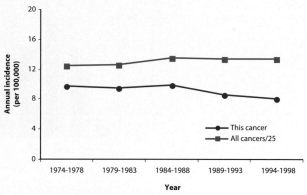

Figure 3: Age-adjusted incidence rate by age and race/ethnicity.

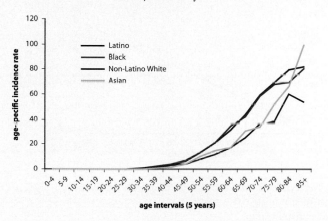

Figure 4: Age-adjusted incidence rate by social class.

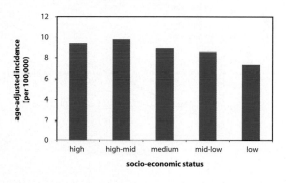

Figure 5: Distribution of the relative risk values for all census tracts.

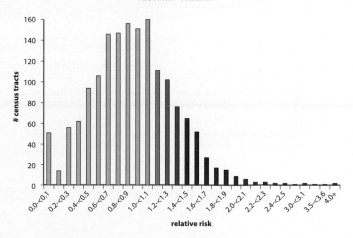

Figure 6: Census tracts by the number of cases per tract.

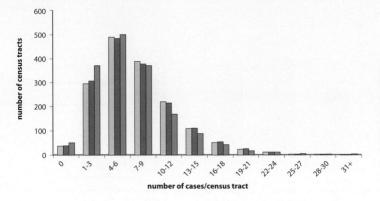

Figure 7a and b: Census tracts at high risk by the number of cases. (a) Unadjusted and (b) adjusted for social class.

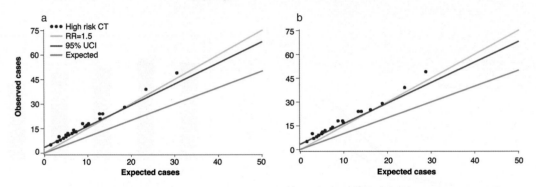

Figure 8: Risk over the period for high-risk census tracts relative to all census tracts.

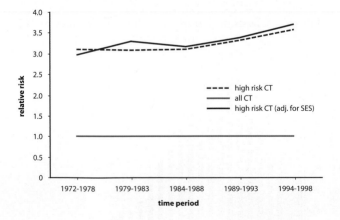

Figure 9: Map of census tracts at high risk.

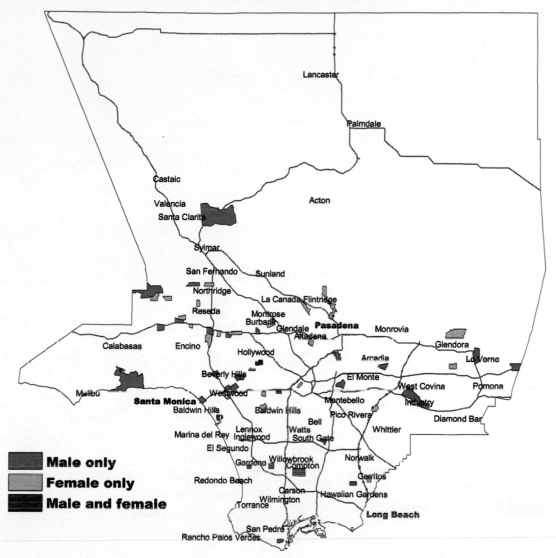

Figure 10: Male-female correlation between the relative risks for high-risk census tracts.

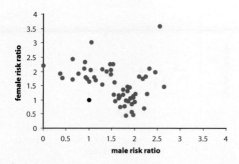

Figure 11: Map of census tracts at high risk, adjusted for social class.

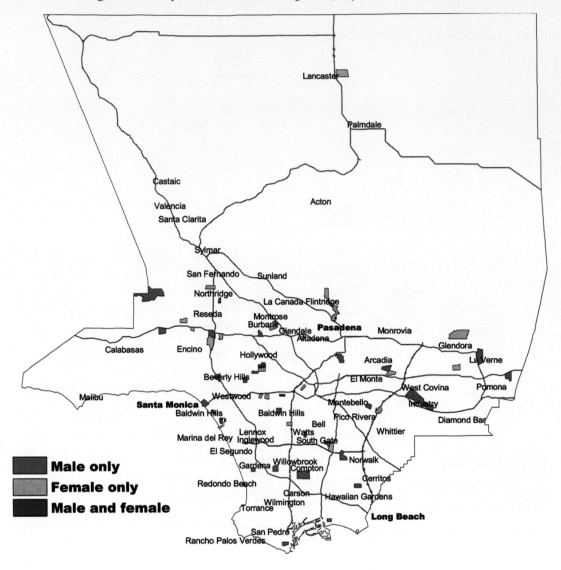

Male only
Female only
Male and female

Figure 12: Male-female correlation between the relative risks for high-risk census tracts, adjusted for social class.

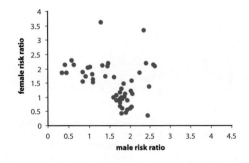

Carcinoma of the Rectum

ICDO-2 Code Anatomic Site: C 19, 20
ICDO-2 Code Histology: 8000–8560, 8570–8573
Age: All
Male Cases: 14250
Female Cases: 11894

Background

See colon carcinoma. Most of the statements made there are also pertinent to carcinoma of the rectum. There exists some evidence that smoking and alcohol consumption, especially heavy beer drinking, play a larger role in the causation of rectal cancer than that of colon cancer generally. In addition, there is no evidence that either physical activity or hormone use, each of which seems to prevent colon cancer, prevent rectal cancer.

Local Pattern

Although incidence also increases greatly with advancing age, rectal cancer occurs with a pattern that differs from that of colon cancer. Rectal malignancies are twice as common among men as women, and occur in Los Angeles County more commonly than in other regions, especially Utah. They are only slightly more common in whites, and they have decreased slightly in frequency with time, both in the county as a whole and among the residents of high-risk census tracts. There is little association with social class. Figure 6 shows a slight nonrandom excess of census tracts with unexpectedly few or unexpectedly many cases. High-risk census tracts are scattered throughout the county before and after adjustment for social class, with few contiguous clusters of census tracts and few census tracts at high risk for both sexes. No census tract stands out on the basis of a particularly high number of excess cases.

Thumbnail Interpretation

The excess risk among men is consistent with the hypothesized relation to beer consumption, but is also consistent with many other possible explanations, from occupational exposure to the hormonal differences between the sexes. This malignancy appears to be evenly distributed among the residents of Los Angeles County, and the pattern of high-risk census tracts does not correspond to any obvious known pattern of exposure.

Figure 1: Age-adjusted incidence rate by place.

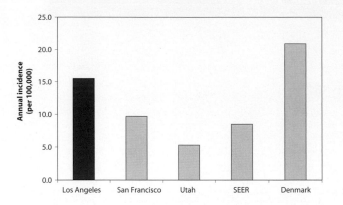

Figure 2: Age-adjusted incidence rate over the period.

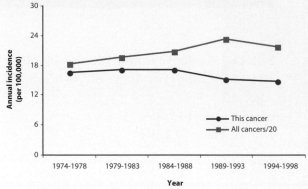

Figure 3: Age-adjusted incidence rate by age and race/ethnicity.

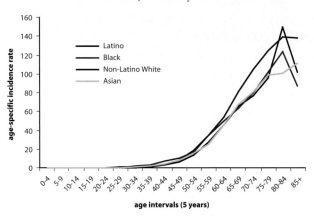

Figure 4: Age-adjusted incidence rate by social class.

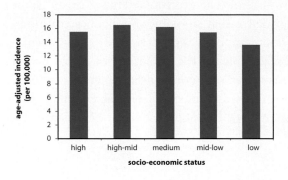

Figure 5: Distribution of the relative risk values for all census tracts.

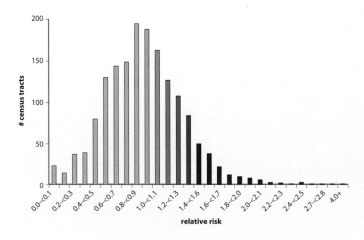

Figure 6: Census tracts by the number of cases per tract.

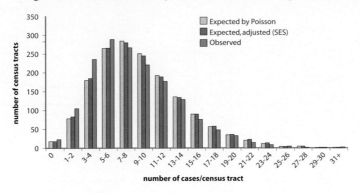

Figure 7a and b: Census tracts at high risk by the number of cases. (a) Unadjusted and (b) adjusted for social class.

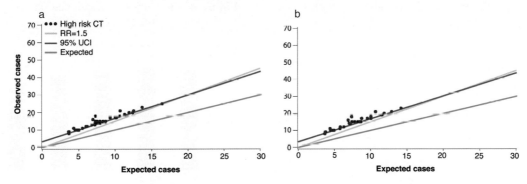

Figure 8: Risk over the period for high-risk census tracts relative to all census tracts.

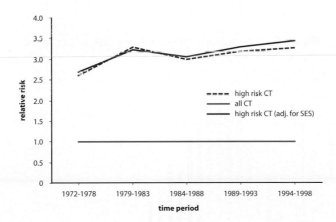

Figure 1: Age-adjusted incidence rate by place.

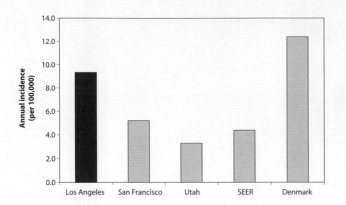

Figure 2: Age-adjusted incidence rate over the period.

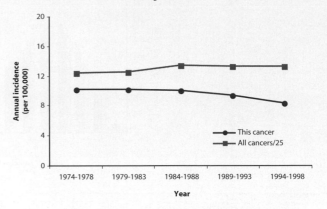

Figure 3: Age-adjusted incidence rate by age and race/ethnicity.

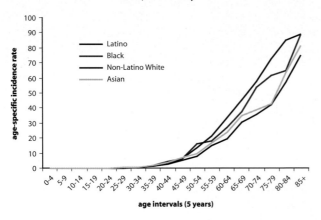

Figure 4: Age-adjusted incidence rate by social class.

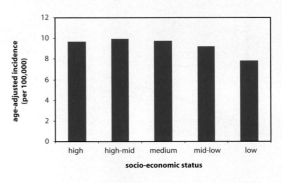

Figure 5: Distribution of the relative risk values for all census tracts.

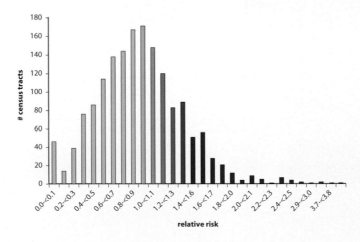

Figure 6: Census tracts by the number of cases per tract.

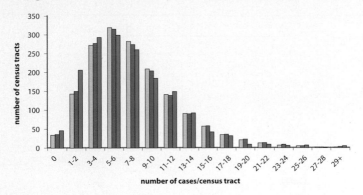

Figure 7a and b: Census tracts at high risk by the number of cases. (a) Unadjusted and (b) adjusted for social class.

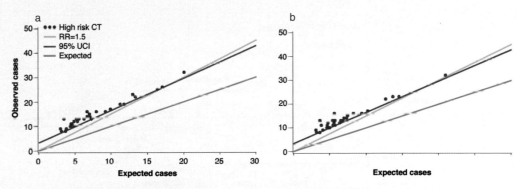

Figure 8: Risk over the period for high-risk census tracts relative to all census tracts.

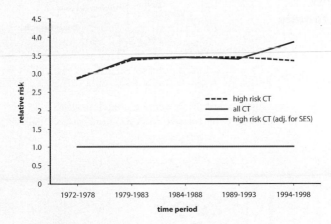

Figure 9: Map of census tracts at high risk.

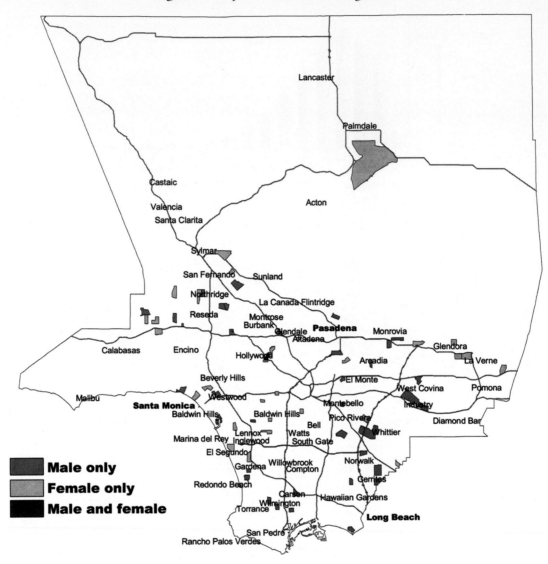

Figure 10: Male-female correlation between the relative risks for high-risk census tracts.

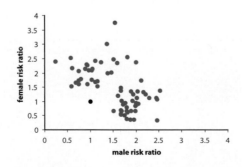

Figure 11: Map of census tracts at high risk, adjusted for social class.

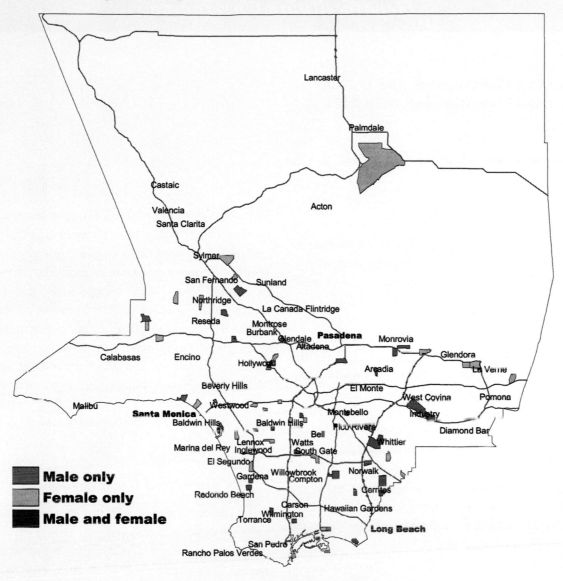

Male only
Female only
Male and female

Figure 12: Male-female correlation between the relative risks for high-risk census tracts, adjusted for social class.

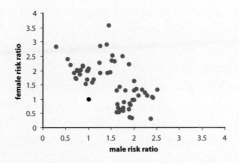

Hepatocellular Carcinoma

ICDO-2 Code Anatomic Site: C 22.0
ICDO-2 Code Histology: 8170–8171
Age: All
Male Cases: 3817
Female Cases: 1473

Background

This carcinoma (also called hepatoma) is one of the most common malignancies in the world, because of the high incidence in East Asia and Africa. It is clearly caused by early infection with the hepatitis B virus, and evidence now indicates it is also caused by infection with the hepatitis C virus, which is usually acquired in adulthood from IV drug use or infected blood products. Genetic factors may render some persons more susceptible. Other known causes include cigarette smoking, alcoholic cirrhosis, and toxins (aflatoxins) produced by fungal contaminants of stored food, especially in developing countries. Oral contraceptives cause non-malignant tumors of the liver, and rarely, carcinomas may develop from these benign growths.

Local Pattern

Carcinoma of the liver is usually a disease of older persons that is more common in Los Angeles County and San Francisco than elsewhere in this country, and is several times more common among men than women. Persons of Asian ancestry are at highest risk, followed by African-Americans and Latinos. This malignancy is increasing in frequency in Los Angeles County, especially among men, although this increase is not occurring among the residents of high-risk census tracts. Hepatoma is more common among persons of lower social class. Figure 6 shows a moderate nonrandom excess of census tracts with unexpectedly few or unexpectedly many cases. The high-risk census tracts are scattered throughout the county, with several aggregates of contiguous census tracts, both before and after adjustment for social class. These groupings are prominent in Chinatown, Koreatown, and the heavily Asian communities in the San Gabriel Valley.

Thumbnail Interpretation

From the known viral and dietary causes of hepatocellular carcinoma, one would expect that this malignancy would appear more often among immigrants from East Asia, and to a lesser degree, Latin America. Both groups are known to have more frequent exposure to hepatitis B virus. In the case of East Asians, this occurs by means of transmission from mother to child. With the exception of the excess among males, the pattern of occurrence in Los Angeles County is completely consistent with that expectation. The reason for the male excess is unknown, and might be the result of differential alcohol and tobacco consumption, or other behavioral differences.

Figure 1: Age-adjusted incidence rate by place.

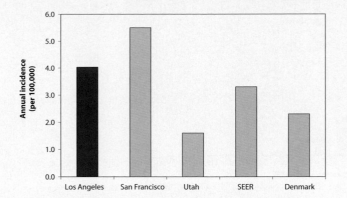

Figure 2: Age-adjusted incidence rate over the period.

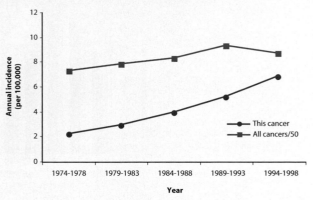

Figure 3: Age-adjusted incidence rate by age and race/ethnicity.

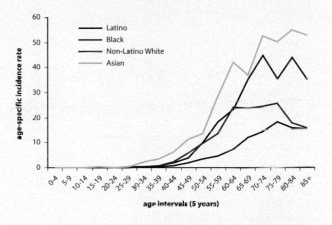

Figure 4: Age-adjusted incidence rate by social class.

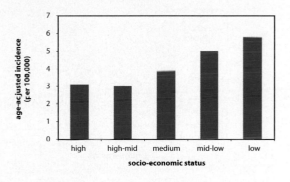

Figure 5: Distribution of the relative risk values for all census tracts.

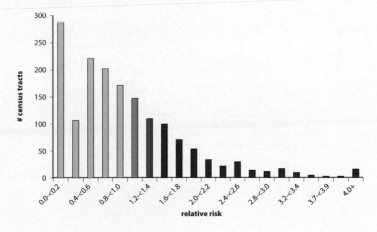

Figure 6: Census tracts by the number of cases per tract.

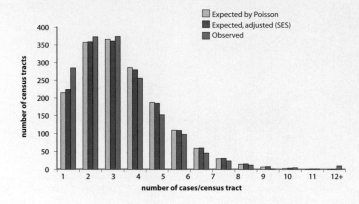

Figure 7a and b: Census tracts at high risk by the number of cases. (a) Unadjusted and (b) adjusted for social class.

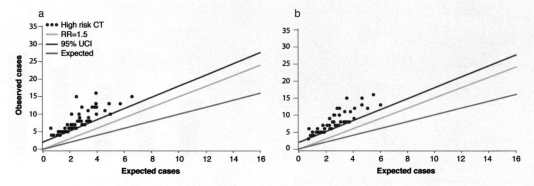

Figure 8: Risk over the period for high-risk census tracts relative to all census tracts.

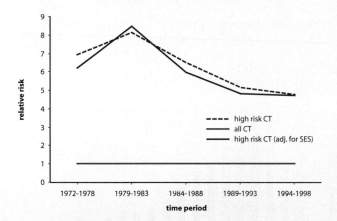

Hepatocellular Carcinoma: Female

Figure 1: Age-adjusted incidence rate by place.

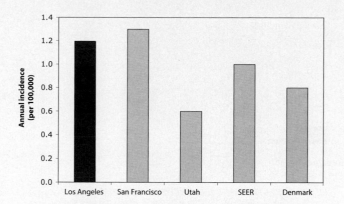

Figure 2: Age-adjusted incidence rate over the period.

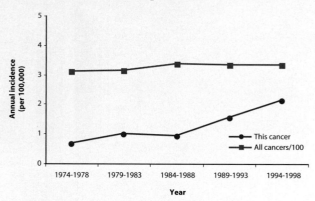

Figure 3: Age-adjusted incidence rate by age and race/ethnicity.

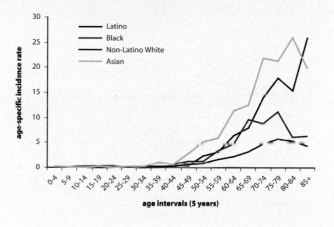

Figure 4: Age-adjusted incidence rate by social class.

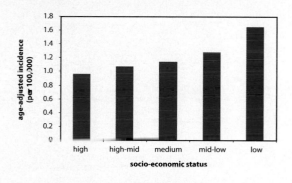

Figure 5: Distribution of the relative risk values for all census tracts.

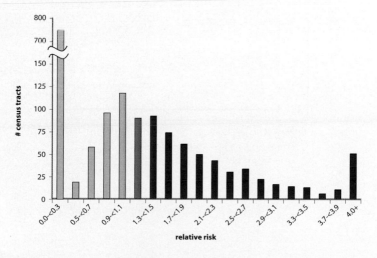

Figure 6: Census tracts by the number of cases per tract.

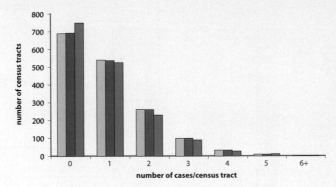

Figure 7a and b: Census tracts at high risk by the number of cases. (a) Unadjusted and (b) adjusted for social class.

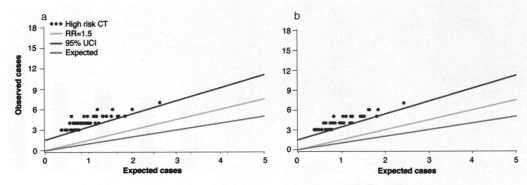

Figure 8: Risk over the period for high-risk census tracts relative to all census tracts.

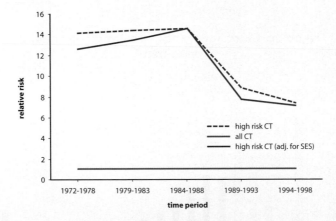

Figure 9: Map of census tracts at high risk.

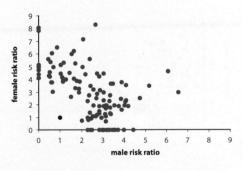

Figure 10: Male-female correlation between the relative risks for high-risk census tracts.

Figure 11: Map of census tracts at high risk, adjusted for social class.

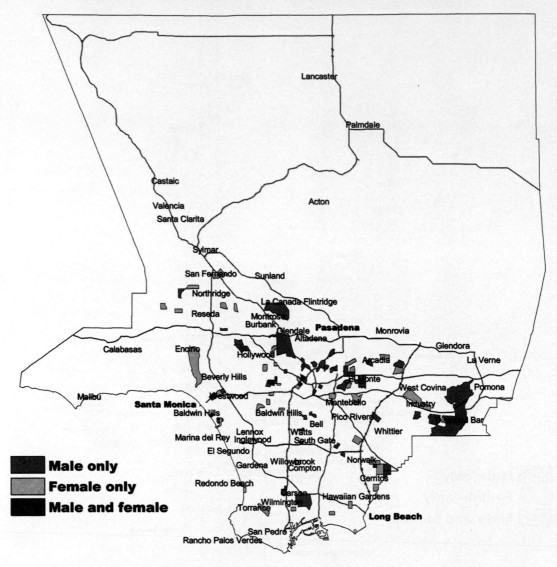

Figure 12: Male-female correlation between the relative risks for high-risk census tracts, adjusted for social class.

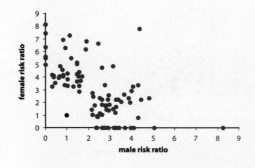

Cholangiocarcinoma

ICDO-2 Code Anatomic Site: C 22
ICDO-2 Code Histology: 8160–8162, 8180
Age: All
Male Cases: 492
Female Cases: 451

Background

Cholangiocarcinomas occur in that portion of the biliary tree still within the liver. Other than the inflammatory effects of liver fluke infestations in Southeast Asia, the causes of these cancers are not known. For unknown reasons, they appear more commonly in patients with ulcerative colitis.

Local Pattern

Cholangiocarcinomas occur slightly more frequently among men than among women, and, except for the higher rate in San Francisco, occur at about the same frequency as in other regions. They are common at advanced ages, and occur at similar rates in persons of all social classes. These malignancies have been increasing in frequency, especially in women, although incidence among the residents of high-risk census tracts has been relatively constant over time. Figure 6 shows a slight nonrandom excess of census tracts with unexpectedly few or unexpectedly many cases. The number of census tracts fulfilling the high-risk criteria is small, and no particular geographic pattern is apparent. No census tract stands out on the basis of a particularly high number of excess cases.

Thumbnail Interpretation

No systematic variation in the occurrence of these malignancies in Los Angeles County is evident, and therefore no local causes are suggested.

Figure 1: Age-adjusted incidence rate by place.

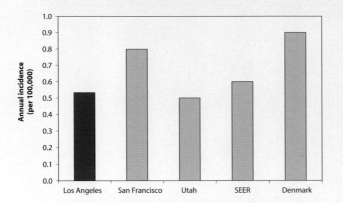

Figure 2: Age-adjusted incidence rate over the period.

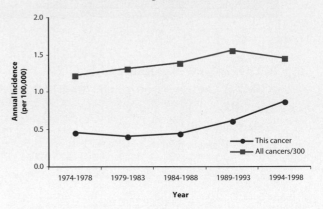

Figure 3: Age-adjusted incidence rate by age and race/ethnicity.

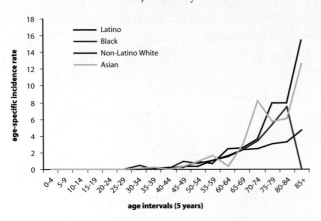

Figure 4: Age-adjusted incidence rate by social class.

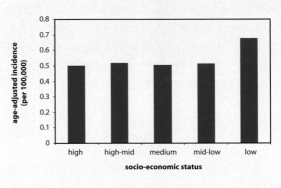

Figure 5: Distribution of the relative risk values for all census tracts.

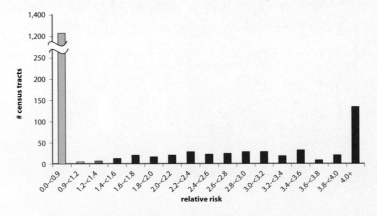

Figure 6: Census tracts by the number of cases per tract.

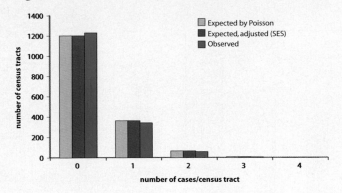

Figure 7a and b: Census tracts at high risk by the number of cases. (a) Unadjusted and (b) adjusted for social class.

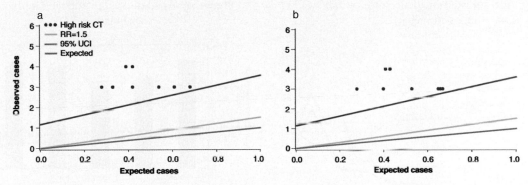

Figure 8: Risk over the period for high-risk census tracts relative to all census tracts.

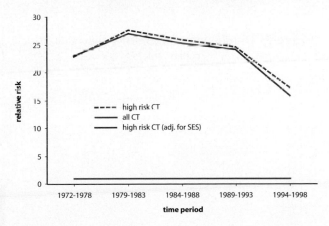

Cholangiocarcinoma: Female

Figure 1: Age-adjusted incidence rate by place.

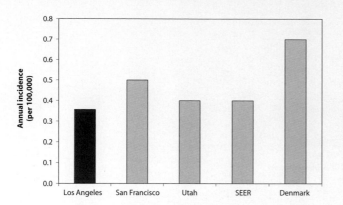

Figure 2: Age-adjusted incidence rate over the period.

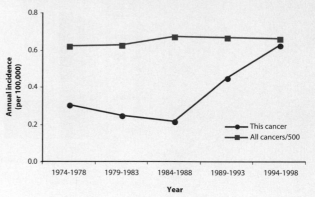

Figure 3: Age-adjusted incidence rate by age and race/ethnicity.

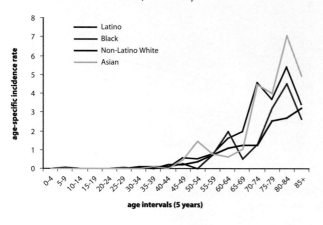

Figure 4: Age-adjusted incidence rate by social class.

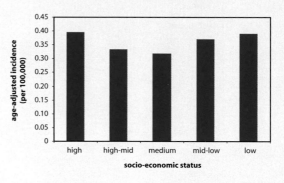

Figure 5: Distribution of the relative risk values for all census tracts.

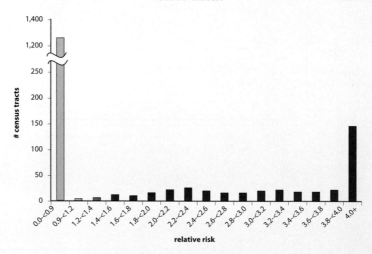

Figure 6: Census tracts by the number of cases per tract.

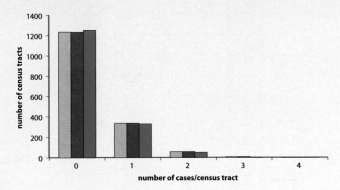

Figure 7a and b: Census tracts at high risk by the number of cases. (a) Unadjusted and (b) adjusted for social class.

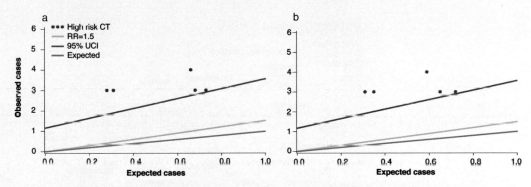

Figure 8: Risk over the period for high-risk census tracts relative to all census tracts.

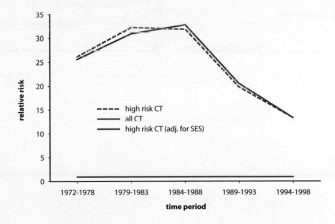

Figure 9: Map of census tracts at high risk.

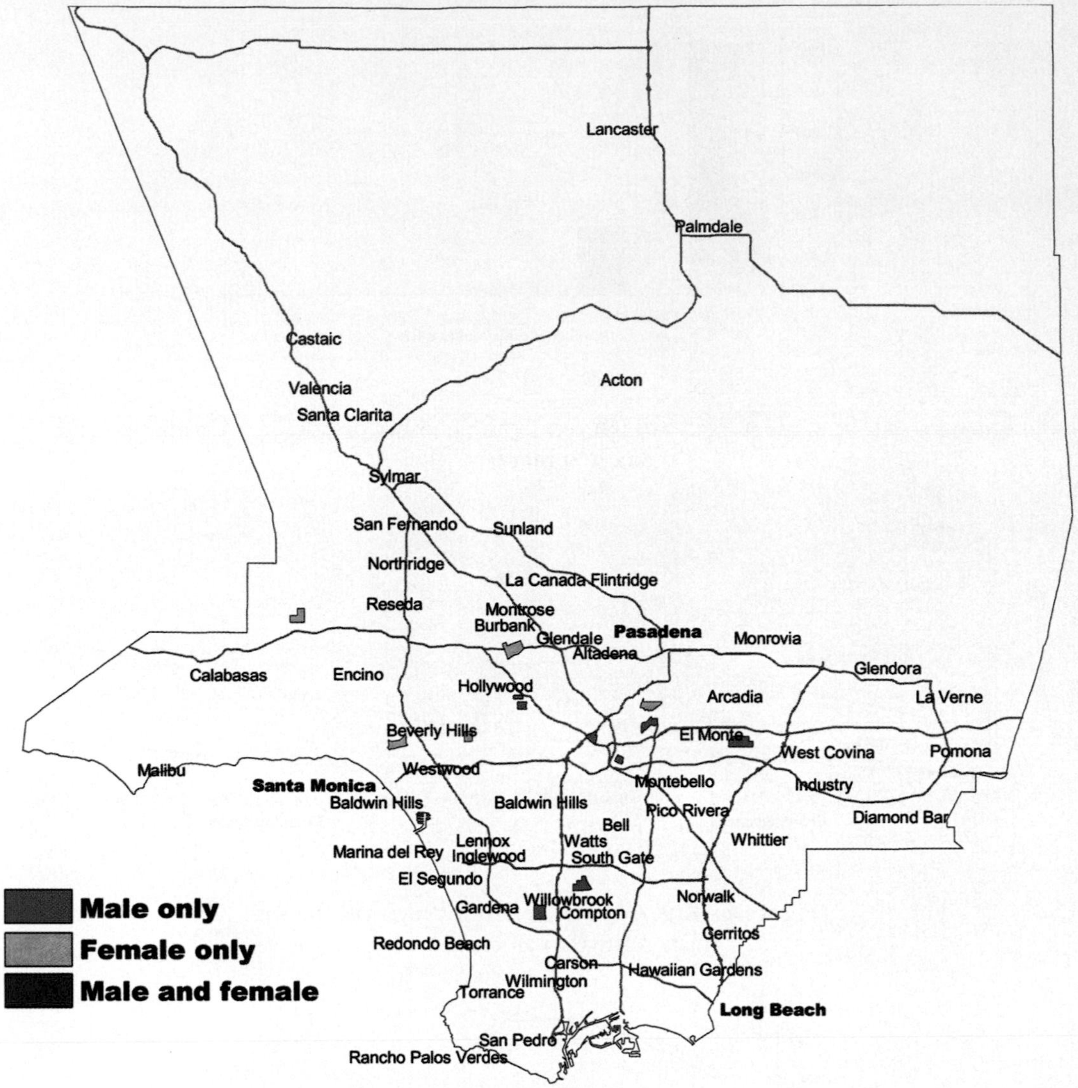

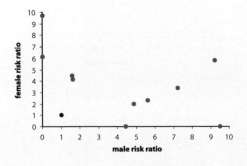

Lancaster

Palmdale

Castaic

Valencia
Santa Clarita

Acton

Sylmar

San Fernando   Sunland

Northridge

La Canada Flintridge

Reseda

Montrose
Burbank
Glendale   **Pasadena**   Monrovia
Altadena

Calabasas   Encino

Glendora

Hollywood

Arcadia   La Verne

Beverly Hills

El Monte   West Covina   Pomona

Malibu   Westwood

Montebello   Industry

Santa Monica   Baldwin Hills   Baldwin Hills   Pico Rivera   Diamond Bar

Bell
Marina del Rey   Lennox   Watts   Whittier
Inglewood   South Gate

El Segundo

Norwalk
Gardena   Willowbrook
Compton

Cerritos

Redondo Beach

Carson   Hawaiian Gardens
Wilmington
Torrance

Long Beach

San Pedro
Rancho Palos Verdes

■ **Male only**
■ **Female only**
■ **Male and female**

Figure 10: Male-female correlation between the relative risks for high-risk census tracts.

female risk ratio

male risk ratio

Figure 11: Map of census tracts at high risk, adjusted for social class.

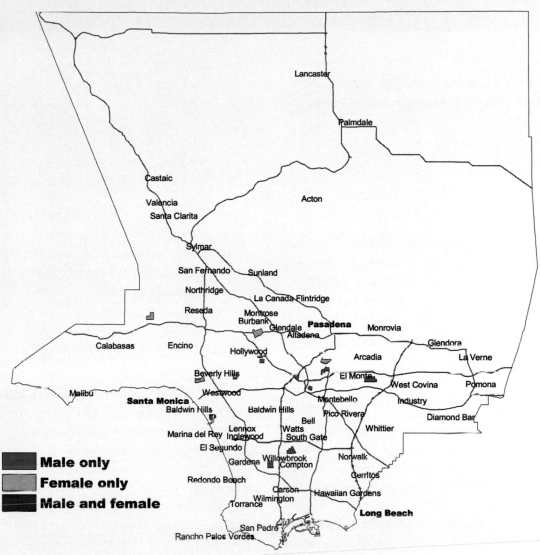

Figure 12: Male-female correlation between the relative risks for high-risk census tracts, adjusted for social class.

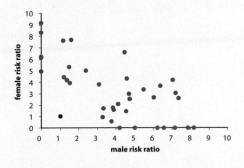

Gallbladder Carcinoma

ICDO-2 Code Anatomic Site: C 23
ICDO-2 Code Histology: 8000–8560, 8570–8573
Age: All
Male Cases: 714
Female Cases: 2092

Background

Cancer of the gallbladder is closely related to the presence of cholesterol-containing gallstones, which in turn are caused by genetic, inflammatory, and dietary factors. Other causes are unknown. It is the most common cancer in many areas of South America.

Local Pattern

This carcinoma occurs among older persons. Women are at higher risk than men, and Latinos are at higher risk than persons of other racial/ethnic groups. Risk in Los Angeles County and the rest of the United States is substantially lower than in Denmark. Risk is higher among persons of lower social class, and is decreasing slightly over time, in Los Angeles County as a whole as well as among the residents of high-risk census tracts. Figure 6 shows a slight nonrandom excess of census tracts with unexpectedly few or unexpectedly many cases. The census tracts at high risk are scattered around the county, with some concentration in East Los Angeles. This geographic pattern is partly modified by adjustment for social class.

Thumbnail Interpretation

Gallstones are more common in women than men, in the residents of Northern Europe than in other Europeans, and in the Western Hemisphere, especially among those with Amerindian genetic heritage, such as Latinos in Los Angeles. On the basis of the known gallstone prevalence, one would predict that a higher than usual risk would occur in Los Angeles, among women, among those of lower social class, and among those residing in Latin-American neighborhoods. These predictions are fulfilled. The geographical distribution in Los Angeles County is largely determined by the prevalence of persons with some Amerindian ancestry.

Figure 1: Age-adjusted incidence rate by place.

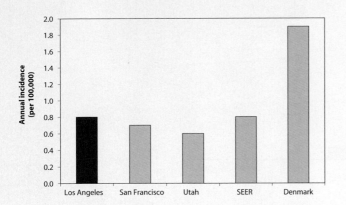

Figure 2: Age-adjusted incidence rate over the period.

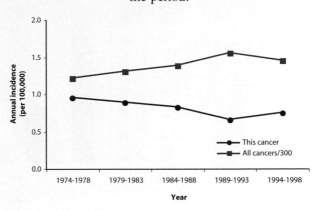

Figure 3: Age-adjusted incidence rate by age and race/ethnicity.

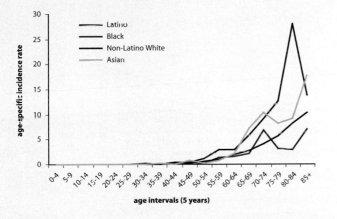

Figure 4: Age-adjusted incidence rate by social class.

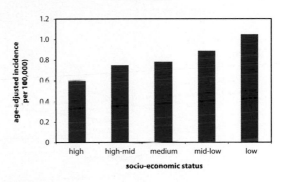

Figure 5: Distribution of the relative risk values for all census tracts.

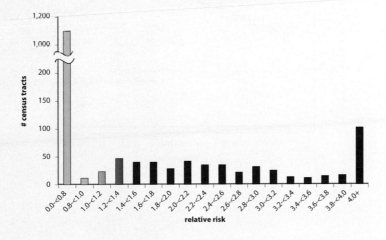

Figure 6: Census tracts by the number of cases per tract.

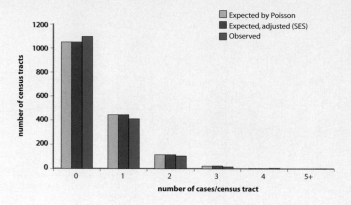

Figure 7a and b: Census tracts at high risk by the number of cases. (a) Unadjusted and (b) adjusted for social class.

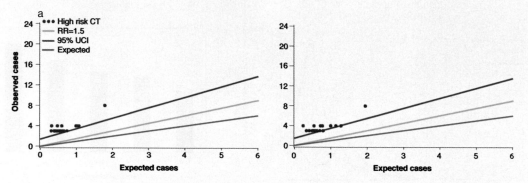

Figure 8: Risk over the period for high-risk census tracts relative to all census tracts.

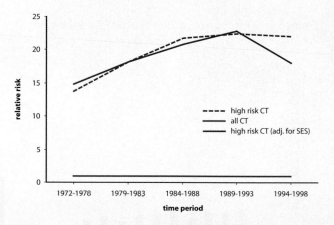

Gallbladder Carcinoma: Female

Figure 1: Age-adjusted incidence rate by place.

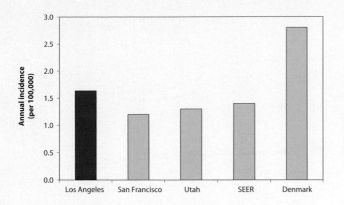

Figure 2: Age-adjusted incidence rate over the period.

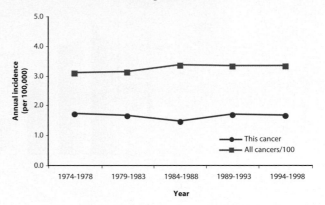

Figure 3: Age-adjusted incidence rate by age and race/ethnicity.

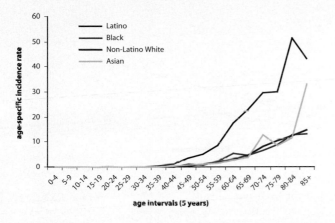

Figure 4: Age-adjusted incidence rate by social class.

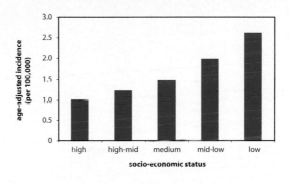

Figure 5: Distribution of the relative risk values for all census tracts.

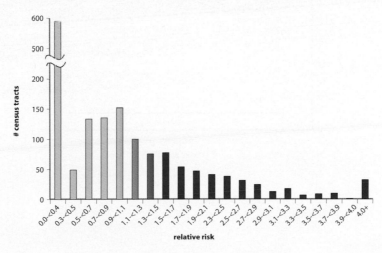

Figure 6: Census tracts by the number of cases per tract.

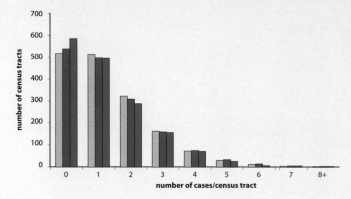

Figure 7a and b: Census tracts at high risk by the number of cases. (a) Unadjusted and (b) adjusted for social class.

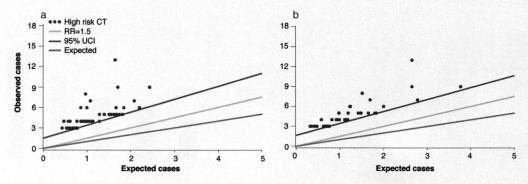

Figure 8: Risk over the period for high-risk census tracts relative to all census tracts.

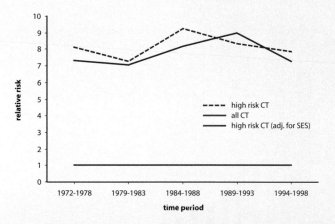

Figure 9: Map of census tracts at high risk.

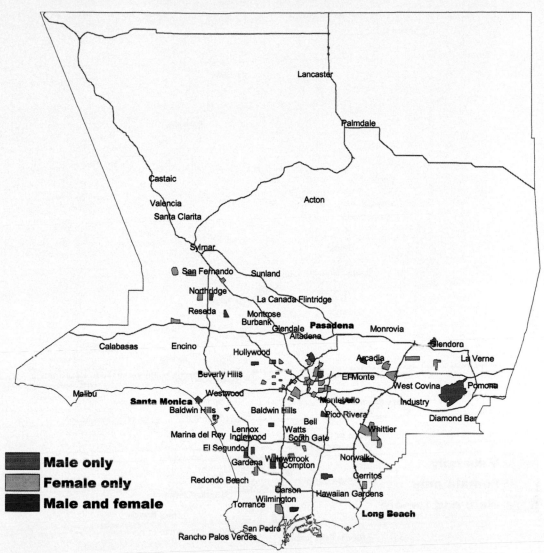

Figure 10: Male-female correlation between the relative risks for high-risk census tracts.

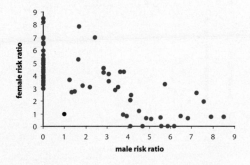

Figure 11: Map of census tracts at high risk, adjusted for social class.

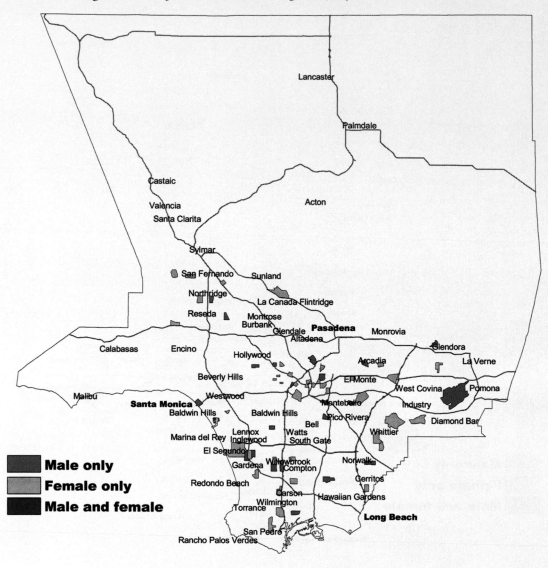

Figure 12: Male-female correlation between the relative risks for high-risk census tracts, adjusted for social class.

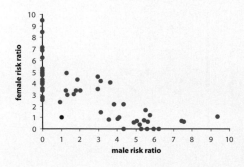

Figure 6: Census tracts by the number of cases per tract.

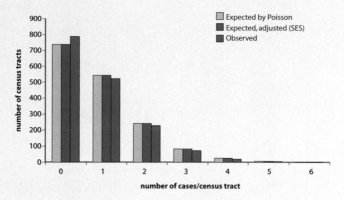

Figure 7a and b: Census tracts at high risk by the number of cases. (a) Unadjusted and (b) adjusted for social class.

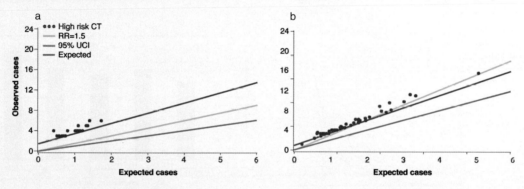

Figure 8: Risk over the period for high-risk census tracts relative to all census tracts.

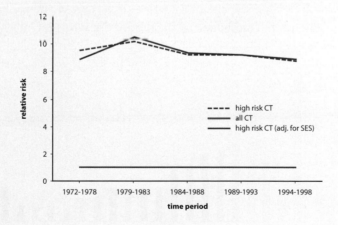

Figure 1: Age-adjusted incidence rate by place.

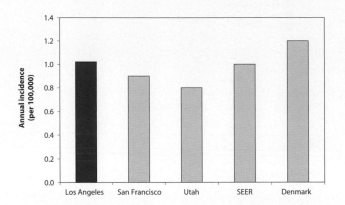

Figure 2: Age-adjusted incidence rate over the period.

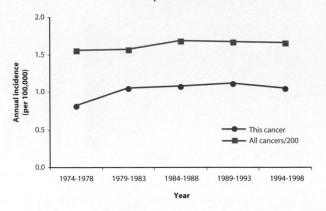

Figure 3: Age-adjusted incidence rate by age and race/ethnicity.

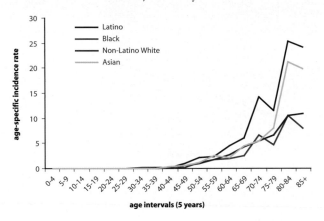

Figure 4: Age-adjusted incidence rate by social class.

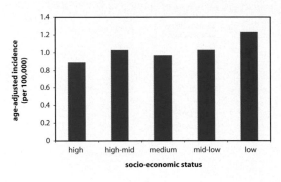

Figure 5: Distribution of the relative risk values for all census tracts.

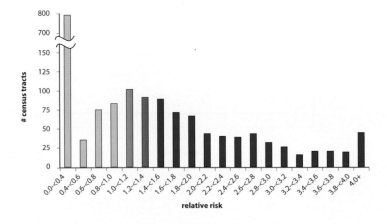

Figure 6: Census tracts by the number of cases per tract.

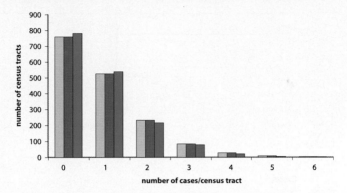

Figure 7a and b: Census tracts at high risk by the number of cases. (a) Unadjusted and (b) adjusted for social class.

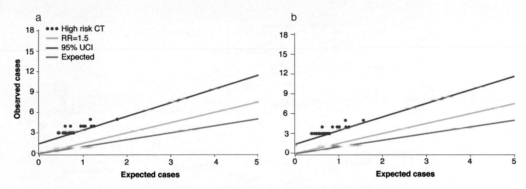

Figure 8: Risk over the period for high-risk census tracts relative to all census tracts.

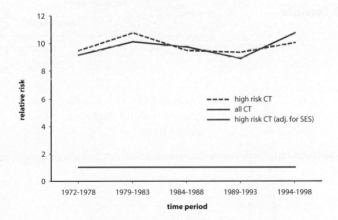

Figure 9: Map of census tracts at high risk.

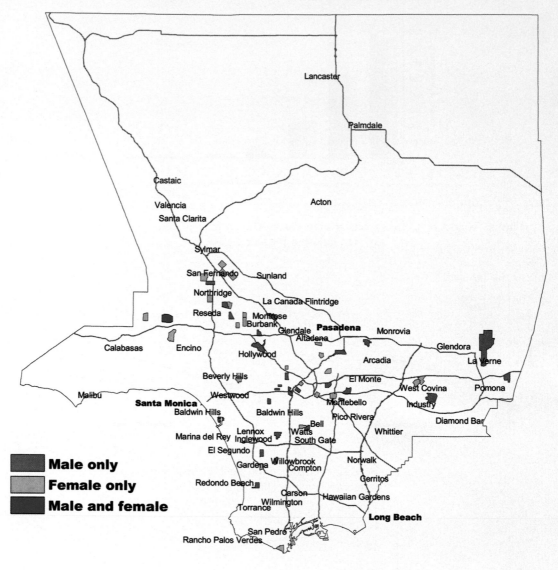

Male only
Female only
Male and female

Figure 10: Male-female correlation between the relative risks for high-risk census tracts.

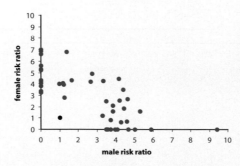

Figure 11: Map of census tracts at high risk, adjusted for social class.

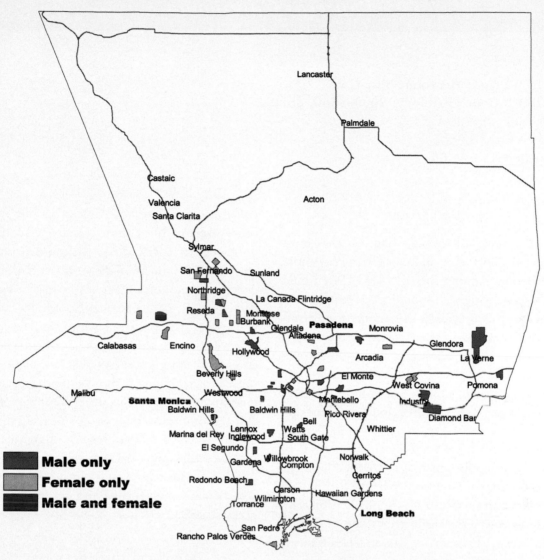

Figure 12: Male-female correlation between the relative risks for high-risk census tracts, adjusted for social class.

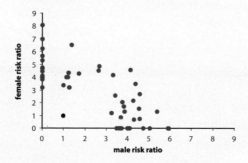

Pancreas Carcinoma

ICDO-2 Code Anatomic Site: C 25
ICDO-2 Code Histology: 8000–8560, 8570–8573
Age: All
Male Cases: 9295
Female Cases: 9760

Background

Pancreas carcinoma, one of the most lethal malignancies, is known to occur more commonly among persons of African and Polynesian origin, although it is not known whether this is a result of genetic or environmental determinants. Cigarette smoking is known to cause pancreas cancer, even though exposure to tobacco constituents occurs only via the blood stream. Although pancreas cancer may cause diabetes, long-standing diabetics may also be at increased risk. Dietary constituents have long been suspected of playing an important role in the development of this cancer. Attention is being focused on leafy green vegetables as potentially protective and on meat, especially well-done or barbecued meat, as potentially causal. Certain forms of pancreatitis, especially the heritable form that recurs in families, greatly increase risk, but are only responsible for a small proportion of cases. Neither high consumption of alcohol nor specific occupational exposures have been clearly linked to pancreatic cancer.

Local Pattern

Pancreas cancer is a disease of old age, and is one and one-half times as common among men as among women. It occurs with roughly equal frequency in Los Angeles County and other parts of the country, and is essentially unrelated to social class. African-Americans are at substantially higher risk and Asian-Americans at substantially lower risk in comparison with members of other groups. This malignancy has occurred with constant incidence over the period in the county as a whole, as well as among the residents of high-risk census tracts. Figure 6 shows a moderate nonrandom excess of census tracts with unexpectedly few or unexpectedly many cases. High-risk census tracts are scattered throughout the county, with only a few sets that are contiguous. No census tract stands out on the basis of a particularly high relative risk and number of excess cases.

Thumbnail Interpretation

One might expect pancreas cancer to occur more commonly in census tracts with predominantly African-American residents, but that is not evident. However, the link between cigarette smoking and pancreas cancer is much weaker than the associations responsible for geographical patterns in the occurrence of lung, larynx, and oral cancers. Because no systematic pattern of geographical occurrence is apparent, no local source of causation can be proposed.

Pancreas Carcinoma: Male

Figure 1: Age-adjusted incidence rate by place.

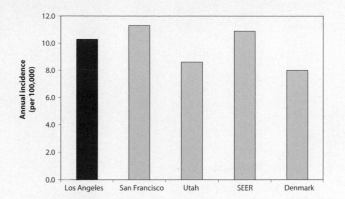

Figure 2: Age-adjusted incidence rate over the period.

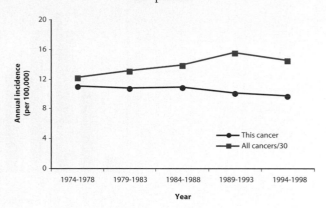

Figure 3: Age-adjusted incidence rate by age and race/ethnicity.

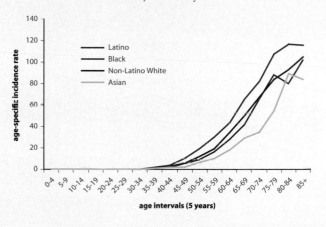

Figure 4: Age-adjusted incidence rate by social class.

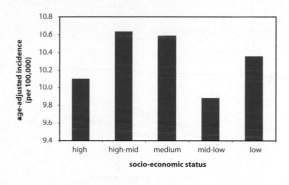

Figure 5: Distribution of the relative risk values for all census tracts.

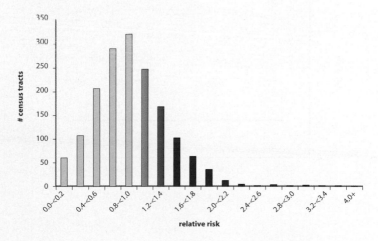

Figure 6: Census tracts by the number of cases per tract.

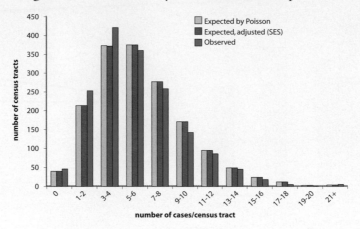

Figure 7a and b: Census tracts at high risk by the number of cases. (a) Unadjusted and (b) adjusted for social class.

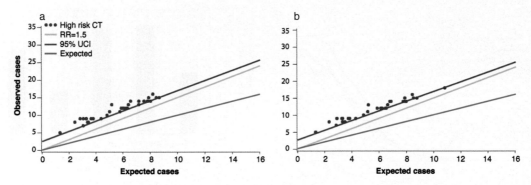

Figure 8: Risk over the period for high-risk census tracts relative to all census tracts.

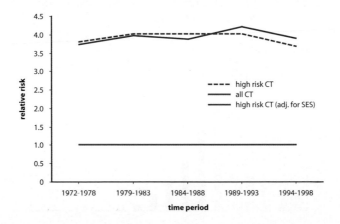

Pancreas Carcinoma: Female

Figure 1: Age-adjusted incidence rate by place.

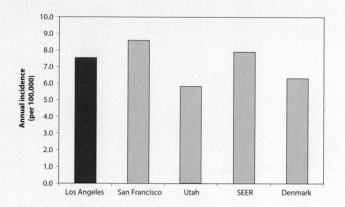

Figure 2: Age-adjusted incidence rate over the period.

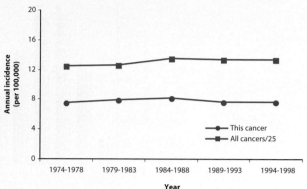

Figure 3: Age-adjusted incidence rate by age and race/ethnicity.

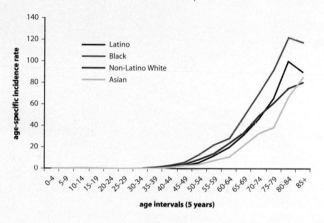

Figure 4: Age-adjusted incidence rate by social class.

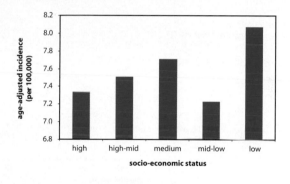

Figure 5: Distribution of the relative risk values for all census tracts.

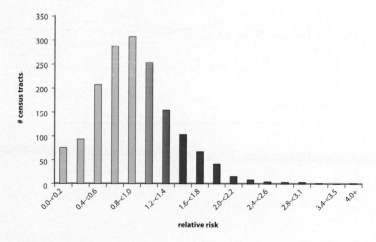

Figure 6: Census tracts by the number of cases per tract.

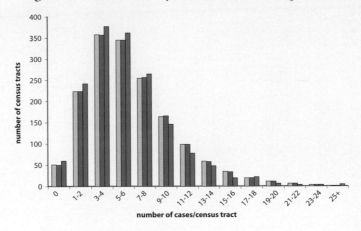

Figure 7a and b: Census tracts at high risk by the number of cases. (a) Unadjusted and (b) adjusted for social class.

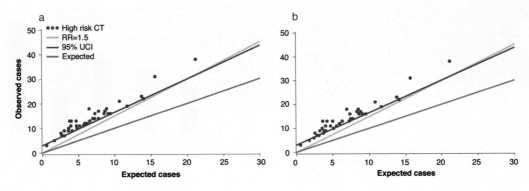

Figure 8: Risk over the period for high-risk census tracts relative to all census tracts.

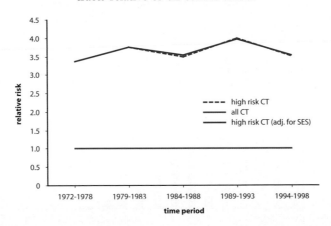

Figure 9: Map of census tracts at high risk.

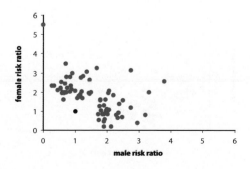

Figure 10: Male-female correlation between the relative risks for high-risk census tracts.

Figure 11: Map of census tracts at high risk, adjusted for social class.

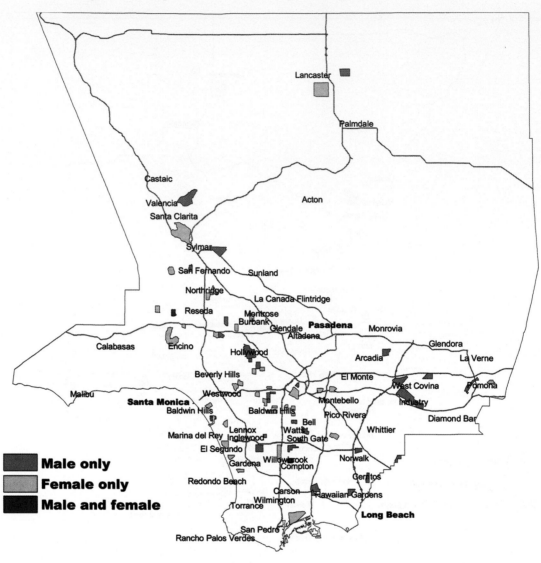

Figure 12: Male-female correlation between the relative risks for high-risk census tracts, adjusted for social class.

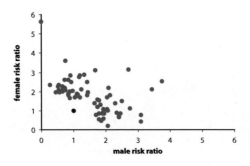

Figure 1: Age-adjusted incidence rate by place.

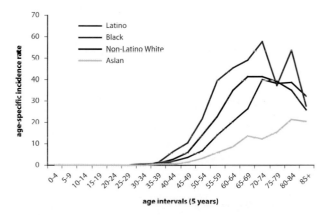

Figure 2: Age-adjusted incidence rate over the period.

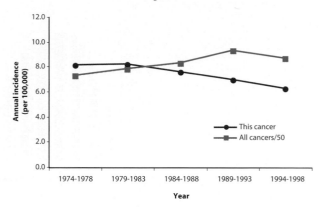

Figure 3: Age-adjusted incidence rate by age and race/ethnicity.

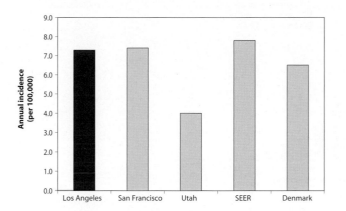

Figure 4: Age-adjusted incidence rate by social class.

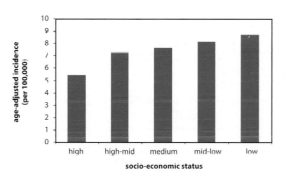

Figure 5: Distribution of the relative risk values for all census tracts.

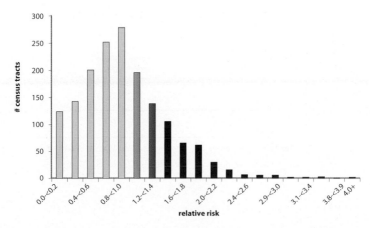

Figure 6: Census tracts by the number of cases per tract.

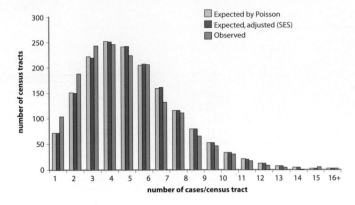

Figure 7a and b: Census tracts at high risk by the number of cases. (a) Unadjusted and (b) adjusted for social class.

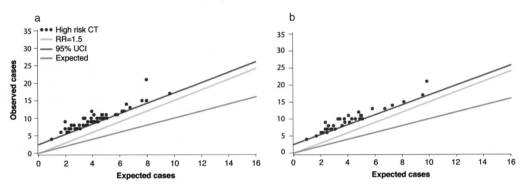

Figure 8: Risk over the period for high-risk census tracts relative to all census tracts.

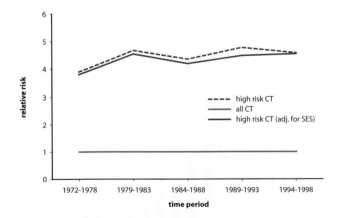

Carcinoma of the Larynx: Female

Figure 1: Age-adjusted incidence rate by place.

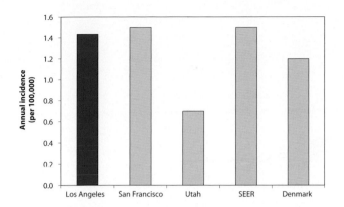

Figure 2: Age-adjusted incidence rate over the period.

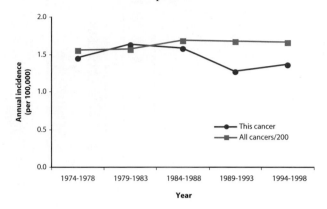

Figure 3: Age-adjusted incidence rate by age and race/ethnicity.

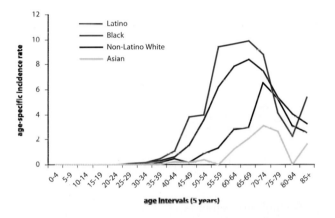

Figure 4: Age-adjusted incidence rate by social class.

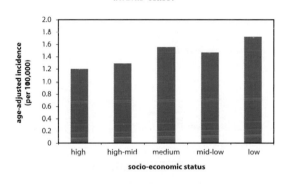

Figure 5: Distribution of the relative risk values for all census tracts.

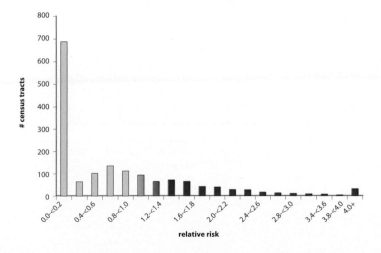

Figure 6: Census tracts by the number of cases per tract.

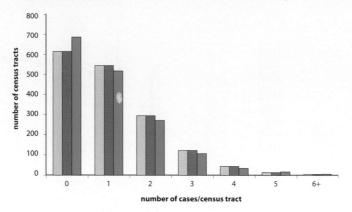

Figure 7a and b: Census tracts at high risk by the number of cases. (a) Unadjusted and (b) adjusted for social class.

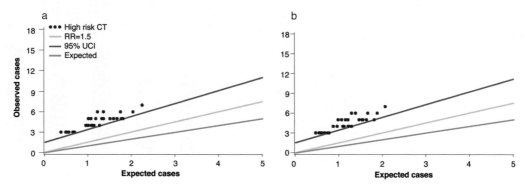

Figure 8: Risk over the period for high-risk census tracts relative to all census tracts.

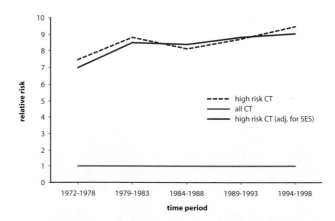

Figure 9: Map of census tracts at high risk.

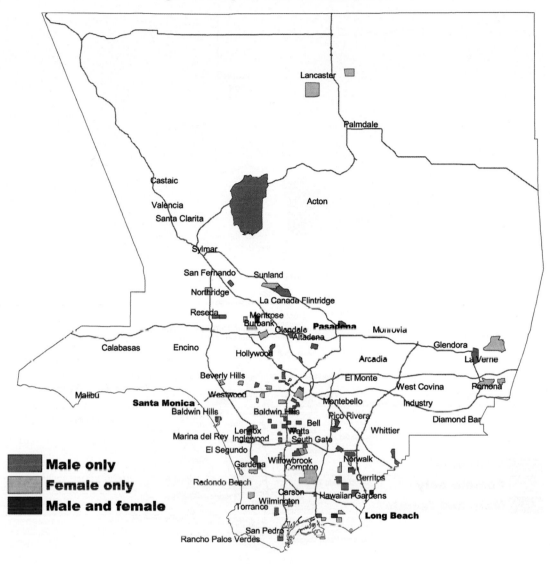

Figure 10: Male-female correlation
between the relative risks for high-risk
census tracts.

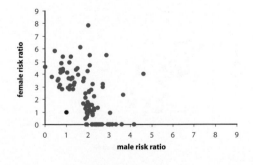

Figure 11: Map of census tracts at high risk, adjusted for social class.

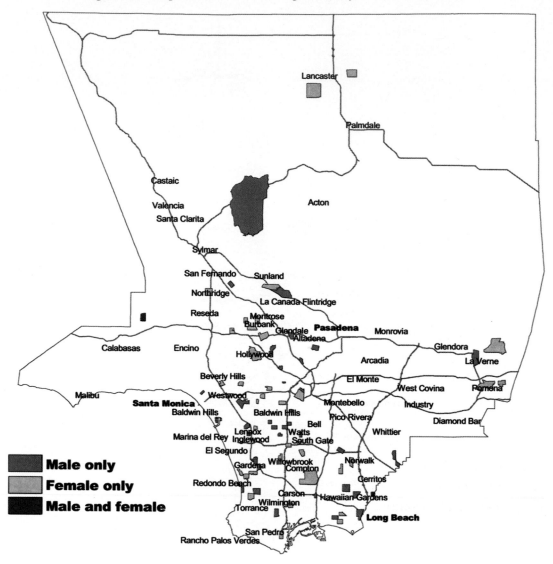

Figure 12: Male-female correlation between the relative risks for high-risk census tracts, adjusted for social class.

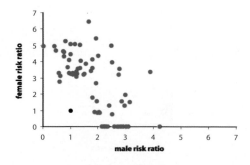

Lung and Bronchus Carcinoma, Total (All Types)

ICDO-2 Code Anatomic Site: C 33, 34
ICDO-2 Code Histology: 8000–8560, 8570–8573
Age: All
Male Cases: 65397
Female Cases: 41398

Background

Lung cancer has usually been studied without distinguishing between the various possible cells of origin. The lung is a complex organ, with a variety of tissues and cell types. As a class, the most important cause of lung cancer by far is cigarette smoking. The degree of causation is modified by the daily amount smoked, the years of smoking, the interval since quitting, and probably even the way cigarettes are smoked. Because of historical patterns of initiating and quitting smoking, the occurrence of lung cancer is closely linked to year of birth. African-Americans, particularly men, have long been known to be at higher risk of lung cancer.

Other causes include pipe and cigar smoking, passive exposure to the smoke coming from the cigarettes of others, and less common causes of lung cancer that one encountered in the workplace. These include arsenic, certain nickel compounds, hexavalent chromium, chloromethyl ethers (BCME or CMME), mustard gas (in production workers), asbestos, and polycyclic hydrocarbons (such as those from coke ovens or diesel exhaust). Radon, a radioactive gas, is emitted from certain geological formations and appears in some mines and even in homes in some parts of the country (risk from radon is increased in the presence of smoking). Lung cancer has resulted from exposure to other forms of ionizing radiation, such as that from outmoded high-dose radiation treatments or the atomic bomb. Other workplace exposures strongly suspected of causing lung cancer are silica, beryllium fumes, acrylonitrile, ferric oxide, sulfuric acid mist, epichlorohydrin, and lead and cadmium dusts. Diseases capable of scarring the lung are also thought to sometimes result in cancer.

Consumption of fruits and vegetables has been reported to decrease the frequency of lung cancer, although the exact nutrient responsible is unclear and may vary. There are also clear genetic factors that increase and decrease susceptibility, sometimes in interaction with dietary items.

Local Pattern

Cancer of the lung and bronchus generally is twice as frequent among men as women, and appears in Los Angeles County as frequently as in other parts of North America and Europe, although Utah enjoys lower rates. African-American men have the highest incidence, but among women, rates for whites are equally high. Rates among men have been falling over the last few decades, whereas rates among women, which were still rising in the first few decades of the period, have more recently leveled off and begun to fall. These trends hold for the county as a whole as well as among the residents of high-risk census tracts. There is a slight tendency for increased risk among men of lower social class and women of higher social class. Figure 6 shows a moderate nonrandom excess of census tracts with unexpectedly few or unexpectedly many cases. Census tracts at high risk, including clusters of contiguous census tracts, appear in South Central Los Angeles, the Antelope Valley, and the upper east side of the county from Glendora to La Verne.

Thumbnail Interpretation

As can be seen from the sections that follow, the prevailing idea that lung cancer is a single condition, explained by smoking alone, is innaccurate. The pattern of occurrence of cancers of the bronchus and lung in Los Angeles County reflects a mixture of different patterns. Nonetheless, as a group, the patterns do conform to the distribution that would be expected on the basis of the known variations in smoking experience, with the highest risk seen in areas that are predominately African-American or of lower social class.

Why some lungs respond to the carcinogens in tobacco smoke with a malignancy of one cell type and others respond with another type is not known, but given the same demographic and geographical target population, subtle environmental differences or genetic factors must be responsible.

Lung and Bronchus Carcinoma, Total (All Types): Male

Figure 1: Age-adjusted incidence rate by place.

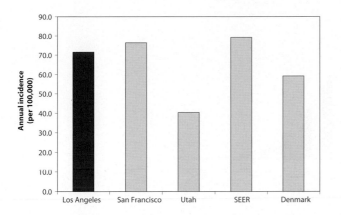

Figure 2: Age-adjusted incidence rate over the period.

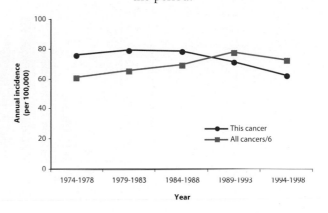

Figure 3: Age-adjusted incidence rate by age and race/ethnicity.

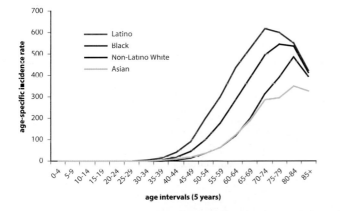

Figure 4: Age-adjusted incidence rate by social class.

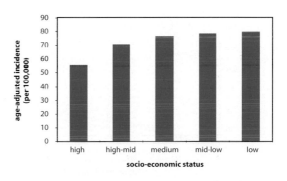

Figure 5: Distribution of the relative risk values for all census tracts.

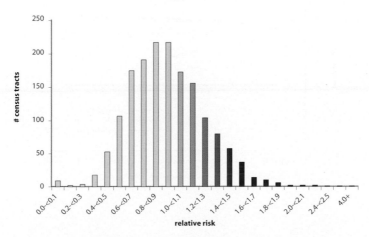

Figure 6: Census tracts by the number of cases per tract.

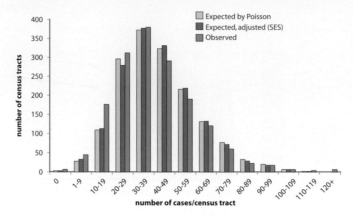

Figure 7a and b: Census tracts at high risk by the number of cases. (a) Unadjusted and (b) adjusted for social class.

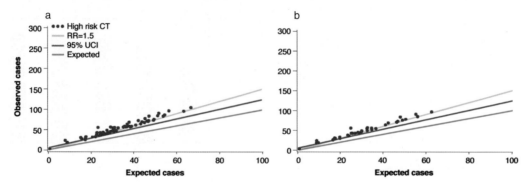

Figure 8: Risk over the period for high-risk census tracts relative to all census tracts.

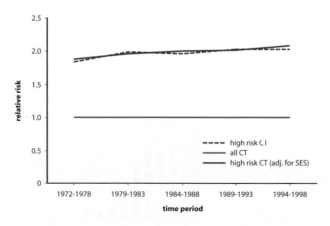

Lung and Bronchus Carcinoma, Total (All Types): Female

Figure 1: Age-adjusted incidence rate by place.

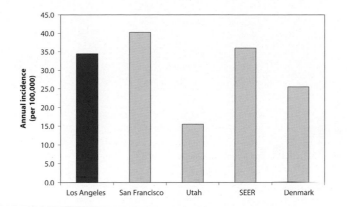

Figure 2: Age-adjusted incidence rate over the period.

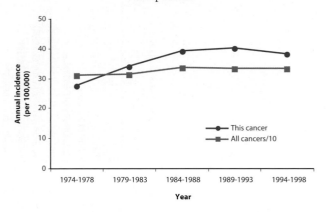

Figure 3: Age-adjusted incidence rate by age and race/ethnicity.

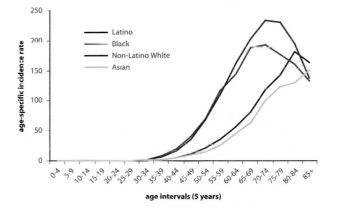

Figure 4: Age-adjusted incidence rate by social class.

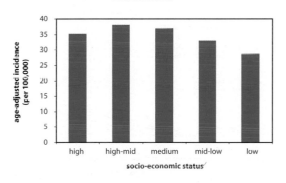

Figure 5: Distribution of the relative risk values for all census tracts.

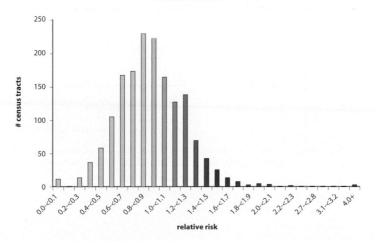

Figure 6: Census tracts by the number of cases per tract.

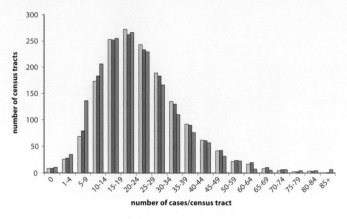

Figure 7a and b: Census tracts at high risk by the number of cases. (a) Unadjusted and (b) adjusted for social class.

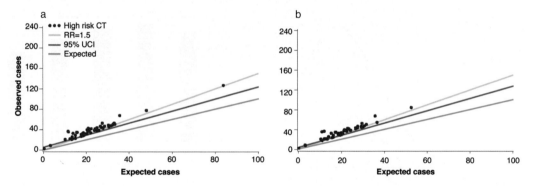

Figure 8: Risk over the period for high-risk census tracts relative to all census tracts.

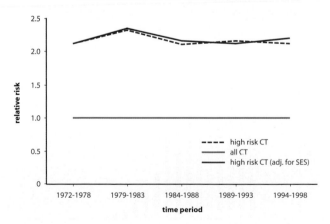

Figure 9: Map of census tracts at high risk.

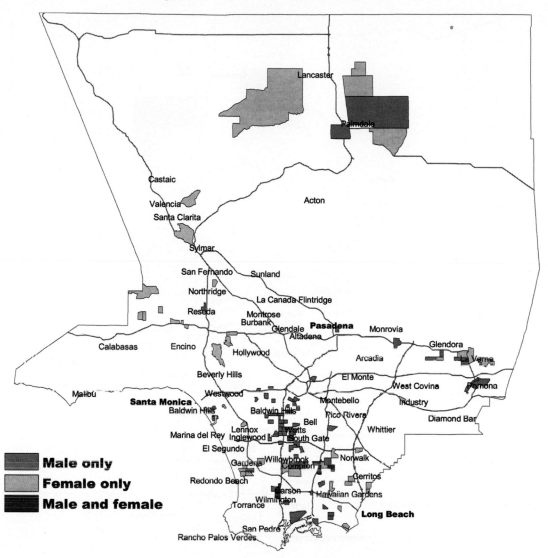

Figure 10: Male-female correlation between the relative risks for high-risk census tracts.

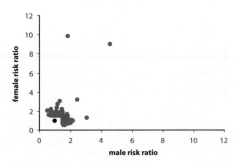

Figure 11: Map of census tracts at high risk, adjusted for social class.

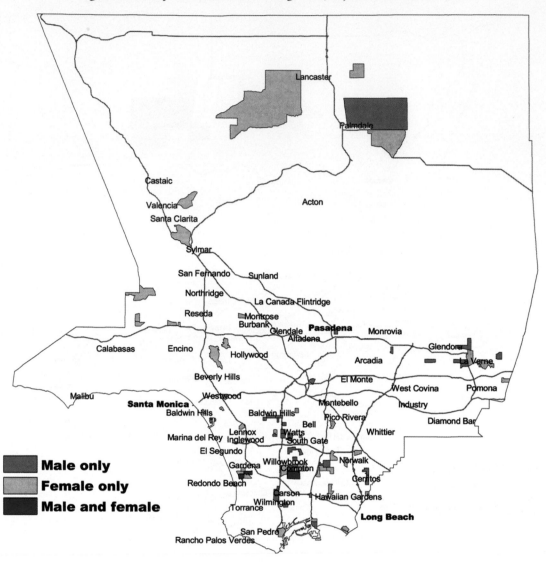

Figure 12: Male-female correlation between the relative risks for high-risk census tracts, adjusted for social class.

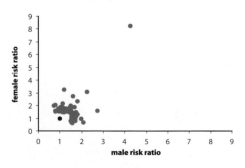

Squamous Cell Carcinoma of the Lung and Bronchus

ICDO-2 Code Anatomic Site: C 33, 34
ICDO-2 Code Histology: 8050–8082
Age: All
Male Cases: 19108
Female Cases: 7378

Background

The most important cause of squamous cell lung cancer by far is cigarette smoking. The degree of causation is modified by the amount smoked, the years of smoking, and the interval since quitting. Because of historical patterns of initiating and quitting smoking, the occurrence of lung cancer is closely linked to year of birth, with persons reaching adulthood before World War II at highest risk.

African-Americans, particularly men, have long been known to be at higher risk of squamous cell lung cancer. More African-Americans than members of other race/ethnicity groups are smokers, but they tend to smoke fewer cigarettes per day. The reason why squamous cell lung cancers are more common among them is still a mystery. While it is possible that they inhale more carcinogens per cigarette, most experts suspect that genetic differences or differences in dietary protection may be partly responsible.

The other causes of lung cancer described in the section pertaining to lung and bronchus cancer, total, are presumed to be true for squamous cell lung cancer, since that has historically been the most common type and only recently has it been possible to distinguish between cell types in studies of sufficient size.

Local Pattern

Rates of squamous cell carcinoma are roughly similar in the populations of Los Angeles County and other parts of the country, except for that of Utah, which has a lower prevalence of smokers and lower rates. Incidence of this malignancy is over twice as high among men as among women. Occurrence is higher among whites and especially African-Americans of both sexes in comparison with those of other racial/ethnic groups. Lower social class men, but not women, experience higher risk. In the county as a whole, risk has decreased among men for more than two decades, but among women the decrease has been apparent for only slightly more than one decade. Among the male residents of high-risk census tracts, incidence is not decreasing, whereas among women the decrease is greater than it is in the county as a whole. Figure 6 shows a moderate nonrandom excess of census tracts with unexpectedly few or unexpectedly many cases, and a substantial number of high-risk census tracts are apparent even after adjusting for social class. Contiguous groups of census tracts and census tracts with high risk for both men and women appear especially in South Central Los Angeles, with a few also in the southeast corner of the county.

Thumbnail Interpretation

As expected, the highest rates are found among African-American men, and in the communities in which they reside. Middle to lower social class whites, who predominate in the high risk communities seen in the southeast of the county, have also traditionally smoked more than those of higher social class.

Squamous Cell Carcinoma of the Lung and Bronchus: Male

Figure 1: Age-adjusted incidence rate by place.

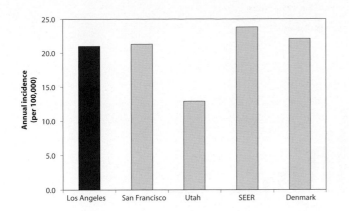

Figure 2: Age-adjusted incidence rate over the period.

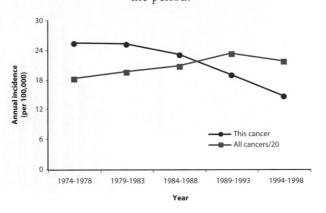

Figure 3: Age-adjusted incidence rate by age and race/ethnicity.

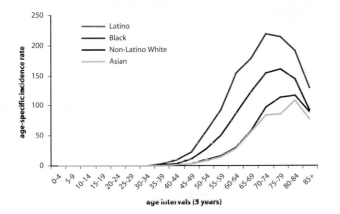

Figure 4: Age-adjusted incidence rate by social class.

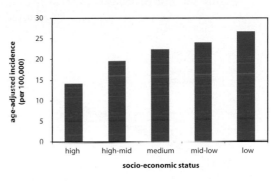

Figure 5: Distribution of the relative risk values for all census tracts.

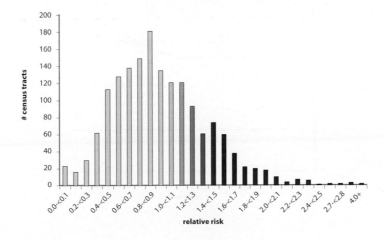

Squamous Cell Carcinoma of the Lung and Bronchus: Male

Figure 6: Census tracts by the number of cases per tract.

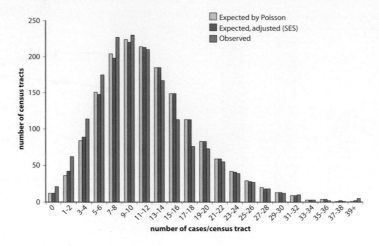

Figure 7a and b: Census tracts at high risk by the number of cases. (a) Unadjusted and (b) adjusted for social class.

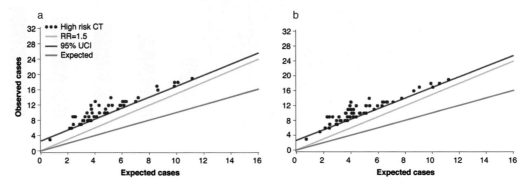

Figure 8: Risk over the period for high-risk census tracts relative to all census tracts.

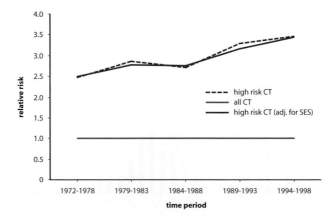

Squamous Cell Carcinoma of the Lung and Bronchus: Female

Figure 1: Age-adjusted incidence rate by place.

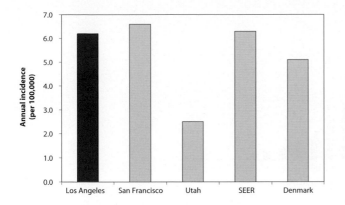

Figure 2: Age-adjusted incidence rate over the period.

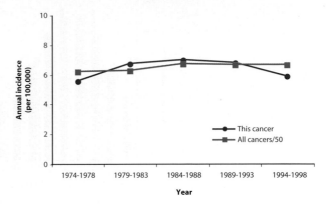

Figure 3: Age-adjusted incidence rate by age and race/ethnicity.

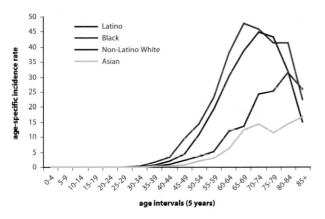

Figure 4: Age-adjusted incidence rate by social class.

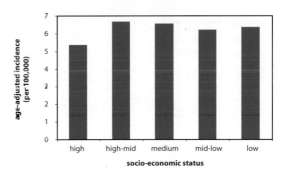

Figure 5: Distribution of the relative risk values for all census tracts.

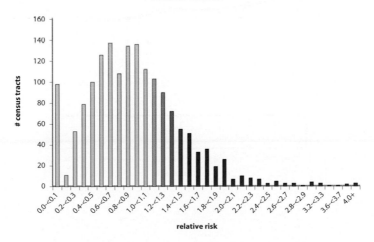

Figure 6: Census tracts by the number of cases per tract.

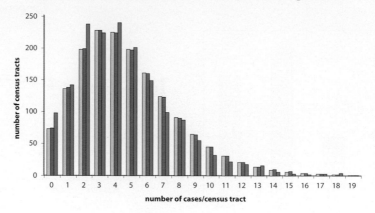

Figure 7a and b: Census tracts at high risk by the number of cases. (a) Unadjusted and (b) adjusted for social class.

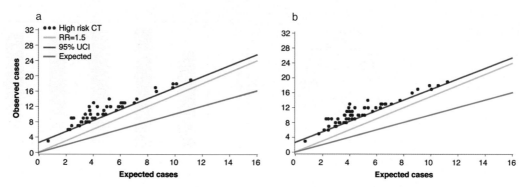

Figure 8: Risk over the period for high-risk census tracts relative to all census tracts.

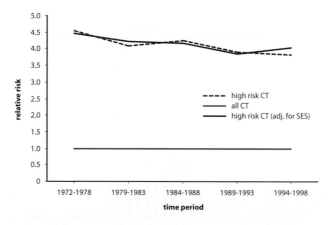

Figure 9: Map of census tracts at high risk.

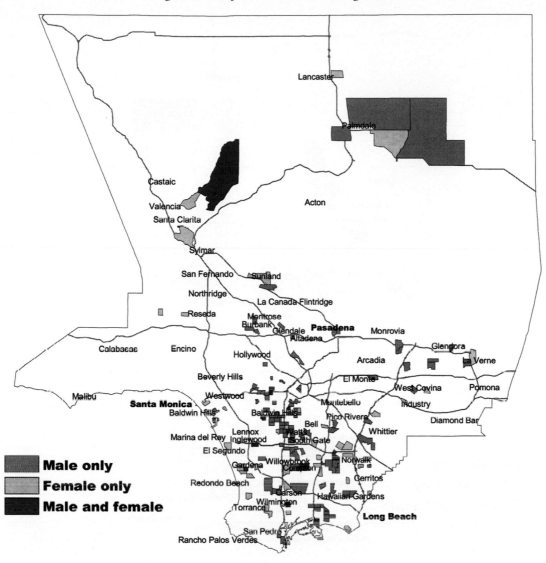

Figure 10: Male-female correlation between the relative risks for high-risk census tracts.

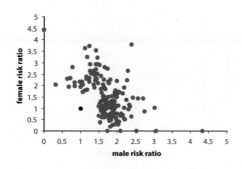

Figure 11: Map of census tracts at high risk, adjusted for social class.

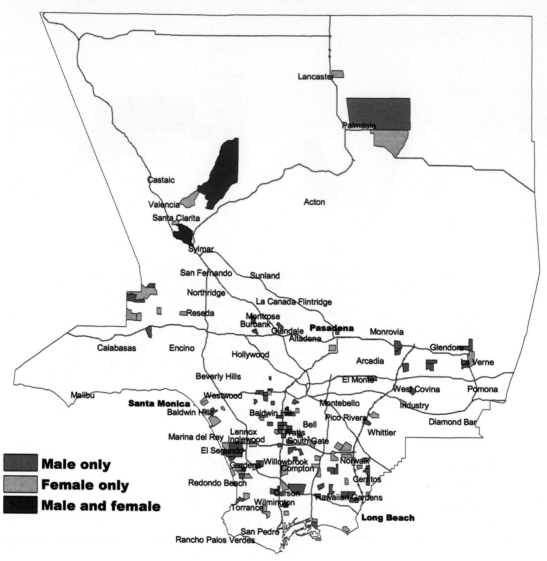

Figure 12: Male-female correlation between the relative risks for high-risk census tracts, adjusted for social class.

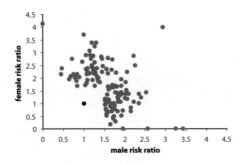

Large Cell Carcinoma of the Lung and Bronchus

ICDO-2 Code Anatomic Site: C 33, 34
ICDO-2 Code Histology: 8012
Age: All
Male Cases: 4466
Female Cases: 2696

Background

Please refer to the discussion of lung and bronchus cancer, total. Large cell carcinomas are thought to be more aggressive (less differentiated) forms of the other cell types of lung cancer, particularly squamous cell carcinoma, and therefore are thought to reflect the same causal experiences.

Local Pattern

This type of lung cancer is about twice as common in men as in women. Incidence is more common in San Francisco and much less common in Utah than in Los Angeles. Occurrence becomes more frequent in middle age, and while the malignancy occurs more often among African American men, the rate among African-American women is similar to that among white women. There is no clear relation-ship with social class. Rates generally have decreased over the past several decades in men, but only in the past decade in women.

Neither the male nor the female decrease appears to have occurred among the residents of high-risk census tracts. Figure 6 shows a moderate nonrandom excess of census tracts with unexpectedly few or unexpectedly many cases. There is a subset of census tracts at very high risk. They are scattered around the county, but appear in contiguous groups in the Antelope Valley, in the San Gabriel Valley, and in South Central Los Angeles, the latter concentration modified after adjustment for social class.

Thumbnail Interpretation

Large cell carcinomas of the lung occur with the same trends, with roughly the same demographic patterns, and in the same mix of neighborhoods that all lung cancers, especially squamous lung cancers, do. One therefore presumes that they have resulted from the same causes, the most important of which is cigarette smoking.

Large Cell Carcinoma of the Lung and Bronchus: Male

Figure 1: Age-adjusted incidence rate by place.

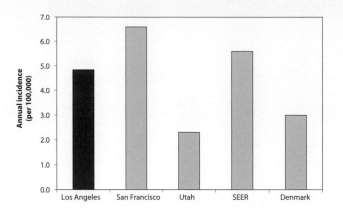

Figure 2: Age-adjusted incidence rate over the period.

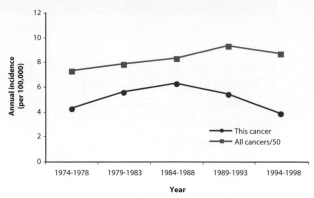

Figure 3: Age-adjusted incidence rate by age and race/ethnicity.

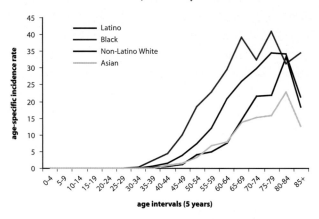

Figure 4: Age-adjusted incidence rate by social class.

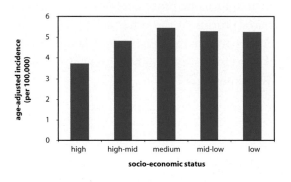

Figure 5: Distribution of the relative risk values for all census tracts.

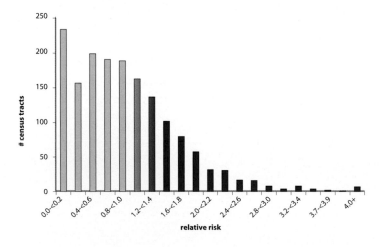

Figure 6: Census tracts by the number of cases per tract.

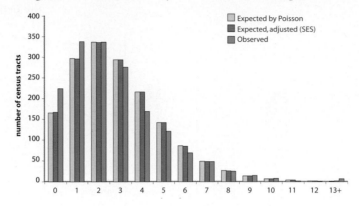

Figure 7a and b: Census tracts at high risk by the number of cases. (a) Unadjusted and (b) adjusted for social class.

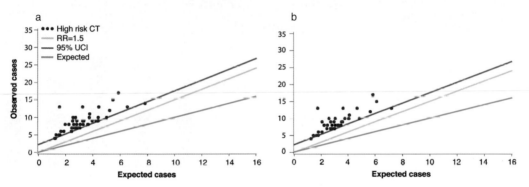

Figure 8: Risk over the period for high-risk census tracts relative to all census tracts.

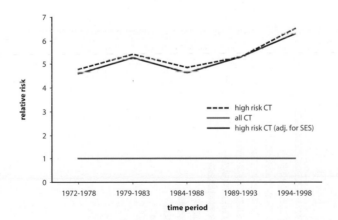

Large Cell Carcinoma of the Lung and Bronchus: Female

Figure 1: Age-adjusted incidence rate by place.

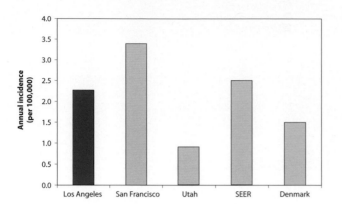

Figure 2: Age-adjusted incidence rate over the period.

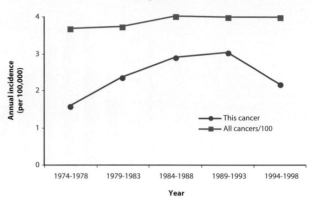

Figure 3: Age-adjusted incidence rate by age and race/ethnicity.

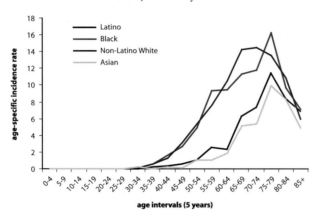

Figure 4: Age-adjusted incidence rate by social class.

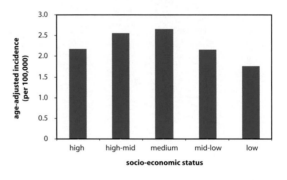

Figure 5: Distribution of the relative risk values for all census tracts.

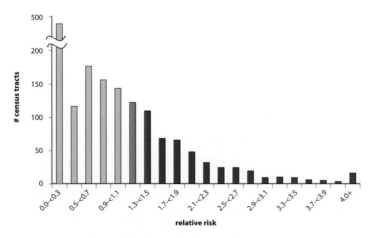

Figure 6: Census tracts by the number of cases per tract.

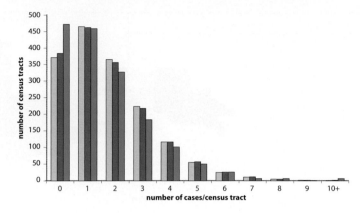

Figure 7a and b: Census tracts at high risk by the number of cases. (a) Unadjusted and (b) adjusted for social class.

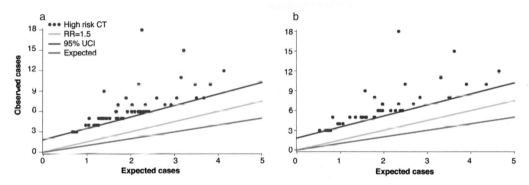

Figure 8: Risk over the period for high-risk census tracts relative to all census tracts.

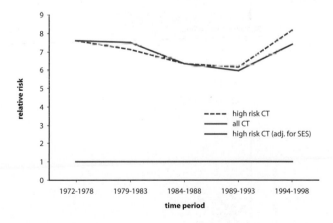

Figure 9: Map of census tracts at high risk.

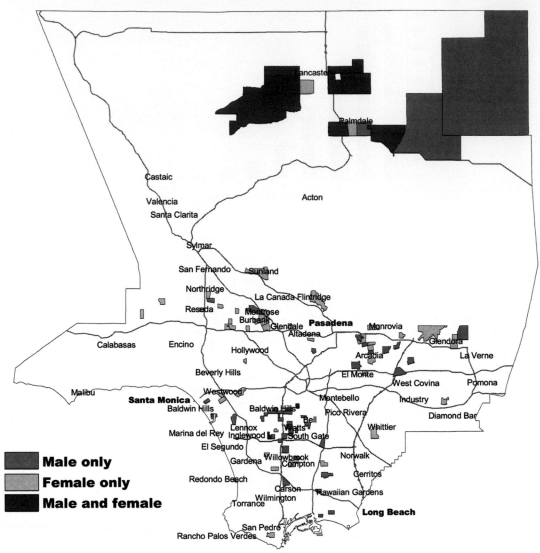

Figure 10: Male-female correlation between the relative risks for high-risk census tracts.

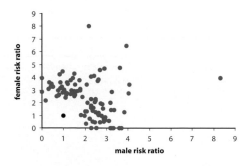

Figure 11: Map of census tracts at high risk, adjusted for social class.

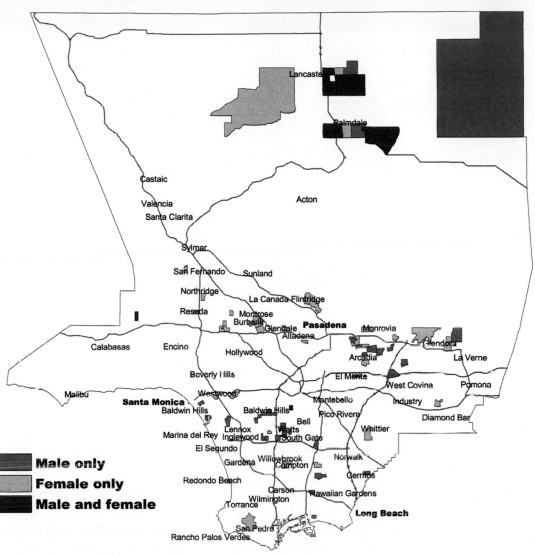

Figure 12: Male-female correlation between the relative risks for high-risk census tracts, adjusted for social class.

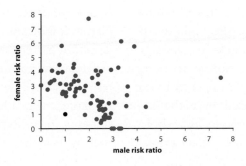

Small Cell Carcinoma of the Lung and Bronchus

ICDO-2 Code Anatomic Site: C 33, 34
ICDO-2 Code Histology: 8040–8045
Age: All
Male Cases: 8684
Female Cases: 6416

Background

Please refer to the discussion of lung and bronchus cancer, total. Small cell carcinomas have cellular characteristics that suggest a relationship to neuroendocrine (carcinoid) tumors, which are included in this book within the multiple endocrine neoplasia category. The appearance of this malignancy is as strongly associated with cigarette smoking as is squamous carcinoma. Small cell carcinoma has also been linked to radon, and to other forms of radiation exposure, and to certain chemicals encountered in the workplace, such as the chloromethyl ethers.

Local Pattern

This form of lung cancer, like other forms, is twice as common among men and less common in Utah than in Los Angeles County or other regions of the country. Risk begins to increase in middle age. Among women, whites are clearly at highest risk, in contrast with squamous carcinoma. Among men, whites and African-Americans share the highest risk position. If anything, Los Angeles residents of middle class, especially men, appear to be at higher risk than those of either upper or lower class. Incidence has decreased over the last decades of the period in the county as a whole, to a greater extent among men than among women. A similar trend has occurred among those residents of high-risk census tracts. Figure 6 shows a strong nonrandom excess of census tracts with unexpectedly few or unexpectedly many cases. While an aggregation of contiguous high-risk census tracts appears in South Central Los Angeles, the heaviest concentration appears in the southeast part of the county, between the 710 and the 605 freeways.

Thumbnail Interpretation

Because the link between this form of lung cancer and cigarette smoking is very strong, about as strong as the link between cigarette smoking and squamous cell lung cancer, the same geographic distribution was expected. It therefore came as a surprise to see that more high-risk census tracts appeared in the blue-collar white communities in the southeastern corner of the county than in the African-American communities of South Central Los Angeles. The reasons for this are unclear. Smoking is such a strong influence that it must be presumed to play an important role in determining the geographical pattern, and therefore the alternative explanations for the contrast in geographic patterns must include differences in the pattern of smoking, the products smoked, or the effect of smoking in combination with other exposures, such as those deriving from the environment.

Figure 1: Age-adjusted incidence rate by place.

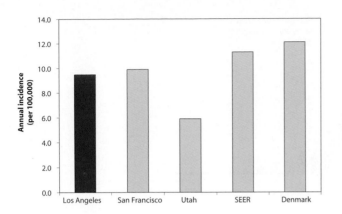

Figure 2: Age-adjusted incidence rate over the period.

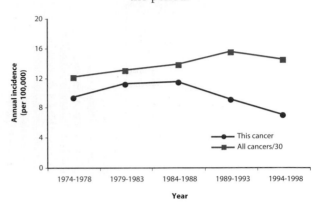

Figure 3: Age-adjusted incidence rate by age and race/ethnicity.

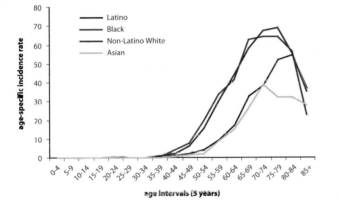

Figure 4: Age-adjusted incidence rate by social class.

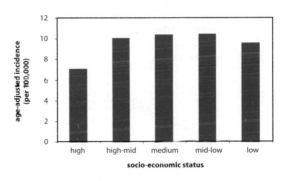

Figure 5: Distribution of the relative risk values for all census tracts.

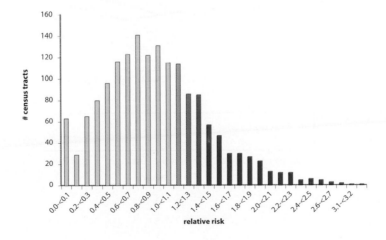

Figure 6: Census tracts by the number of cases per tract.

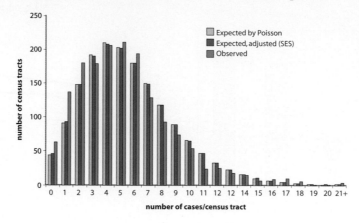

Figure 7a and b: Census tracts at high risk by the number of cases. (a) Unadjusted and (b) adjusted for social class.

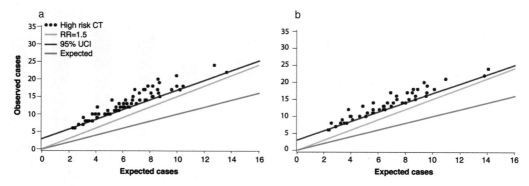

Figure 8: Risk over the period for high-risk census tracts relative to all census tracts.

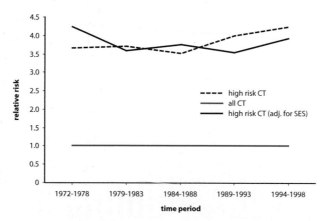

Small Cell Carcinoma of the Lung and Bronchus: Female

Figure 1: Age-adjusted incidence rate by place.

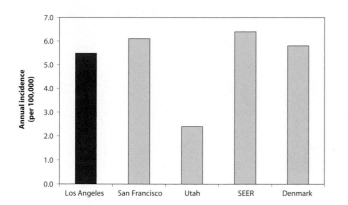

Figure 2: Age-adjusted incidence rate over the period.

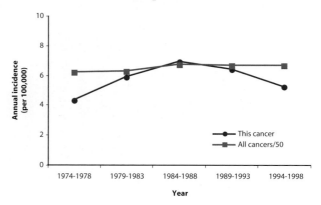

Figure 3: Age-adjusted incidence rate by age and race/ethnicity.

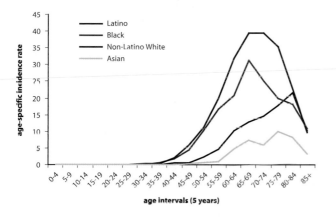

Figure 4: Age-adjusted incidence rate by social class.

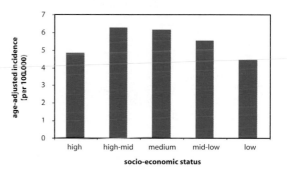

Figure 5: Distribution of the relative risk values for all census tracts.

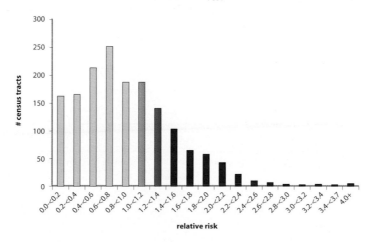

Figure 6: Census tracts by the number of cases per tract.

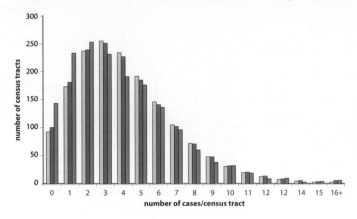

Figure 7a and b: Census tracts at high risk by the number of cases. (a) Unadjusted and (b) adjusted for social class.

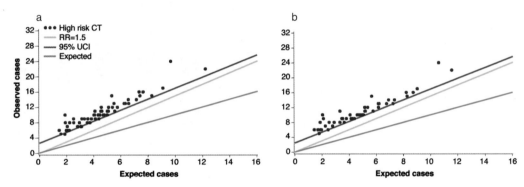

Figure 8: Risk over the period for high-risk census tracts relative to all census tracts.

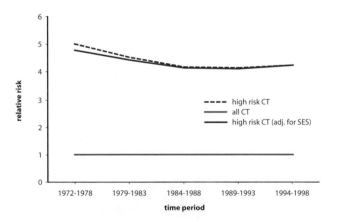

Figure 9: Map of census tracts at high risk.

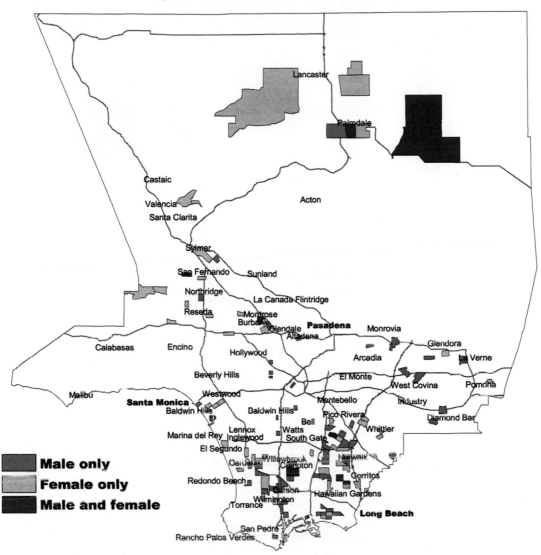

Figure 10: Male-female correlation between the relative risks for high-risk census tracts.

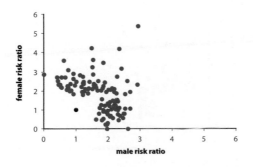

Figure 11: Map of census tracts at high risk, adjusted for social class.

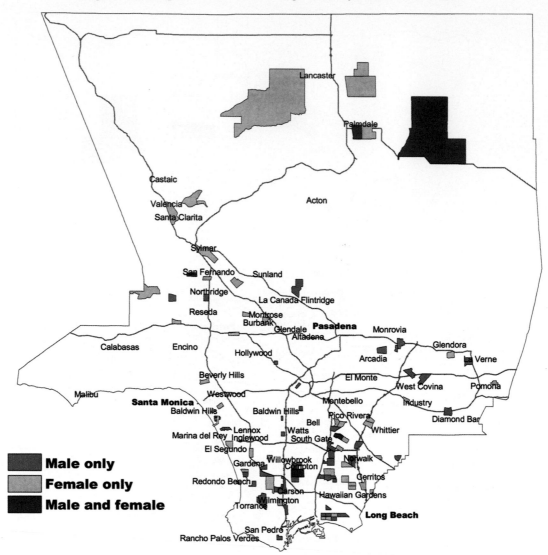

Figure 12: Male-female correlation between the relative risks for high-risk census tracts, adjusted for social class.

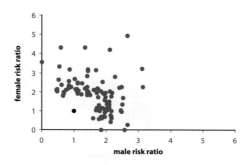

Adenocarcinoma of the Lung and Bronchus

ICDO-2 Code Anatomic Site: C 33, 34
ICDO-2 Code Histology: 8140–8247, 8260–8560, 8570–8573
Age: All
Male Cases: 16517
Female Cases: 13390

Background

Cigarette smoking adenocarcinoma of the lung, as it does other forms of lung cancer, although the causal association is weaker than it is for squamous cell or small cell lung cancer. It has been linked in particular to filter cigarettes, possibly because adenocarcinomas appear at the outer ends of the bronchioles (the terminal sacs or alveoli), and to get a satisfactory dose of smoke from a filter cigarette requires more vigorous inhalation, thus possibly distributing the smoke more widely. Now the most common form of lung cancer, adenocarcinoma has long been known to be especially common in women, particularly in China and Israel, and for unknown reasons to be generally increasing, not decreasing, in world-wide frequency. The oil vapors produced by cooking with unrefined rapeseed oil have been thought partly responsible for the higher rates among women in China. Adenocarcinomas also constitute the majority of cases found in association with scars from previous lung disease.

Local Pattern

Like other forms of lung cancer, adenocarcinoma has been less common in Utah, and slightly more common in San Francisco, than in Los Angeles County. Incidence is higher in men than in women, particularly in African-Americans and particularly with advancing age. Unlike squamous cell carcinoma, this cancer has not decreased in incidence in recent decades, and it has been increasing among women, both those in high-risk tracts and those in the county as a whole, it has been increasing. Also among women, but not men, those of higher social class are at higher risk. Figure 6 shows a strong nonrandom excess of census tracts with unexpectedly few or unexpectedly many cases. While some census tracts at high-risk for males are evident in the south of the county, the most concentrated area of risk is seen in the affluent census tracts of the upper west side, and these appear almost exclusively on the basis of risk to women. This pattern is only partly eliminated by adjustment for social class.

Thumbnail Interpretation

Since this form of lung cancer is known to be increasing among women, and to be more common among those with higher income and education, a different geographical distribution was predicted. Nonetheless, the reason for the striking geographical and demographic distribution is a matter for speculation, and as an observation it definitely requires attention. Since the high-risk among female residents of the upper west side of the city cannot be explained by social class alone, the pattern suggests that wealthy women in that area have different smoking habits, or some other different exposure, than wealthy women in other parts of the country.

Adenocarcinoma of the Lung and Bronchus: Male

Figure 1: Age-adjusted incidence rate by place.

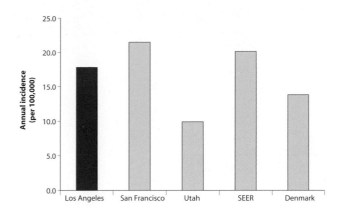

Figure 2: Age-adjusted incidence rate over the period.

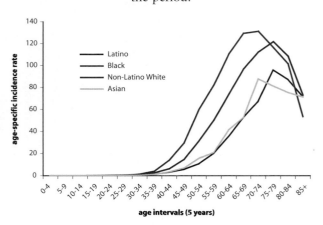

Figure 3: Age-adjusted incidence rate by age and race/ethnicity.

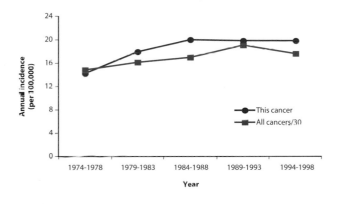

Figure 4: Age-adjusted incidence rate by social class.

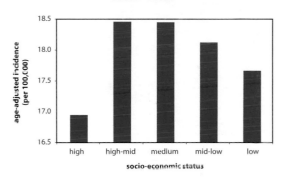

Figure 5: Distribution of the relative risk values for all census tracts.

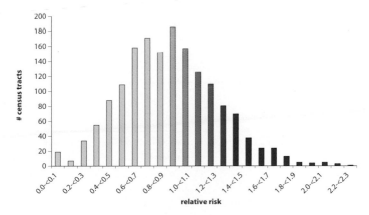

Figure 6: Census tracts by the number of cases per tract.

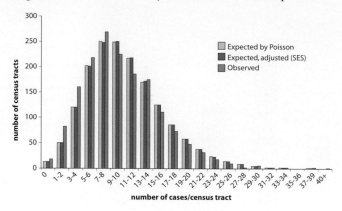

Figure 7a and b: Census tracts at high risk by the number of cases. (a) Unadjusted and (b) adjusted for social class.

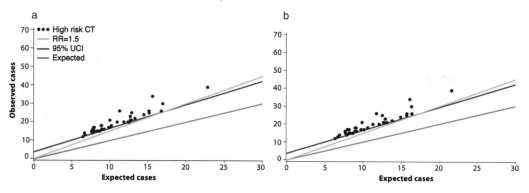

Figure 8: Risk over the period for high-risk census tracts relative to all census tracts.

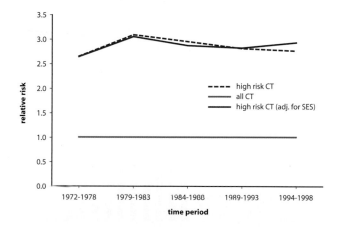

Adenocarcinoma of the Lung and Bronchus: Female

Figure 1: Age-adjusted incidence rate by place.

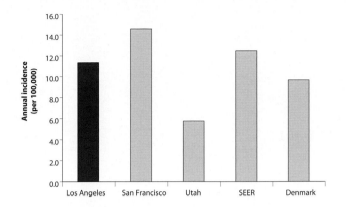

Figure 2: Age-adjusted incidence rate over the period.

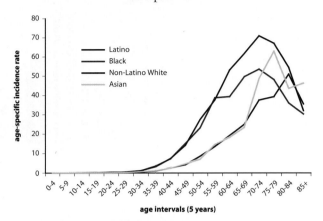

Figure 3: Age-adjusted incidence rate by age and race/ethnicity.

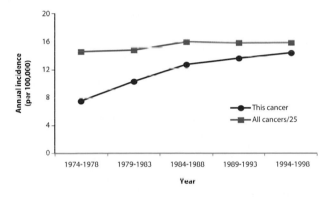

Figure 4: Age-adjusted incidence rate by social class.

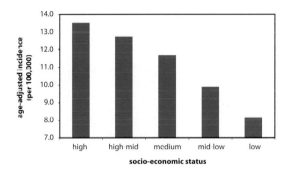

Figure 5: Distribution of the relative risk values for all census tracts.

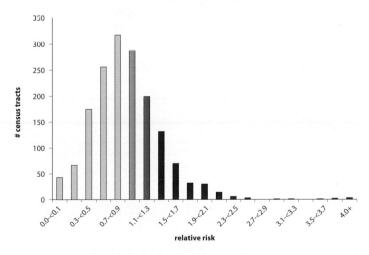

Figure 6: Census tracts by the number of cases per tract.

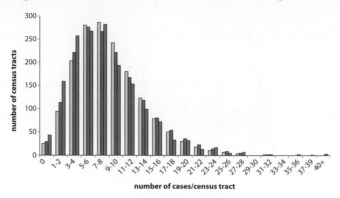

Figure 7a and b: Census tracts at high risk by the number of cases. (a) Unadjusted and (b) adjusted for social class.

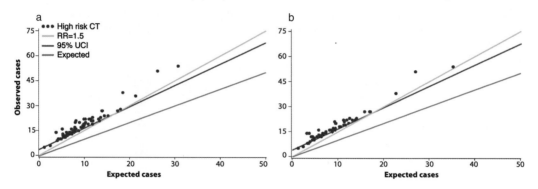

Figure 8: Risk over the period for high-risk census tracts relative to all census tracts.

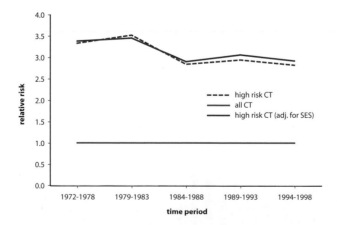

Figure 9: Map of census tracts at high risk.

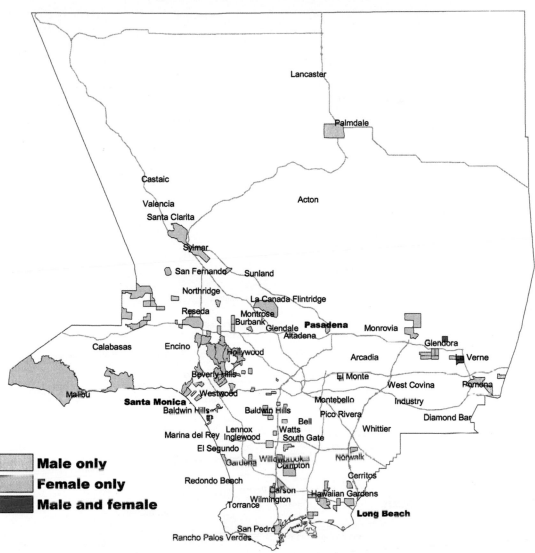

Figure 10: Male-female correlation between the relative risks for high-risk census tracts.

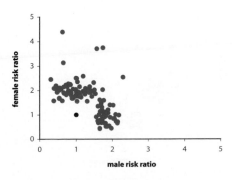

Figure 11: Map of census tracts at high risk, adjusted for social class.

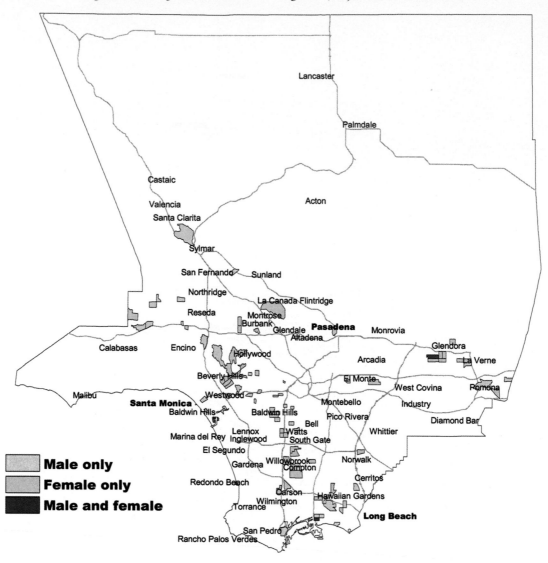

Figure 12: Male-female correlation between the relative risks for high-risk census tracts, adjusted for social class.

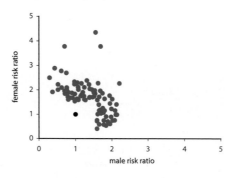

Bronchioloalveolar Carcinoma of the Lung and Bronchus

ICDO-2 Code Anatomic Site: C 33, 34
ICDO-2 Code Histology: 8250–8251
Age: All
Male Cases: 1945
Female Cases: 2409

Background

This uncommon carcinoma is believed to derive from the cells lining the terminal bronchioles and the alveoli, sacs at the end from which the oxygen is transferred from the air to the blood. It appears in many degrees of severity, from benign-appearing groups of cells to aggressive malignancy. There appears to be substantial variation in appearance as well as behavior, the malignancy sometimes appearing as a single large mass, and sometimes as multiple foci of malignant cells. It is the form of lung carcinoma most difficult to distinguish from pulmonary metastases, malignancies originating in other organs that have spread to the lung.

Local Pattern

This malignancy is also less common in Utah and Denmark, than in Los Angeles County. Whites are at highest risk among women and older men and risk is roughly equal between men and women. Also unlike adenocarcinoma of the lung, both men and women of higher social class are at higher risk of this malignancy. In the county as a whole, and in high-risk census tracts, incidence has increased slightly over time in women, but not men. Figure 6 shows only a slight nonrandom excess of census tracts with unexpectedly few or unexpectedly many cases. The census tracts at high risk, predominately on the basis of female cases, concentrate in the same upper west side area as do those at high-risk of adenocarcinoma, but with fewer high-risk tracts, and with a larger proportion that disappear after adjustment for social class.

Thumbnail Interpretation

While the lower risk in Utah suggests a link to smoking, the similar risk patterns among men and women, the high social class gradient, and the modest geographical concentration suggest that other determinants are important. The pattern does not suggest a specific hypothesis.

Figure 1: Age-adjusted incidence rate by place.

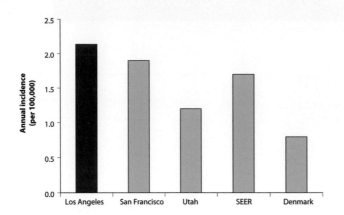

Figure 2: Age-adjusted incidence rate over the period.

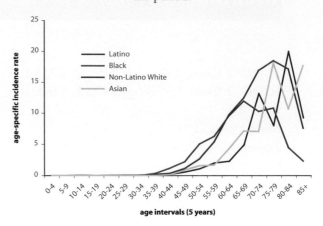

Figure 3: Age-adjusted incidence rate by age and race/ethnicity.

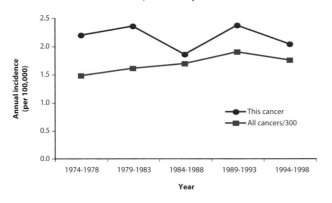

Figure 4: Age-adjusted incidence rate by social class.

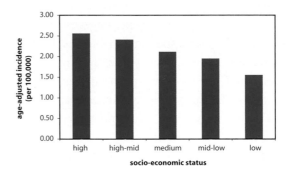

Figure 5: Distribution of the relative risk values for all census tracts.

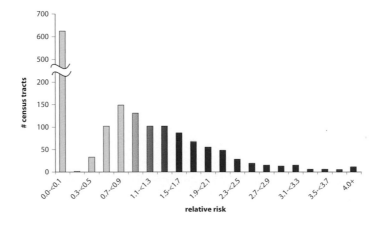

Figure 6: Census tracts by the number of cases per tract.

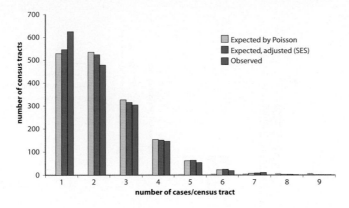

Figure 7a and b: Census tracts at high risk by the number of cases. (a) Unadjusted and (b) adjusted for social class.

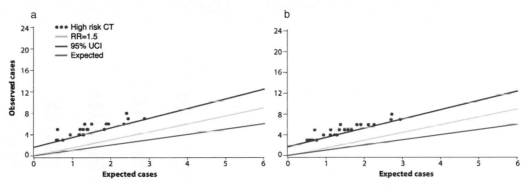

Figure 8: Risk over the period for high-risk census tracts relative to all census tracts.

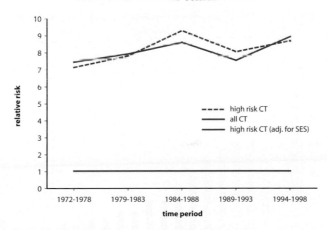

Figure 1: Age-adjusted incidence rate by place.

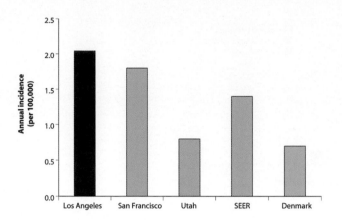

Figure 2: Age-adjusted incidence rate over the period.

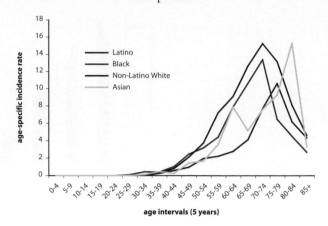

Figure 3: Age-adjusted incidence rate by age and race/ethnicity.

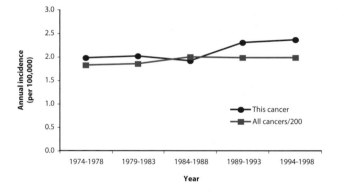

Figure 4: Age-adjusted incidence rate by social class.

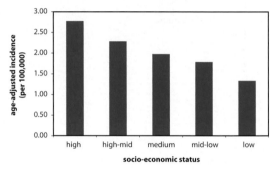

Figure 5: Distribution of the relative risk values for all census tracts.

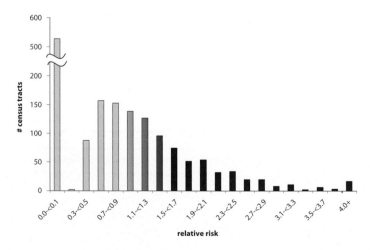

Figure 6: Census tracts by the number of cases per tract.

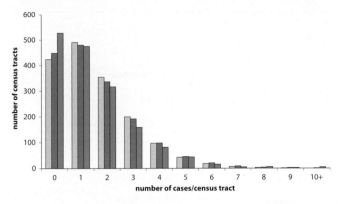

Figure 7a and b: Census tracts at high risk by the number of cases. (a) Unadjusted and (b) adjusted for social class.

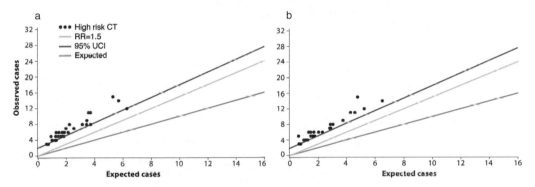

Figure 8: Risk over the period for high-risk census tracts relative to all census tracts.

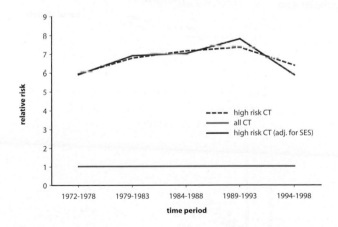

Figure 9: Map of census tracts at high risk.

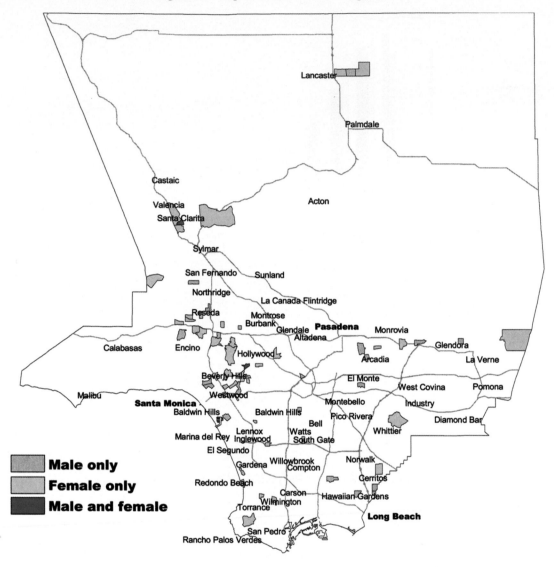

Figure 10: Male-female correlation between the relative risks for high-risk census tracts.

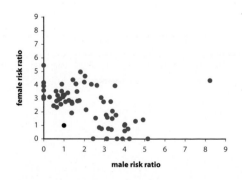

Figure 11: Map of census tracts at high risk, adjusted for social class.

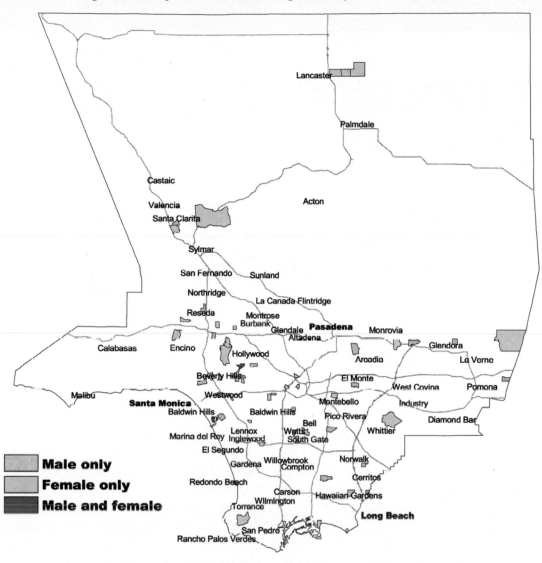

Figure 12: Male-female correlation between the relative risks for high-risk census tracts, adjusted for social class.

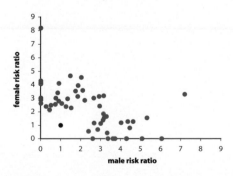

Undifferentiated Carcinoma of the Lung and Bronchus

ICDO-2 Code Anatomic Site: C 33, 34
ICDO-2 Code Histology: 8020–8034
Age: All
Male Cases: 1436
Female Cases: 836

Background

These lung cancers are comprised of cells thought by the pathologists to be very primitive, without features distinguishing them as cells of pulmonary origin. They could be viewed as tumor cells that have become more nonspecific as a result of uncontrolled growth, or as the cells of a malignancy deriving from the least-specialized cells of the lung, stem cells.

Local Pattern

These malignancies are also less common in Utah than elsewhere. They occur with roughly equal frequency in African-Americans and whites, but are much more common among men than among women. Incidence is more common among men of lower social class and women of middle class. In the county as a whole, but not in high-risk census tracts, the frequency of diagnoses has dropped dramatically over time, especially in men. Figure 6 shows little if any nonrandom excess of census tracts with unexpectedly few or unexpectedly many cases. No nonrandom geographic pattern of occurrence is evident in the maps.

Thumbnail Interpretation

The number of cases coded to this malignancy has probably decreased because pathologists have developed more sophisticated methods, and have increasingly classified lung malignancies that show undifferentiated features to other histologic types, especially adenocarcinoma. Otherwise, no systematic pattern of geographical occurrence is apparent, and therefore no local source of causation can be proposed.

Undifferentiated Carcinoma of the Lung and Bronchus: Male

Figure 1: Age-adjusted incidence rate by place.

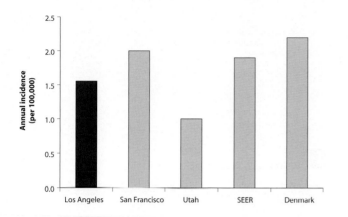

Figure 2: Age-adjusted incidence rate over the period.

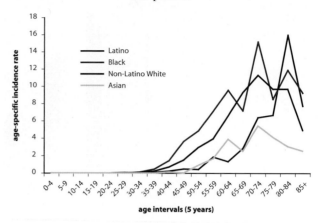

Figure 3: Age-adjusted incidence rate by age and race/ethnicity.

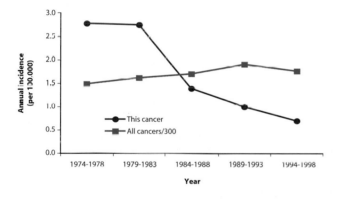

Figure 4: Age-adjusted incidence rate by social class.

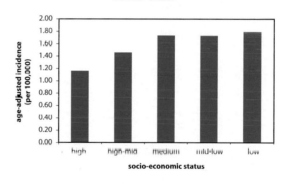

Figure 5: Distribution of the relative risk values for all census tracts.

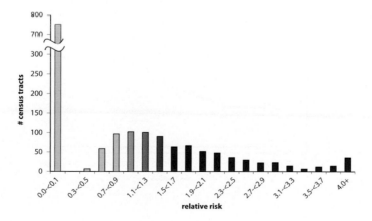

Figure 6: Census tracts by the number of cases per tract.

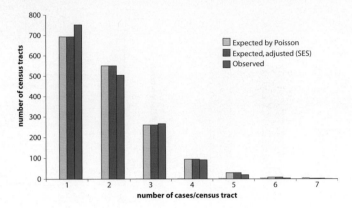

Figure 7a and b: Census tracts at high risk by the number of cases. (a) Unadjusted and (b) adjusted for social class.

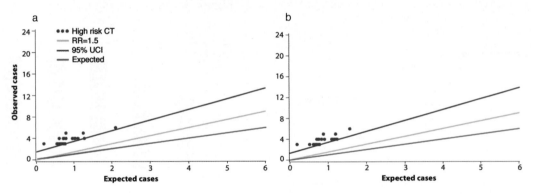

Figure 8: Risk over the period for high-risk census tracts relative to all census tracts.

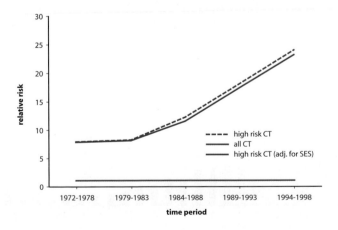

Figure 1: Age-adjusted incidence rate by place.

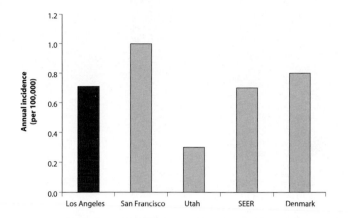

Figure 2: Age-adjusted incidence rate over the period.

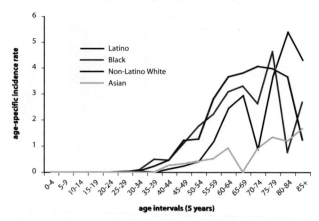

Figure 3: Age-adjusted incidence rate by age and race/ethnicity.

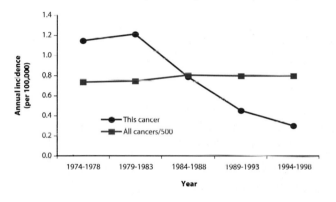

Figure 4: Age-adjusted incidence rate by social class.

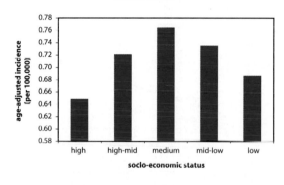

Figure 5: Distribution of the relative risk values for all census tracts.

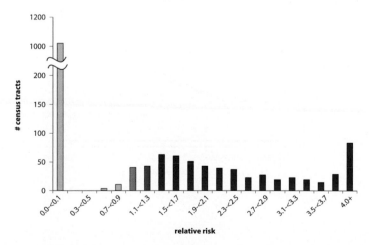

Figure 6: Census tracts by the number of cases per tract.

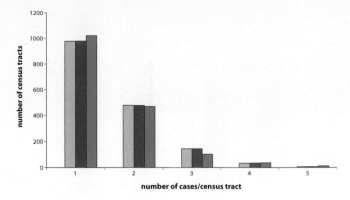

Figure 7a and b: Census tracts at high risk by the number of cases. (a) Unadjusted and (b) adjusted for social class.

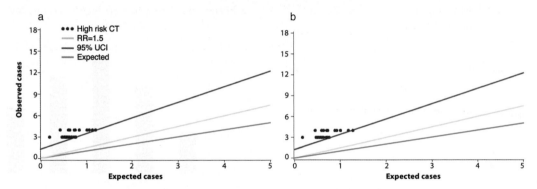

Figure 8: Risk over the period for high-risk census tracts relative to all census tracts.

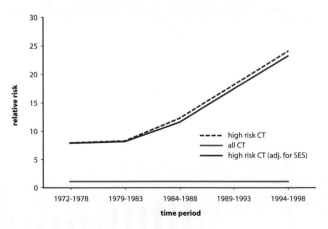

Figure 9: Map of census tracts at high risk.

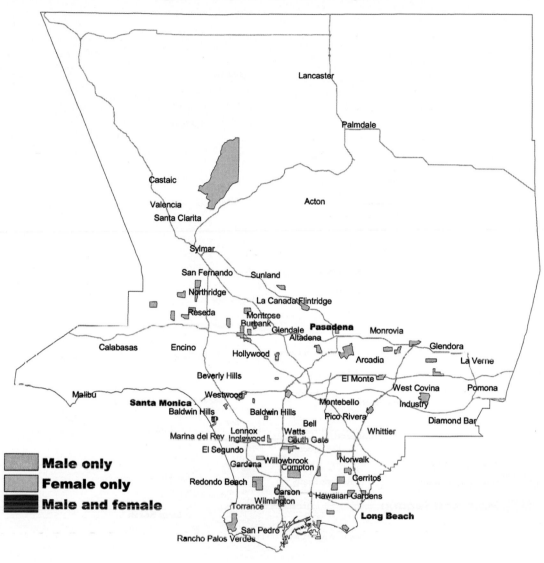

Figure 10: Male-female correlation between the relative risks for high-risk census tracts.

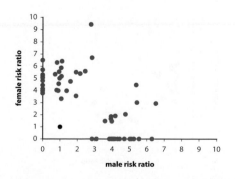

Figure 11: Map of census tracts at high risk, adjusted for social class.

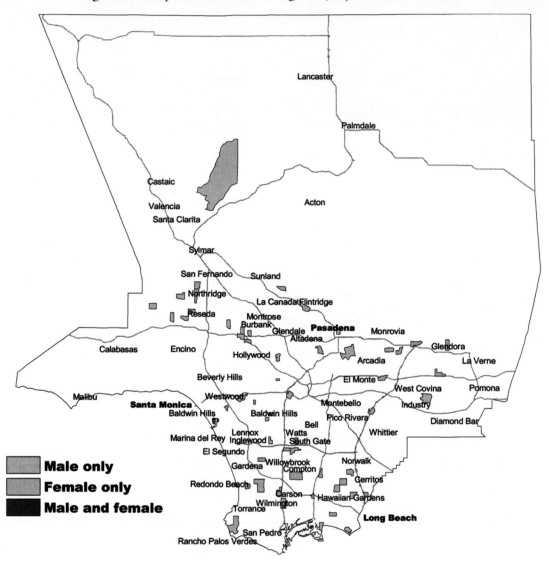

Figure 12: Male-female correlation between the relative risks for high-risk census tracts, adjusted for social class.

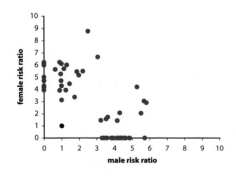

Carcinoma of the Lung and Bronchus, Not Otherwise Classified

ICDO-2 Code Anatomic Site: C 33, 34
ICDO-2 Code Histology: 8000–8011
Age: All
Male Cases: 13230
Female Cases: 8271

Background

Please refer to the discussion of lung and bronchus cancer, total. This category of lung cancers consists of cancers that were described in the medical records only in very general terms, in many cases because no biopsy was performed and the diagnosis was based on clinical grounds. As a group they would be expected to have the same causes as the total set of lung cancers.

Local Pattern

Lung cancers of unclassified histology, like other lung cancers were particularly uncommon in Utah, but also uncommon in Denmark. Risk to men is substantially higher than risk to women, and whereas among those under 70, African-Americans are at highest risk; rates among whites become highest after that age. Men of lower social class and women of middle class are at higher risk. Rates of incidence have been stable among men, but have increased somewhat among women, both in the county as a whole, and in high-risk census tracts. Figure 6 shows a strong nonrandom excess of census tracts with unexpectedly few or unexpectedly many cases. Census tracts at high-risk appear throughout the southern part of the county, with another area of aggregation south of the mountains in the northeastern part of the Los Angeles basin.

Thumbnail Interpretation

This group of lung cancers shows a pattern of occurrence consistent with the occurrence of squamous and small cell lung cancers combined, presumably because smoking is largely responsible.

Figure 1: Age-adjusted incidence rate by place.

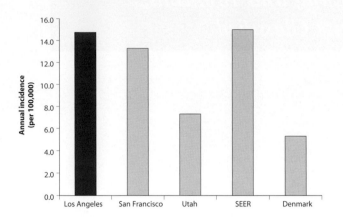

Figure 2: Age-adjusted incidence rate over the period.

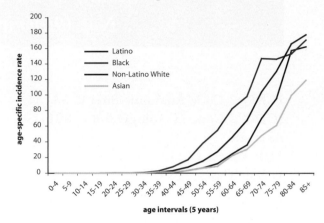

Figure 3: Age-adjusted incidence rate by age and race/ethnicity.

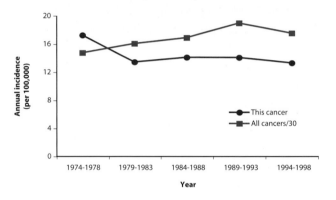

Figure 4: Age-adjusted incidence rate by social class.

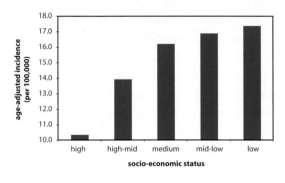

Figure 5: Distribution of the relative risk values for all census tracts.

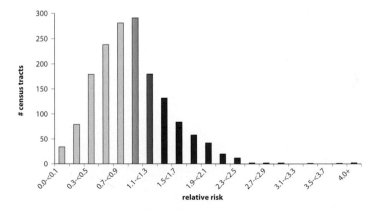

Figure 6: Census tracts by the number of cases per tract.

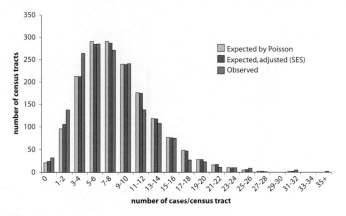

Figure 7a and b: Census tracts at high risk by the number of cases. (a) Unadjusted and (b) adjusted for social class.

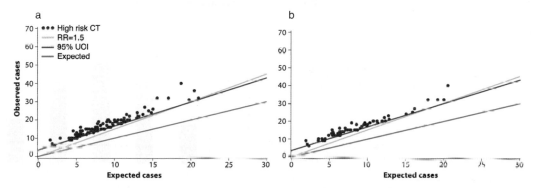

Figure 8: Risk over the period for high-risk census tracts relative to all census tracts.

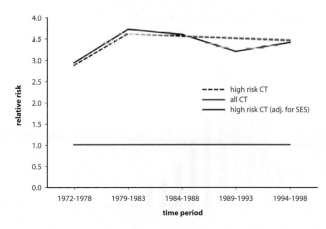

Figure 1: Age-adjusted incidence rate by place.

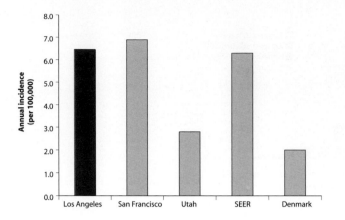

Figure 2: Age-adjusted incidence rate over the period.

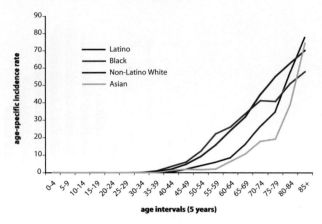

Figure 3: Age-adjusted incidence rate by age and race/ethnicity.

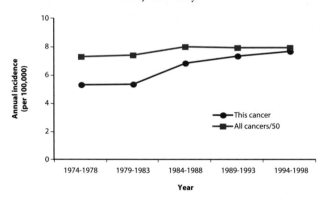

Figure 4: Age-adjusted incidence rate by social class.

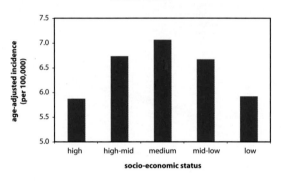

Figure 5: Distribution of the relative risk values for all census tracts.

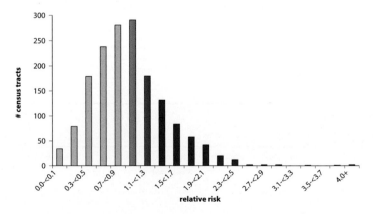

Figure 6: Census tracts by the number of cases per tract.

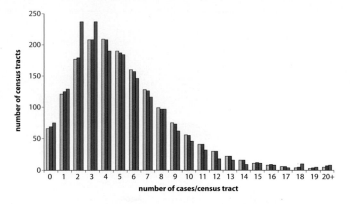

Figure 7a and b: Census tracts at high risk by the number of cases. (a) Unadjusted and (b) adjusted for social class.

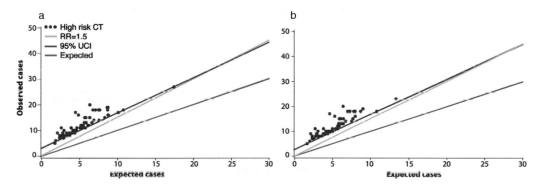

Figure 8: Risk over the period for high-risk census tracts relative to all census tracts.

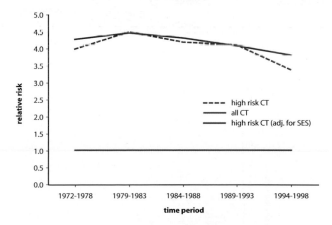

Figure 9: Map of census tracts at high risk.

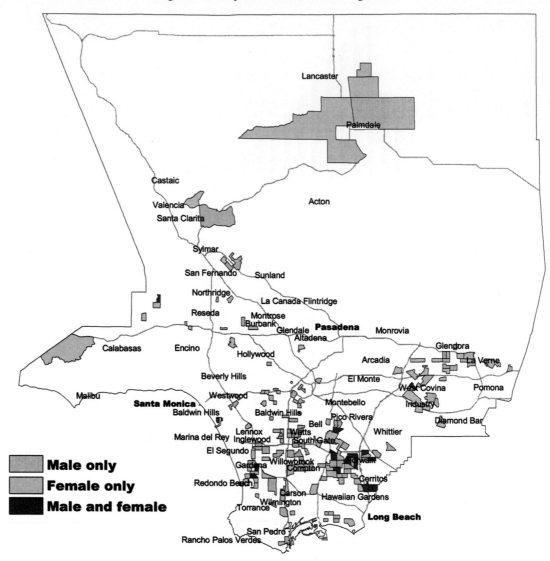

Figure 10: Male-female correlation between the relative risks for high-risk census tracts.

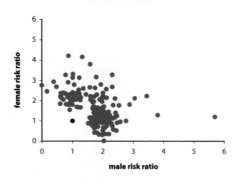

Figure 11: Map of census tracts at high risk, adjusted for social class.

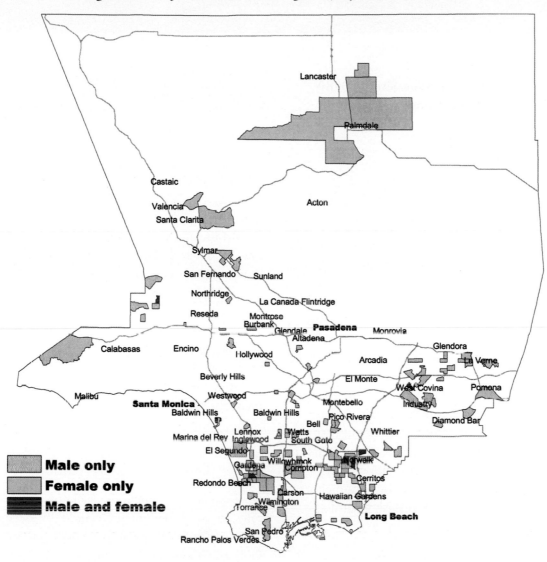

Figure 12: Male-female correlation between the relative risks for high-risk census tracts, adjusted for social class.

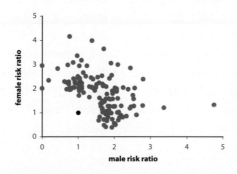

Mesothelioma

ICDO-2 Code Anatomic Site: C 0–80
ICDO-2 Code Histology: 9050–9055
Age: All
Male Cases: 1095
Female Cases: 318

Background

The most important cause of mesothelioma is exposure to various forms of asbestos in the workplace. Smoking does not appear to increase risk, although asbestos may amplify the effect of smoking in the etiology of certain forms of lung cancer. Rarely, household exposure to asbestos has been responsible, because of asbestos carried home in the clothing of working family members, or because of proximity to an asbestos mine or mill. Clusters of mesothelioma have occurred because of the use of asbestos minerals as ingredients in whitewash. A closely related mineral, erionite, has been found responsible for mesotheliomas occurring in an area of Turkish Anatolia.

Local Pattern

Mesothelioma occurs many times more commonly in men than women, and is more common in San Francisco than in Los Angeles. Incidence begins to rise in middle age, and among men, cases among whites and Latinos are more common than those among persons of other groups. Incidence has increased with time in the county as a whole, although male residents of high-risk census tracts experienced higher incidence early in the time period.

Persons of middle class have been more commonly affected than persons of either high or low social class. Figure 6 shows a moderate nonrandom excess of census tracts with unexpectedly few or unexpectedly many cases. Some census tracts are at very high risk, especially among men. High-risk census tracts, almost exclusively based on risk to men, are most common in the southeast of the county, north of Long Beach, before and after adjustment for social class.

Thumbnail Interpretation

Because mesothelioma is known historically to be a malignancy caused by occupational exposure, it would be expected to occur more often among those who have been blue-collar workers. Exposure to asbestos among shipyard workers took place in Long Beach during and just after World War II, it is reasonable to expect that mesothelioma would more commonly occur among the white and Latino men who once held those jobs. Because the census tracts showing high-risk for men are distributed in a semi-circle around Long Beach, it seems possible that many of these census tracts are at high-risk because malignancies have occurred among such men.

Figure 1: Age-adjusted incidence rate by place.

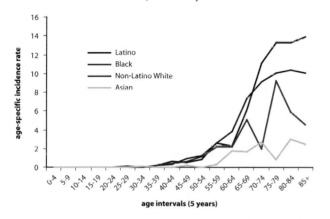

Figure 2: Age-adjusted incidence rate over the period.

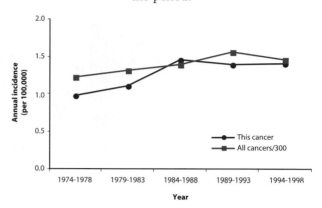

Figure 3: Age-adjusted incidence rate by age and race/ethnicity.

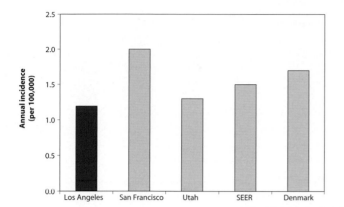

Figure 4: Age-adjusted incidence rate by social class.

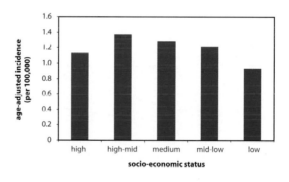

Figure 5: Distribution of the relative risk values for all census tracts.

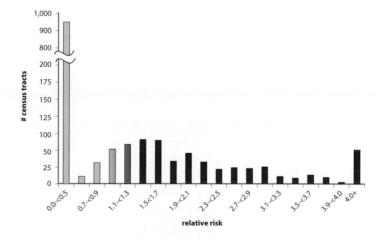

Figure 6: Census tracts by the number of cases per tract.

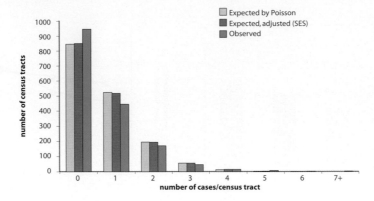

Figure 7a and b: Census tracts at high risk by the number of cases. (a) Unadjusted and (b) adjusted for social class.

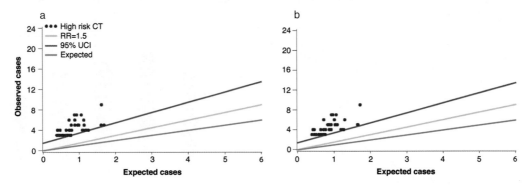

Figure 8: Risk over the period for high-risk census tracts relative to all census tracts.

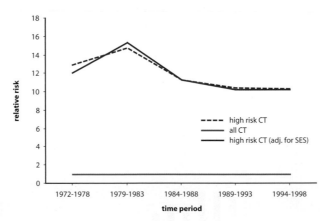

Figure 1: Age-adjusted incidence rate by place.

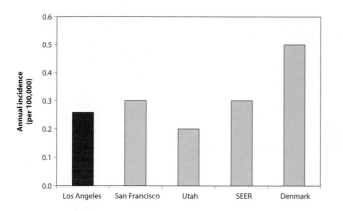

Figure 2: Age-adjusted incidence rate over the period.

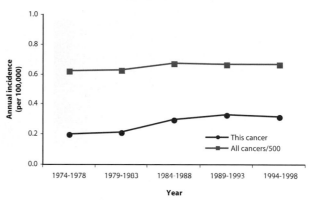

Figure 3: Age-adjusted incidence rate by age and race/ethnicity.

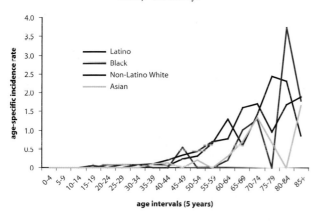

Figure 4: Age-adjusted incidence rate by social class.

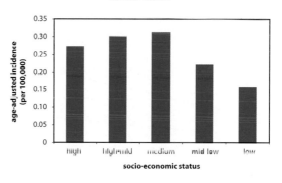

Figure 5: Distribution of the relative risk values for all census tracts.

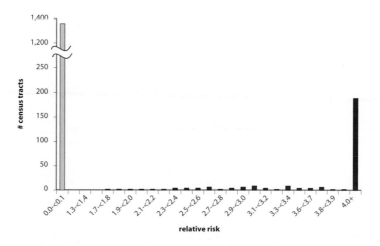

Figure 6: Census tracts by the number of cases per tract.

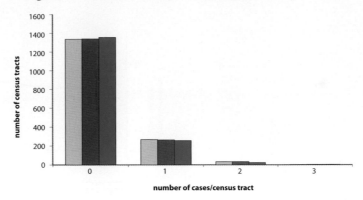

Figure 7a and b: Census tracts at high risk by the number of cases. (a) Unadjusted and (b) adjusted for social class.

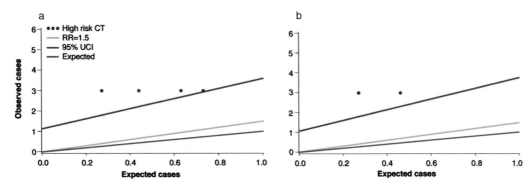

Figure 8: Risk over the period for high-risk census tracts relative to all census tracts.

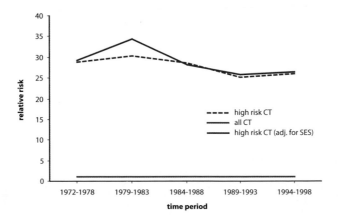

Figure 9: Map of census tracts at high risk.

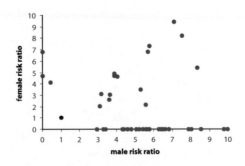

Figure 10: Male-female correlation between the relative risks for high-risk census tracts.

Figure 11: Map of census tracts at high risk, adjusted for social class.

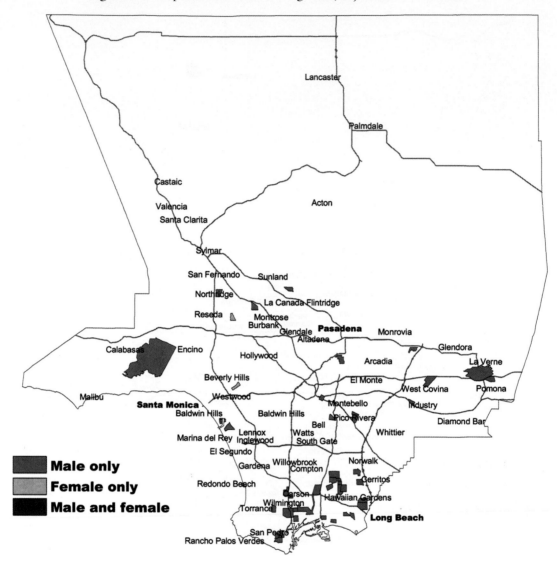

Figure 12: Male-female correlation between the relative risks for high-risk census tracts, adjusted for social class.

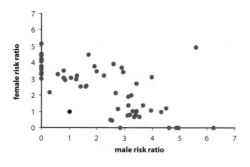

Soft Tissue Sarcoma

ICDO-2 Code Anatomic Site: C 0–80
ICDO-2 Code Histology: 8800–8814, 8830–8833, 8840–8841, 8850–8861,
8890–8897, 8900–8920, 8930–8933, 8963–8991, 9020–9044, 9251, 9261–9330
Age: All
Male Cases: 4531
Female Cases: 6127

Background

This set of malignancies consists of a very large number of different but related tumors deriving from the structural cells of the body. They are lumped together because each individual group is rather rare, and because the known causes tend to be common to all subgroups. Sarcomas are known to occur after high-dose radiotherapy, and after immunosuppression, such as that produced by the drugs given to facilitate transplantation. Persons with any of several rare genetic syndromes are at higher risk of sarcomas. Long-term or intense exposure to certain herbicides or related chemicals (including the dioxins) has been suspected of causing these malignancies. As a group, soft tissue sarcomas occur somewhat more commonly among postmenopausal women, especially African-American women who are known to be at higher risk of a smooth muscle sarcoma of the uterus.

Local Pattern

Sarcomas occur slightly more often among females than among males, but with equal frequency in Los Angeles County and other parts of the country. They occur at all ages, but incidence begins to increase in young adulthood, particularly in women. Among such women, African-Americans are at higher risk and Asian-Americans at lower risk. There is a generally slightly higher rate among those of higher social class. Risk has increased slightly over time in the County as a whole, less so among the residents of high-risk census tracts. Figure 6 shows a slight nonrandom excess of census tracts with unexpectedly few or unexpectedly many cases. The census tracts fulfilling the high-risk criteria appear scattered throughout the county before and after adjustment for social class, with only a few contiguous census tracts at high risk. No census tract stands out on the basis of a particularly high relative risk and number of excess cases.

Thumbnail Interpretation

No systematic pattern of geographical occurrence is apparent, and therefore no local source of causation can be proposed.

Figure 1: Age-adjusted incidence rate by place.

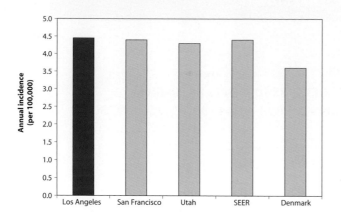

Figure 2: Age-adjusted incidence rate over the period.

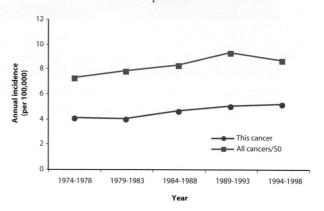

Figure 3: Age-adjusted incidence rate by age and race/ethnicity.

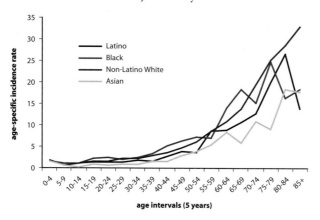

Figure 4: Age-adjusted incidence rate by social class.

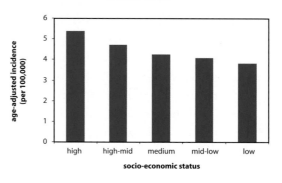

Figure 5: Distribution of the relative risk values for all census tracts.

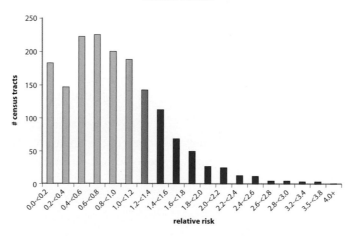

Figure 6: Census tracts by the number of cases per tract.

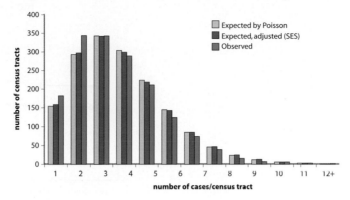

Figure 7a and b: Census tracts at high risk by the number of cases. (a) Unadjusted and (b) adjusted for social class.

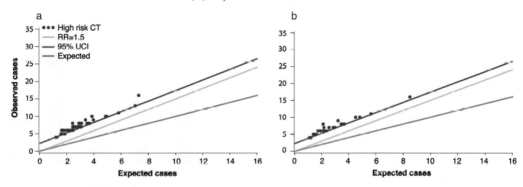

Figure 8: Risk over the period for high-risk census tracts relative to all census tracts.

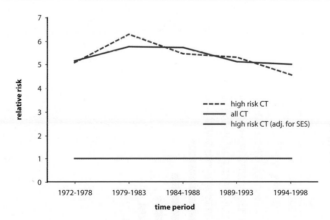

Figure 1: Age-adjusted incidence rate by place.

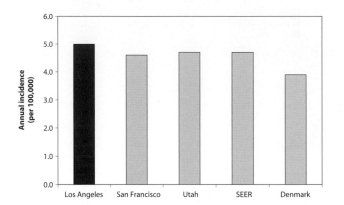

Figure 2: Age-adjusted incidence rate over the period.

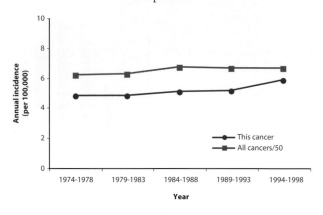

Figure 3: Age-adjusted incidence rate by age and race/ethnicity.

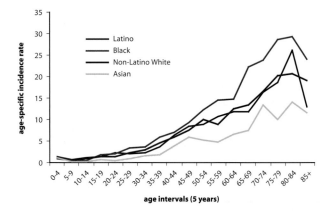

Figure 4: Age-adjusted incidence rate by social class.

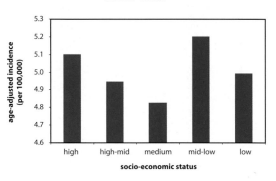

Figure 5: Distribution of the relative risk values for all census tracts.

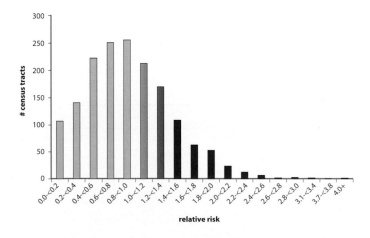

Figure 6: Census tracts by the number of cases per tract.

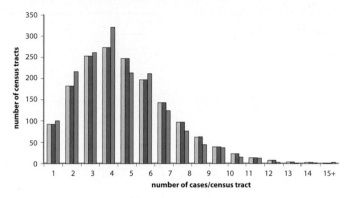

Figure 7a and b: Census tracts at high risk by the number of cases. (a) Unadjusted and (b) adjusted for social class.

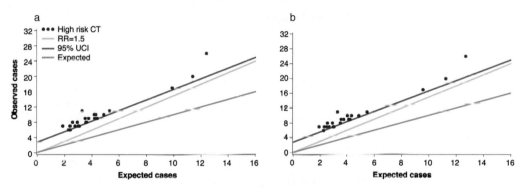

Figure 8: Risk over the period for high-risk census tracts relative to all census tracts.

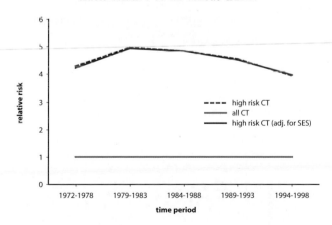

Figure 9: Map of census tracts at high risk.

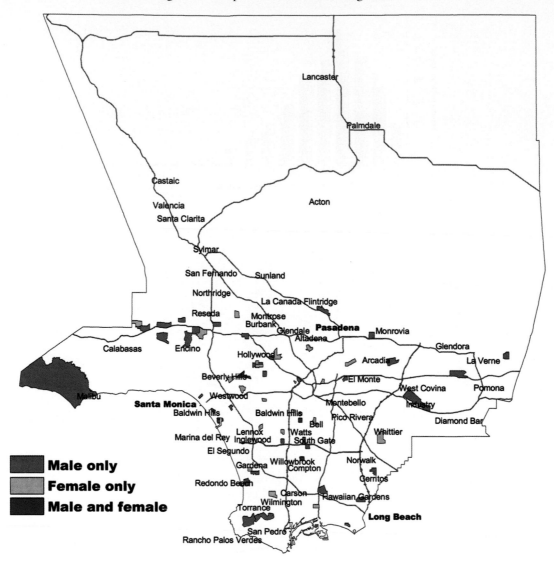

Figure 10: Male-female correlation between the relative risks for high-risk census tracts.

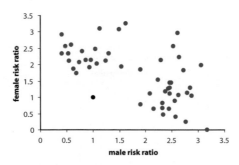

Figure 11: Map of census tracts at high risk, adjusted for social class.

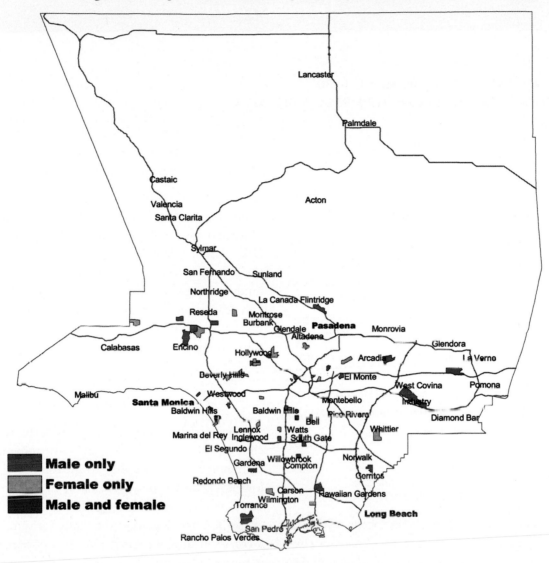

Figure 12: Male-female correlation between the relative risks for high-risk census tracts, adjusted for social class.

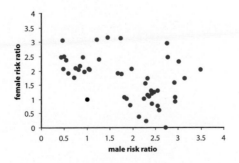

Angiosarcoma

ICDO-2 Code Anatomic Site: C 0–80
ICDO-2 Code Histology: 9120–9134, 9141–9175
Age: All
Male Cases: 224
Female Cases: 242

Background

Angiosarcomas are malignancies of the blood vessels that can occur anywhere in the body. Angiosarcomas of the liver have been caused by thorotrast (a radioactive diagnostic agent) or by exposure to the chemicals vinyl chloride or arsenic in the workplace. Women treated by radiation for breast cancer sometimes develop angiosarcomas occurring in the breast or in the arm on the same side as the treated breast, probably because of damage to blocked lymphatic vessels.

Local Pattern

These rare malignancies occur with the same frequency in Los Angeles County as elsewhere, and show no great predilection for persons by race. Although incidence begins rising in young adulthood, especially among women, the malignancies are more common in older persons and those of relatively high social class. Risk has remained constant over time in the county as a whole. Figure 6 shows no non-random excess of census tracts with unexpectedly few or unexpectedly many cases. Only one census tract fulfilled the high-risk criteria, presumably on the basis of chance.

Thumbnail Interpretation

Despite the potential relationship to vinyl chloride or arsenic, which are environmental contaminants that are easy to measure at very low levels and widely present in most urban areas, including Los Angeles County, the exposure is too low to result in any measurable increase in angiosarcoma risk. No systematic pattern of geographical occurrence is apparent, and therefore no local source of causation can be proposed.

Angiosarcoma: Male

Figure 1: Age-adjusted incidence rate by place.

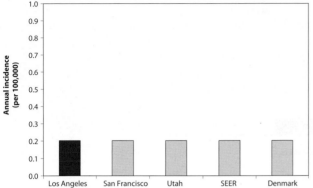

Figure 2: Age-adjusted incidence rate over the period.

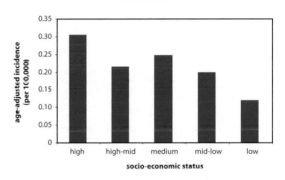

Figure 3: Age-adjusted incidence rate by age and race/ethnicity.

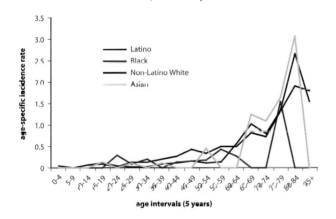

Figure 4: Age-adjusted incidence rate by social class.

Figure 5: Distribution of the relative risk values for all census tracts.

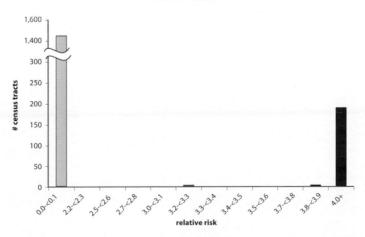

Figure 6: Census tracts by the number of cases per tract.

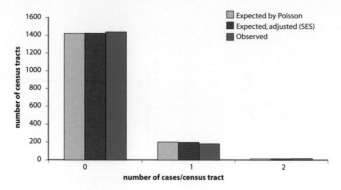

Figure 7a and b: Census tracts at high risk by the number of cases. (a) Unadjusted and (b) adjusted for social class.

There were no high risk census tracts

Figure 8: Risk over the period for high-risk census tracts relative to all census tracts.

There were no high risk census tracts

Figure 1: Age-adjusted incidence rate by place.

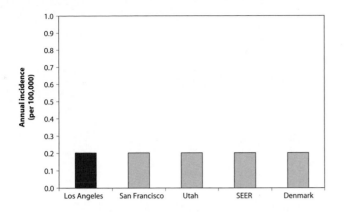

Figure 2: Age-adjusted incidence rate over the period.

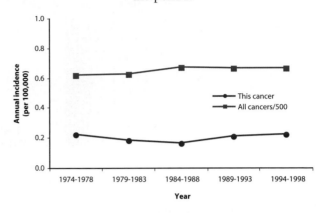

Figure 3: Age-adjusted incidence rate by age and race/ethnicity.

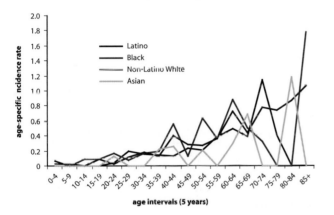

Figure 4: Age-adjusted incidence rate by social class.

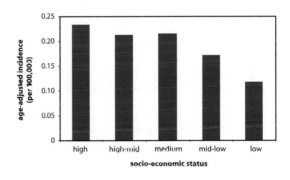

Figure 5: Distribution of the relative risk values for all census tracts.

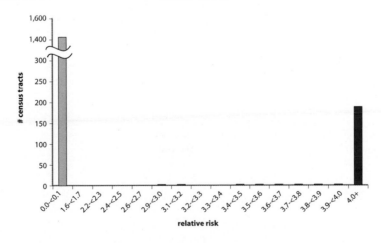

Figure 6: Census tracts by the number of cases per tract.

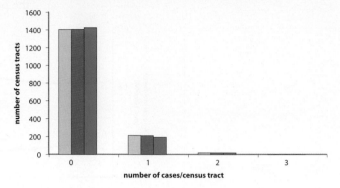

Figure 7a and b: Census tracts at high risk by the number of cases. (a) Unadjusted and (b) adjusted for social class.

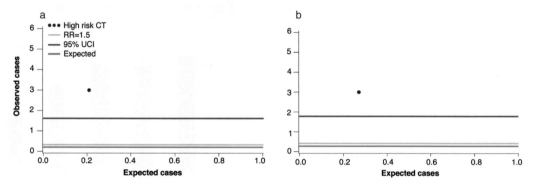

Figure 8: Risk over the period for high-risk census tracts relative to all census tracts.

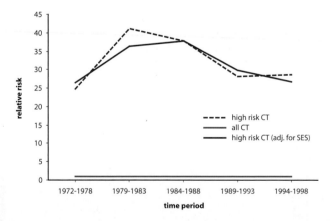

Figure 9: Map of census tracts at high risk.

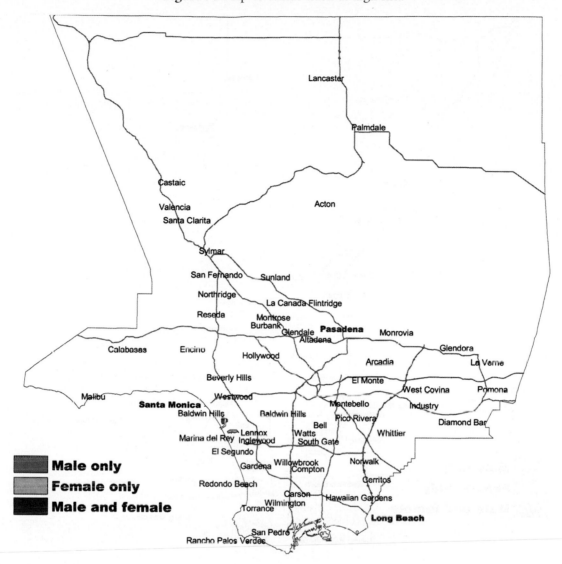

Figure 10: Male-female correlation between the relative risks for high-risk census tracts.

There were no high risk census tracts

Figure 11: Map of census tracts at high risk, adjusted for social class.

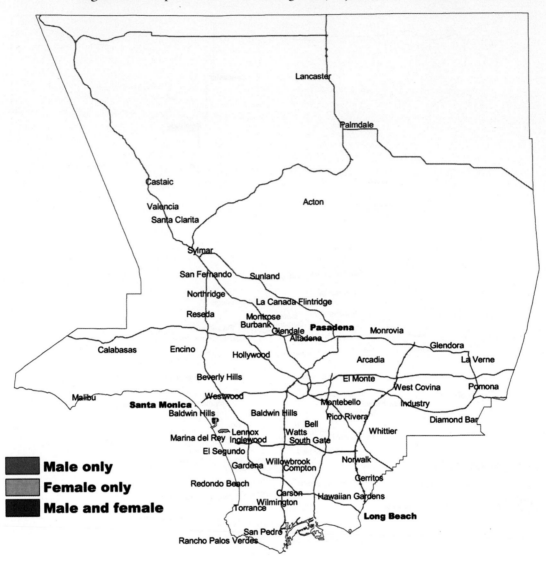

Figure 12: Male-female correlation between the relative risks for high-risk census tracts, adjusted for social class.

There were no high risk census tracts

Kaposi Sarcoma

ICDO-2 Code Anatomic Site: C 0–80
ICDO-2 Code Histology: 9140
Age: All
Male Cases: 6893
Female Cases: 255

Background

This malignancy of small blood vessels, once quite rare, is caused by a specific virus (human herpes virus 8) and now occurs mainly in the setting of AIDS when immunological competence is destroyed. A small number of Kaposi sarcomas, especially in East Africans but also in persons from central and southern Europe, occur without any obvious immunological abnormality other than that (in East Africa) related to malaria.

Local Pattern

Kaposi sarcoma is twenty times more common among men than women in Los Angeles County, and is even more common among men in San Francisco. In the AIDS era it has been a disease of young and middle-aged men, and the cases in women have mostly been older. Asians are at lower risk. In women occurrence in the county as a whole has been relatively constant over time, but in men this malignancy began increasing in the middle 1970s, reached a peak in about 1990, and has since begun to decrease, especially among residents of high-risk census tracts. Figure 6 shows an extreme nonrandom excess of census tracts with very few or very many cases.

Geographically, many high-risk census tracts form contiguous blocks in a region extending from West Hollywood to Silver Lake, with other contiguous groups in Long Beach and Marina del Rey.

Thumbnail Interpretation

Incidence of Kaposi sarcoma among men increased greatly over the period as the AIDS epidemic progressed, but with effective and available therapy for HIV infection, incidence has subsequently decreased, as have other conditions associated with late-stage AIDS. As would be predicted by the frequency of AIDS generally, incidence of this malignancy is more common among white and African-Americans, and is more common in Los Angeles and especially in San Francisco. Within Los Angeles County, the malignancy is concentrated in those regions of the city with populations of gay men especially vulnerable to AIDS. These appear in a band from West Hollywood across to Silver Lake, in Marina del Rey, and in the Signal Hill and Belmont Shores districts of Long Beach. Among women, a hint of the same distribution is present, in addition to the seemingly random distribution consistent with the pre-AIDS occurrence of Kaposi sarcoma.

Figure 1: Age-adjusted incidence rate by place.

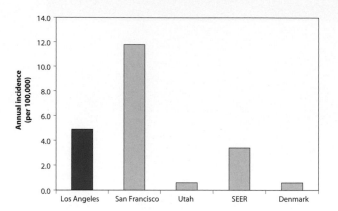

Figure 2: Age-adjusted incidence rate over the period.

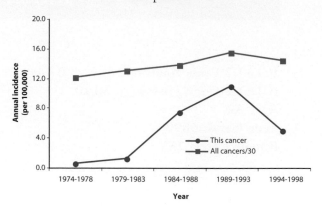

Figure 3: Age-adjusted incidence rate by age and race/ethnicity.

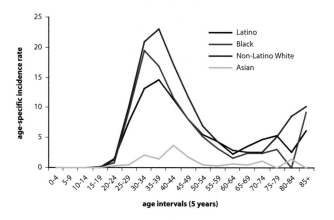

Figure 4: Age-adjusted incidence rate by social class.

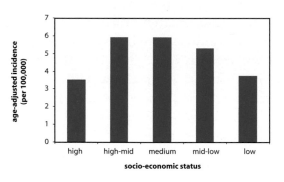

Figure 5: Distribution of the relative risk values for all census tracts.

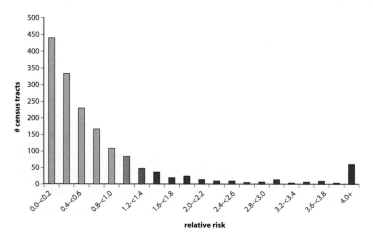

Figure 6: Census tracts by the number of cases per tract.

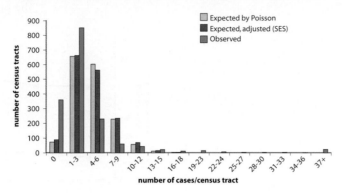

Figure 7a and b: Census tracts at high risk by the number of cases. (a) Unadjusted and (b) adjusted for social class.

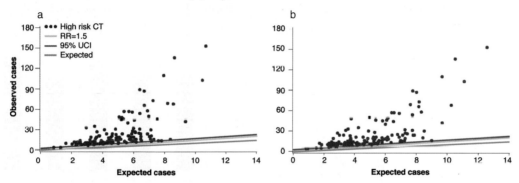

Figure 8: Risk over the period for high-risk census tracts relative to all census tracts.

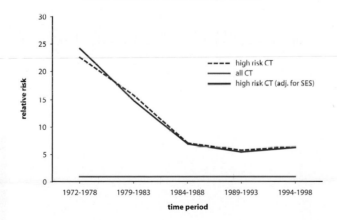

Kaposi Sarcoma: Female

Figure 1: Age-adjusted incidence rate by place.

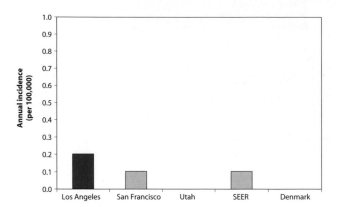

Figure 2: Age-adjusted incidence rate over the period.

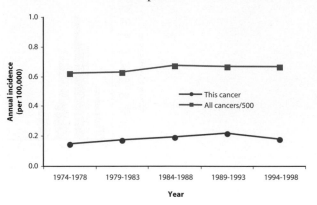

Figure 3: Age-adjusted incidence rate by age and race/ethnicity.

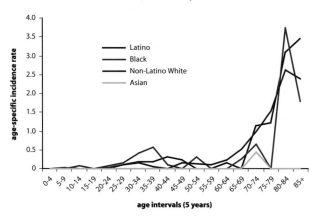

Figure 4: Age-adjusted incidence rate by social class.

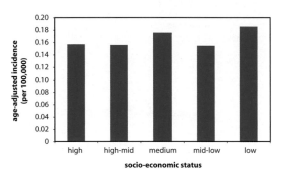

Figure 5: Distribution of the relative risk values for all census tracts.

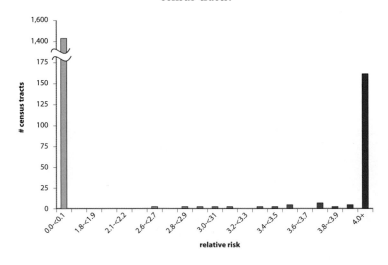

Figure 6: Census tracts by the number of cases per tract.

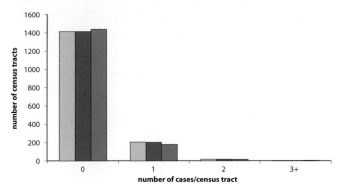

number of cases/census tract

Figure 7a and b: Census tracts at high risk by the number of cases. (a) Unadjusted and (b) adjusted for social class.

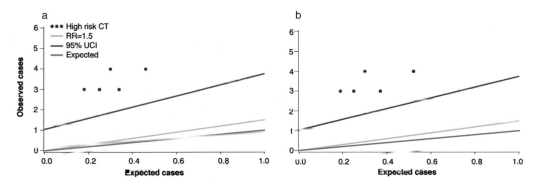

Figure 8: Risk over the period for high-risk census tracts relative to all census tracts.

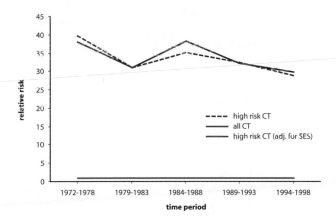

Figure 9: Map of census tracts at high risk.

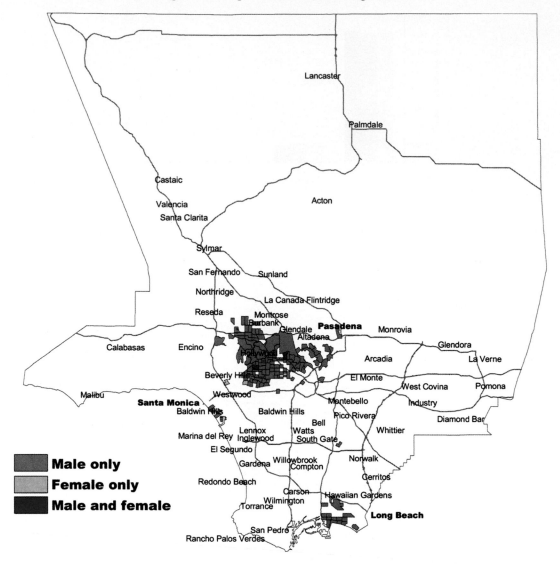

Figure 10: Male-female correlation between the relative risks for high-risk census tracts.

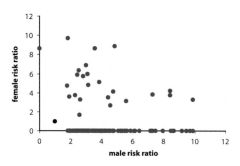

Kaposi Sarcoma

Figure 11: Map of census tracts at high risk, adjusted for social class.

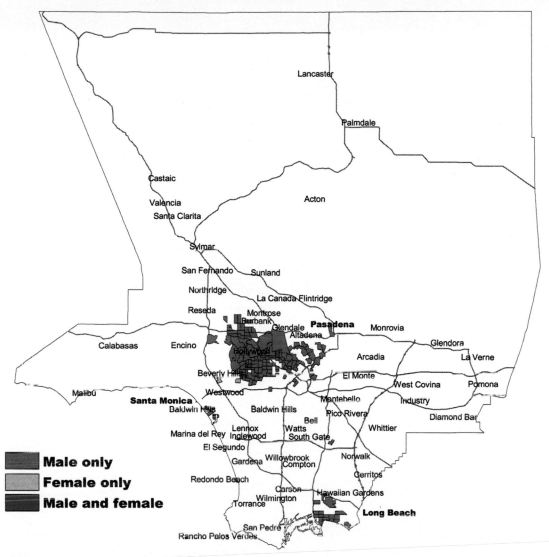

Figure 12: Male-female correlation between the relative risks for high-risk census tracts, adjusted for social class.

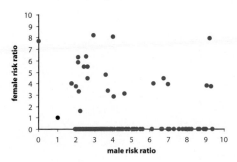

Osteosarcoma

ICDO-2 Code Anatomic Site: C 0–80
ICDO-2 Code Histology: 9180–9240
Age: All
Male Cases: 665
Female Cases: 615

Background

These bone cancers have few known causes. They occasionally appear after high-dose radiation treatment for other conditions and after treatment with certain chemotherapeutic agents. They occur in bones affected by a condition of bone called Paget's disease (not the same Paget's disease as in the breast) and by any of a number of other inherited conditions of bone. Osteosarcomas also occur in the members of families at genetic risk of multiple different forms of cancer.

Local Pattern

Osteosarcomas are only slightly more common in men than women, and appear with equal frequency in Los Angeles County and other parts of the country. They occur in all races and in both children and adults, although among Asian-Americans, older adults appear to be at lower risk. Risk does not vary by social class, and has been constant over time in the county as a whole, although residents of high-risk census tracts were at higher risk in the earlier part of the period covered. Figure 6 shows a slight nonrandom excess of census tracts with unexpectedly few or unexpectedly many cases. Census tracts fulfilling the high-risk criteria appear scattered throughout the county before and after adjustment for social class. No census tract stands out on the basis of a particularly high number of excess cases.

Thumbnail Interpretation

The reasons for the nonrandom pattern of osteosarcomas in relation to sex and age are unknown. No systematic pattern of geographical occurrence is apparent, and therefore no local source of causation can be proposed.

Figure 1: Age-adjusted incidence rate by place.

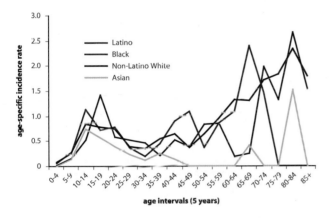

Figure 2: Age-adjusted incidence rate over the period.

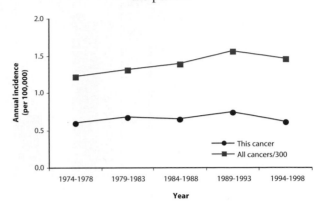

Figure 3: Age-adjusted incidence rate by age and race/ethnicity.

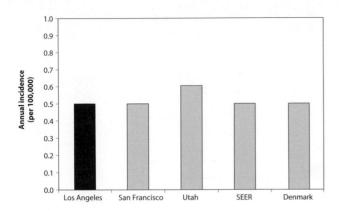

Figure 4: Age-adjusted incidence rate by social class.

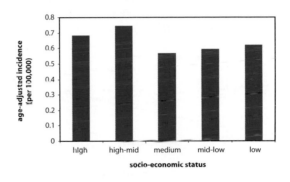

Figure 5: Distribution of the relative risk values for all census tracts.

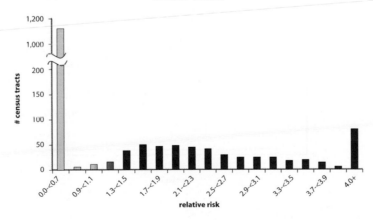

Figure 6: Census tracts by the number of cases per tract.

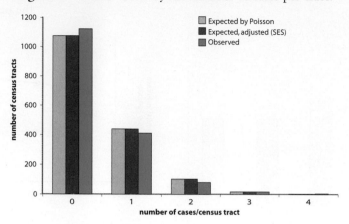

Figure 7a and b: Census tracts at high risk by the number of cases. (a) Unadjusted and (b) adjusted for social class.

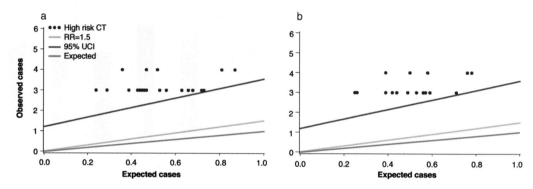

Figure 8: Risk over the period for high-risk census tracts relative to all census tracts.

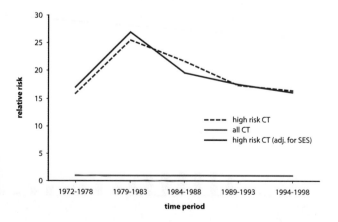

Figure 1: Age-adjusted incidence rate by place.

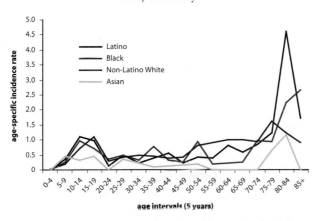

Figure 2: Age-adjusted incidence rate over the period.

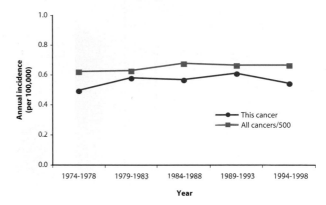

Figure 3: Age-adjusted incidence rate by age and race/ethnicity.

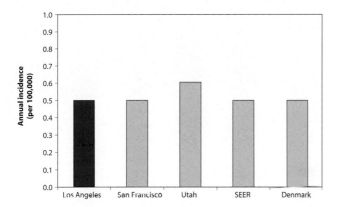

Figure 4: Age-adjusted incidence rate by social class.

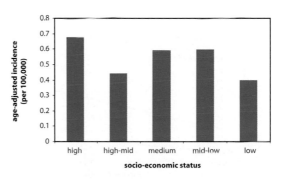

Figure 5: Distribution of the relative risk values for all census tracts.

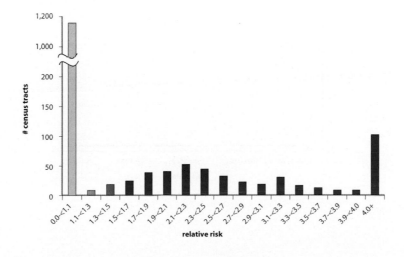

Figure 6: Census tracts by the number of cases per tract.

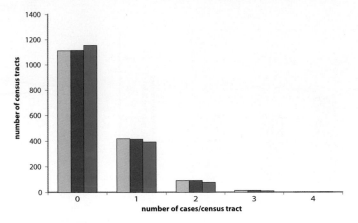

Figure 7a and b: Census tracts at high risk by the number of cases. (a) Unadjusted and (b) adjusted for social class.

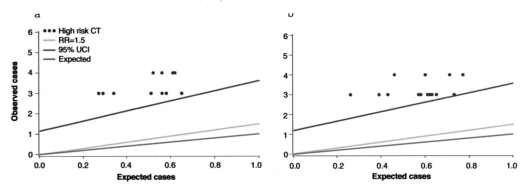

Figure 8: Risk over the period for high-risk census tracts relative to all census tracts.

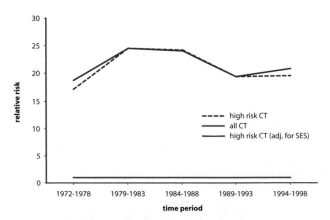

Figure 9: Map of census tracts at high risk.

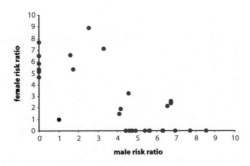

Figure 10: Male-female correlation between the relative risks for high-risk census tracts.

Figure 11: Map of census tracts at high risk, adjusted for social class.

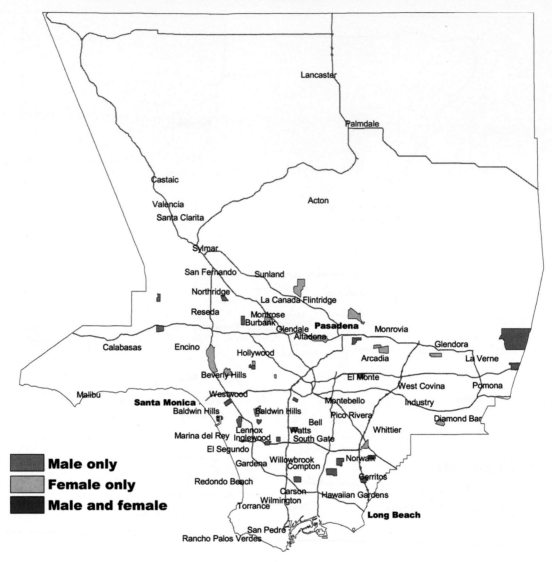

Male only

Female only

Male and female

Figure 12: Male-female correlation between the relative risks for high-risk census tracts, adjusted for social class.

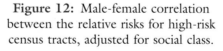

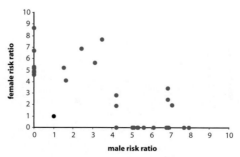

Ewing's Sarcoma

ICDO-2 Code Anatomic Site: C 0–80
ICDO-2 Code Histology: 9260
Age: All
Male Cases: 167
Female Cases: 113

Background

There are no known external causes of this form of bone cancer. A small number of cases are linked to genetic abnormalities.

Local Pattern

This malignancy occurs with remarkable consistency in registry populations across Europe and the United States, including Los Angeles County. Few cases occur after age 30. While African-Americans and Asian-Americans are at substantially lower risk than whites or Latinos, there is still a slight tendency toward higher risk among males of lower social class. Incidence has been relatively constant over the period. Figure 6 shows no nonrandom excess of census tracts with unexpectedly few or unexpectedly many cases. No census tracts fulfilled the high-risk criteria for either males or females.

Thumbnail Interpretation

Despite the racial/ethnic disparity, no systematic pattern of geographical occurrence is apparent, and therefore no local source of causation can be proposed.

Figure 1: Age-adjusted incidence rate by place.

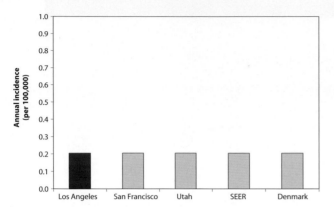

Figure 2: Age-adjusted incidence rate over the period.

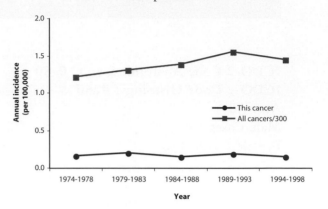

Figure 3: Age-adjusted incidence rate by age and race/ethnicity.

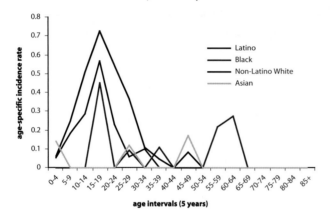

Figure 4: Age-adjusted incidence rate by social class.

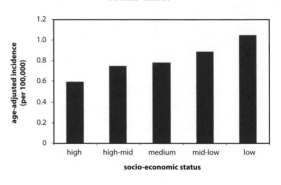

Figure 5: Distribution of the relative risk values for all census tracts.

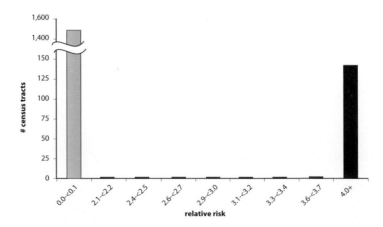

Figure 6: Census tracts by the number of cases per tract.

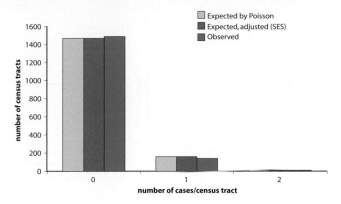

Figure 7a and b: Census tracts at high risk by the number of cases. (a) Unadjusted and (b) adjusted for social class.

There were no high risk census tracts

Figure 8: Risk over the period for high-risk census tracts relative to all census tracts.

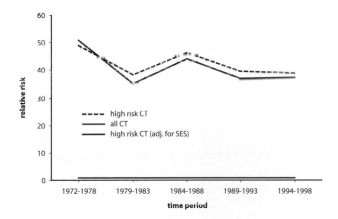

Figure 1: Age-adjusted incidence rate by place.

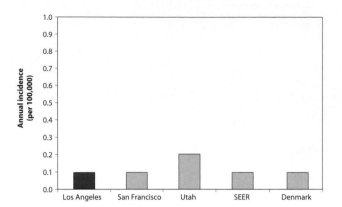

Figure 2: Age-adjusted incidence rate over the period.

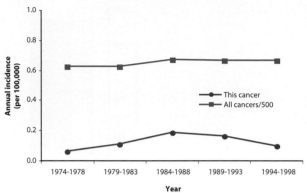

Figure 3: Age-adjusted incidence rate by age and race/ethnicity.

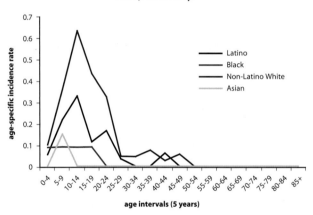

Figure 4: Age-adjusted incidence rate by social class.

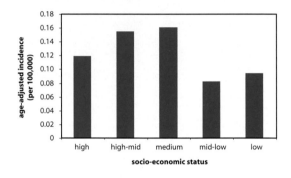

Figure 5: Distribution of the relative risk values for all census tracts.

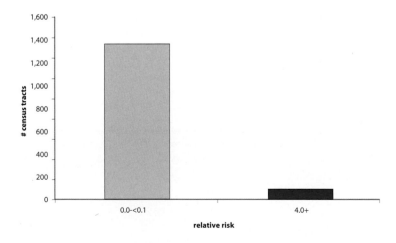

Figure 6: Census tracts by the number of cases per tract.

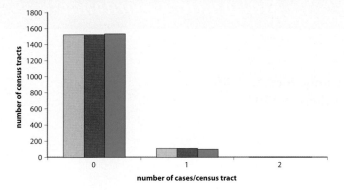

Figure 7a and b: Census tracts at high risk by the number of cases. (a) Unadjusted and (b) adjusted for social class.

There were no high risk census tracts

Figure 8: Risk over the period for high-risk census tracts relative to all census tracts.

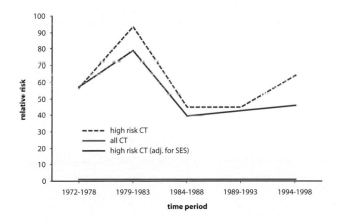

Malignant Chordoma

ICDO-2 Code Anatomic Site: C 0–80
ICDO-2 Code Histology: 9370
Age: All
Male Cases: 101
Female Cases: 67

Background

The causes of this rare malignancy are not known.

Local Pattern

This very rare malignancy occurs with equal frequency among men and women in Los Angeles County and in other areas, with no racial/ethnic or consistent social class gradients. It has occurred with a constant rate over the period, at a level so low that even a single observed case produces a very high relative risk. Figure 6 shows no nonrandom excess of census tracts with unexpectedly few or unexpectedly many cases. A single census tract fulfilled the high-risk criteria on the basis of male cases.

Thumbnail Interpretation

No systematic pattern of occurrence is apparent, and therefore no local source of causation can be proposed.

Figure 1: Age-adjusted incidence rate by place.

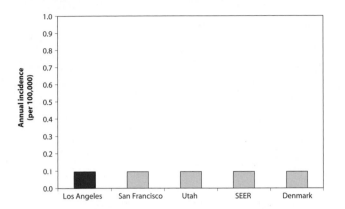

Figure 2: Age-adjusted incidence rate over the period.

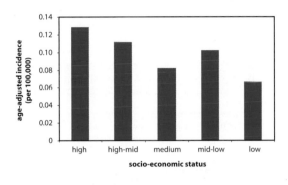

Figure 3: Age-adjusted incidence rate by age and race/ethnicity.

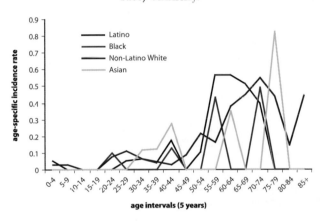

Figure 4: Age-adjusted incidence rate by social class.

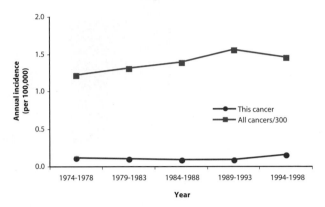

Figure 5: Distribution of the relative risk values for all census tracts.

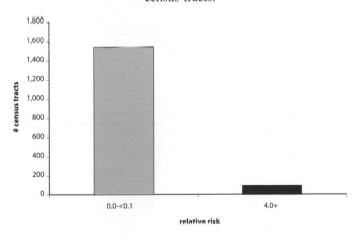

Figure 6: Census tracts by the number of cases per tract.

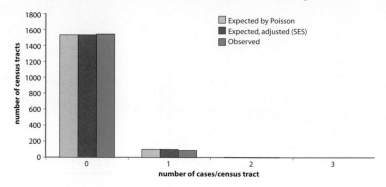

Figure 7a and b: Census tracts at high risk by the number of cases. (a) Unadjusted and (b) adjusted for social class.

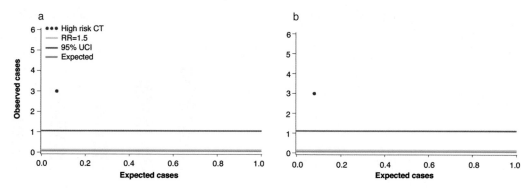

Figure 8: Risk over the period for high-risk census tracts relative to all census tracts.

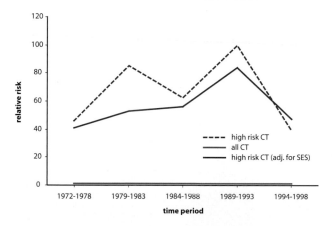

Figure 1: Age-adjusted incidence rate by place.

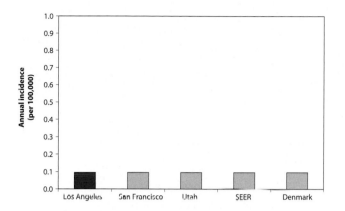

Figure 2: Age-adjusted incidence rate over the period.

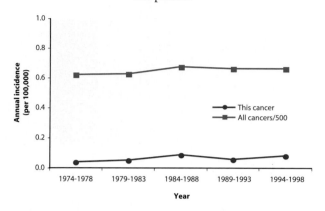

Figure 3: Age-adjusted incidence rate by age and race/ethnicity.

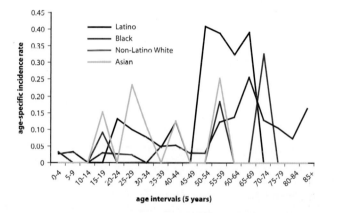

Figure 4: Age-adjusted incidence rate by social class.

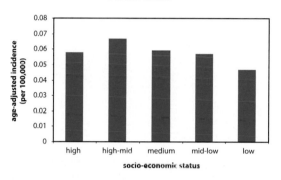

Figure 5: Distribution of the relative risk values for all census tracts.

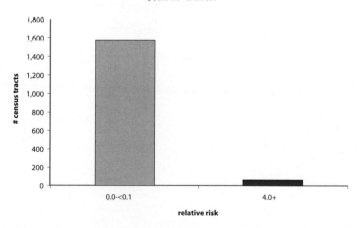

Figure 6: Census tracts by the number of cases per tract.

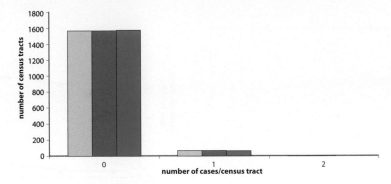

Figure 7a and b: Census tracts at high risk by the number of cases. (a) Unadjusted and (b) adjusted for social class.

There were no high risk census tracts

Figure 8: Risk over the period for high-risk census tracts relative to all census tracts.

There were no high risk census tracts

Figure 9: Map of census tracts at high risk.

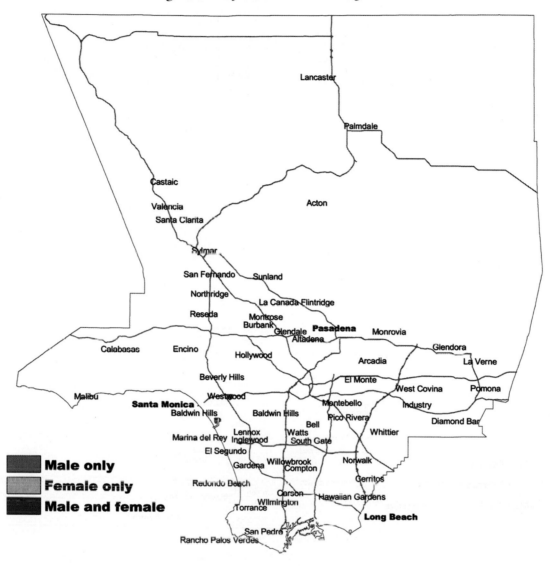

Male only

Female only

Male and female

Figure 10: Male-female correlation between the relative risks for high-risk census tracts.

There were no high risk census tracts

Figure 11: Map of census tracts at high risk, adjusted for social class.

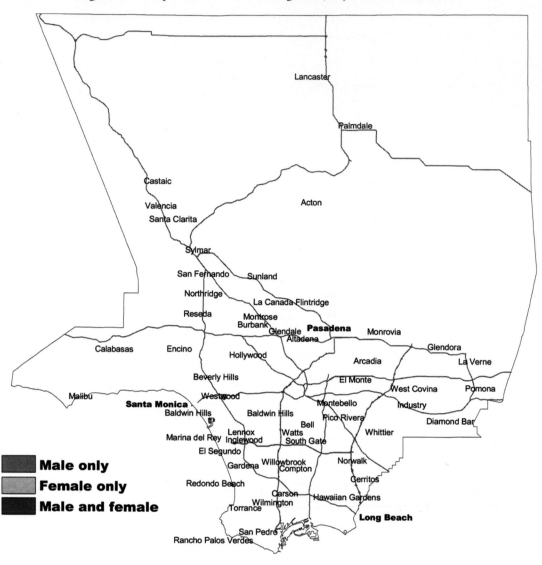

Male only
Female only
Male and female

Figure 12: Male-female correlation between the relative risks for high-risk census tracts, adjusted for social class.

There were no high risk census tracts

Malignant Melanoma

ICDO-2 Code Anatomic Site: C 0–80
ICDO-2 Code Histology: 8720–8780
Age: All
Male Cases: 12412
Female Cases: 10616

Background

Melanoma is caused by solar radiation, but the details (such as the nature and timing of the exposure and the dose required) are not entirely clear. Other things being equal, these malignancies are more common in persons living nearer to the equator. Exposure to sunlight in the first few decades of life appears to be much more important than later exposure, even for tumors that appear in later life. It is unclear whether simple exposure is responsible or whether exposure heavy enough to cause sunburn is needed. The members of races having dark skin color are at very low risk, and melanomas are much more common in redheads and in others with very light skin color. The role of repeated low level exposure, sufficient to produce a tan but not a burn, is not clearly protective, although outdoor workers do tend to get fewer, not more, melanomas than indoor workers. Melanomas are especially common among persons with many large ordinary moles on their skin. They also sometimes occur in several individuals in the same family, and in families with a hereditary pattern of unusual moles. Specific genes are associated with the risk in some of these families.

Local Pattern

Malignant melanoma is about as common in sunny Los Angeles County, despite the polyglot population, as it is in the uniformly white populations served by more northerly registries. In Los Angeles County, whites are at much higher risk than the members of other race/ethnicity groups, and incidence is very strongly linked to higher social class. The malignancy is again about half as common among men as women. Incidence begins to increase in the third decade of life, more sharply and earlier in women, and more gradually and later in men. Incidence among men, but not among women, has increased throughout the period in the county as a whole. There may even be a slight decrease among women in high-risk census tracts over the period. Figure 6 shows a substantial nonrandom excess of census tracts with unexpectedly few or unexpectedly many cases. Many census tracts fulfill the high-risk criteria and many are at very high risk. The unadjusted map of such census tracts at high risk shows many contiguous census tracts with populations of high social class, both on the west and the north side of the Los Angeles basin. Moreover, a very large number of high-risk tracts are at high risk for both men and women. After adjustment, most of these census tracts disappear from the map, leaving a scattering of census tracts, but still representing the high social class regions of the county.

Thumbnail Interpretation

The occurrence of malignant melanoma is determined by exposure to solar radiation, especially in the first several decades of life, and is reduced, as would be expected, by darker skin color. Whites in Southern California are therefore at higher risk, and because the baby-boomer generation has habitually accumulated more sun exposure with less protection than previous generations, incidence has increased as they have progressed into the higher risk ages. Risk among women increases earlier and regresses earlier than risk among men, presumably following, after a long latency, the trend in sun exposure over age. The occurrence of this malignancy is strongly linked to social class, for unknown reasons. The striking geographical distribution of the disease is confounded by the racial/ethnic geography of Los Angeles County, but generally reflects this social class gradient. This is apparent because the high social class pattern of occurrence largely disappears after adjustment.

Figure 1: Age-adjusted incidence rate by place.

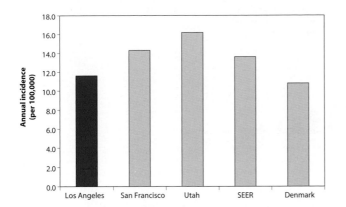

Figure 2: Age-adjusted incidence rate over the period.

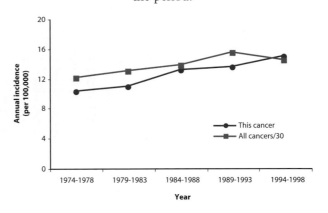

Figure 3: Age-adjusted incidence rate by age and race/ethnicity.

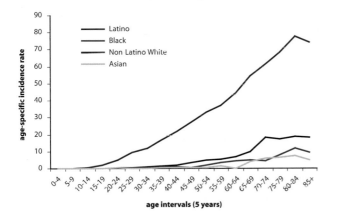

Figure 4: Age-adjusted incidence rate by social class.

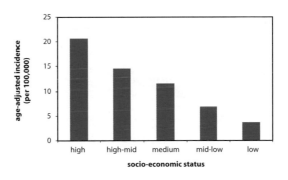

Figure 5: Distribution of the relative risk values for all census tracts.

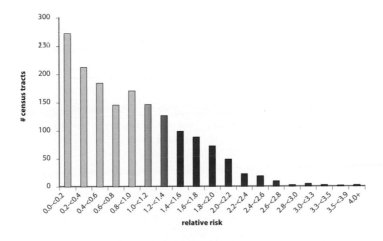

Figure 6: Census tracts by the number of cases per tract.

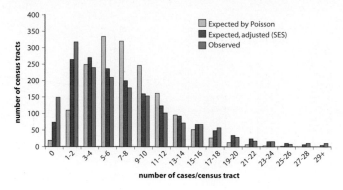

Figure 7a and b: Census tracts at high risk by the number of cases. (a) Unadjusted and (b) adjusted for social class.

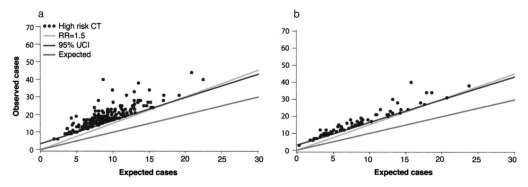

Figure 8: Risk over the period for high-risk census tracts relative to all census tracts.

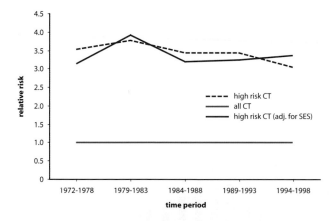

Malignant Melanoma: Female

Figure 1: Age-adjusted incidence rate by place.

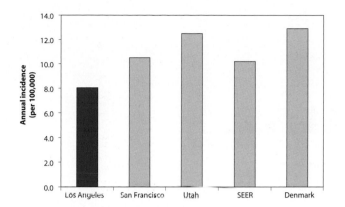

Figure 2: Age-adjusted incidence rate over the period.

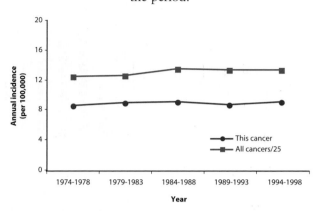

Figure 3: Age-adjusted incidence rate by age and race/ethnicity.

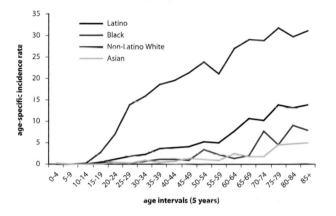

Figure 4: Age-adjusted incidence rate by social class.

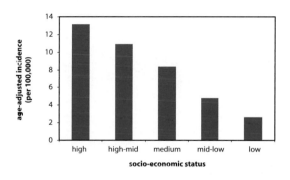

Figure 5: Distribution of the relative risk values for all census tracts.

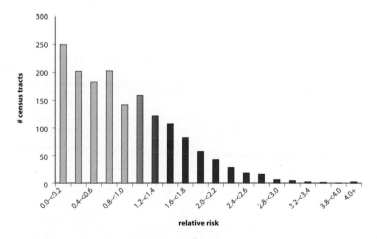

Figure 6: Census tracts by the number of cases per tract.

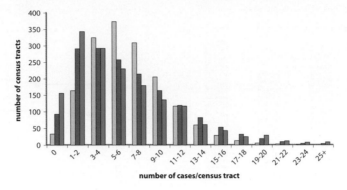

Figure 7a and b: Census tracts at high risk by the number of cases. (a) Unadjusted and (b) adjusted for social class.

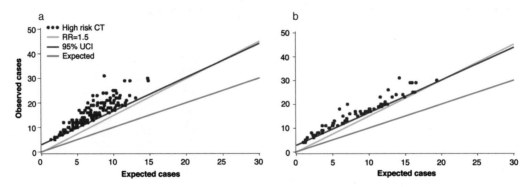

Figure 8: Risk over the period for high-risk census tracts relative to all census tracts.

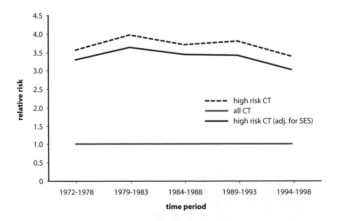

Figure 9: Map of census tracts at high risk.

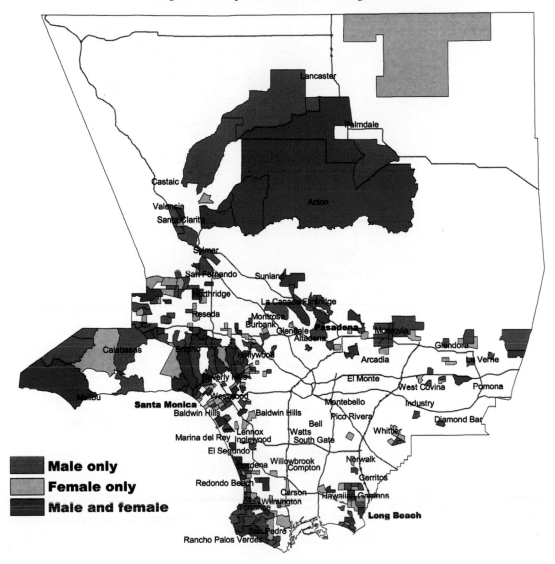

Figure 10: Male-female correlation between the relative risks for high-risk census tracts.

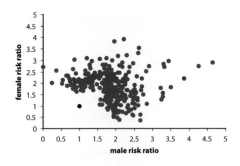

Figure 11: Map of census tracts at high risk, adjusted for social class.

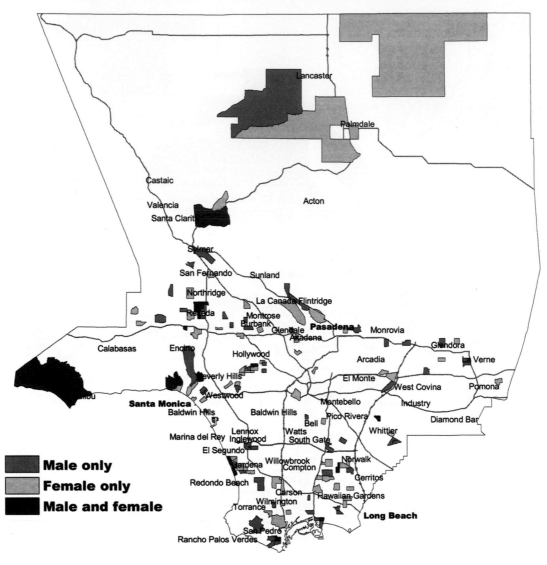

Figure 12: Male-female correlation between the relative risks for high-risk census tracts, adjusted for social class.

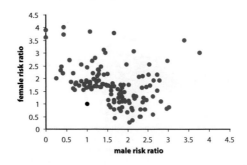

Breast Carcinoma

ICDO-2 Code Anatomic Site: C 50
ICDO-2 Code Histology: 8000–8560, 8570–8573
Age: All
Male Cases: 803
Female Cases: 113066

Background

Breast cancer is the most common cancer of women in the United States as well as in other economically developed countries. Most characteristics common among women with breast cancer suggest the presence of over-abundant female sex hormones. These include early age at menarche (first menstrual period), early menstrual regularity, few or no pregnancies, late first full-term pregnancy, little or no breastfeeding, and late age at menopause. Height, associated with early menarche, also is linked to higher risk. A modest excess risk also results from the long-term use of menopausal hormonal replacement therapy. Alcohol use, which may cause liver damage and slow the elimination of hormones, results in increased risk. Older women who are obese are at higher risk, presumably because fat cells are an important source of postmenopausal estrogens. Regular physical activity, which may delay menarche, diminish menstrual regularity, and reduce obesity, appears to reduce risk. High consumption of soy products early in life also may lower risk, probably because soy compounds compete with estrogens for receptors on cells.

Exposure to ionizing radiation early in life has been shown to increase risk, and for that reason young women treated for other diseases by radiation to the chest have higher risk. Another clear risk factor is a history of the proliferative type of benign breast disease. Genetic factors are important, and although much is known about a few rare high-risk genes, such as BRCA1 and BRCA2, the genetic factors that result in most kinds of hereditary breast cancer have not been identified. Ashkenazi Jewish women are at higher risk of breast cancer, probably due to a combination of genetic and nongenetic factors.

Although it is presumed that hormonal factors are also responsible for male breast cancer, the evidence to date is insufficient for certainty.

Local Pattern

Breast cancer is ten times as common among women as among men. Incidence is high in Los Angeles and all other American and European communities. Risk to white and African-American women is higher than that to Latinas and Asian-American women (the same pattern holds for risk to men). Women, but not men, of higher social class are at higher risk. Incidence among women, but not among men, has increased slightly with time in the county as a whole, but not among residents of high-risk census tracts. Female risk begins increasing in young adulthood and peaks after age 65 in each ethnic group. Figure 6 shows a substantial nonrandom excess of census tracts with unexpectedly few or unexpectedly many cases. Geographically, the census tracts at high risk for both male and female breast cancer are concentrated on the upper west side of the Los Angeles basin, in the wealthy census tracts in Beverly Hills, Encino, Brentwood, and the Hollywood Hills. Adjustment for social class completely eliminates this geographical pattern among women, and decreases it among men.

Thumbnail Interpretation

The pattern of female breast cancer in Los Angeles County is overwhelmingly the pattern of social class, largely because high social class implies early menarche, fewer children, and early first full-term pregnancy. The increase in risk over time has probably occurred in part because of more complete detection of less aggressive tumors as mammography has become more widespread, and in part from the modern shift toward fewer children and later first childbirth. After adjustment for social class, no systematic pattern of occurrence is apparent, and therefore no local source of causation can be proposed.

The determinants of male breast cancer incidence are less clear, and the reason for the concentration of high-risk census tracts in the same area of Los Angeles County, relatively unmodified by social class adjustment, is unknown.

Figure 1: Age-adjusted incidence rate by place.

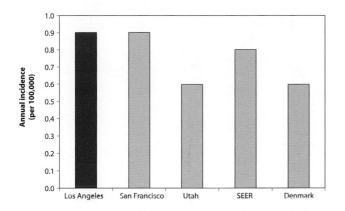

Figure 2: Age-adjusted incidence rate over the period.

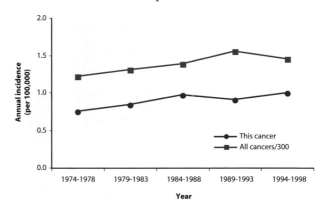

Figure 3: Age-adjusted incidence rate by age and race/ethnicity.

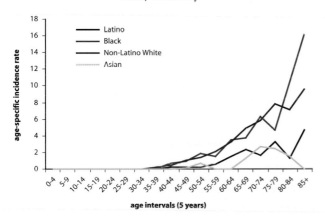

Figure 4: Age-adjusted incidence rate by social class.

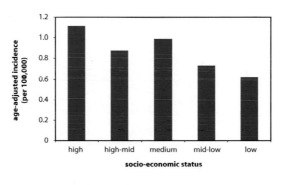

Figure 5: Distribution of the relative risk values for all census tracts.

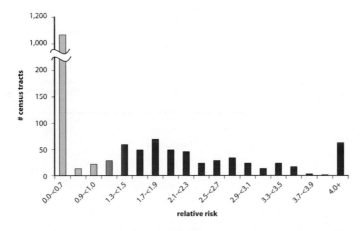

Figure 6: Census tracts by the number of cases per tract.

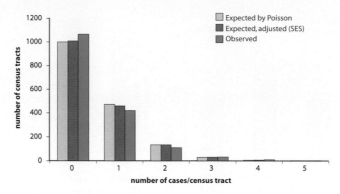

Figure 7a and b: Census tracts at high risk by the number of cases. (a) Unadjusted and (b) adjusted for social class.

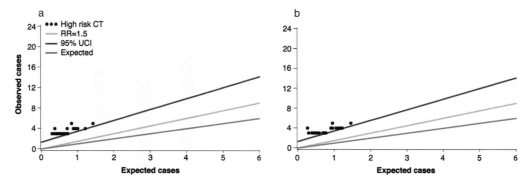

Figure 8: Risk over the period for high-risk census tracts relative to all census tracts.

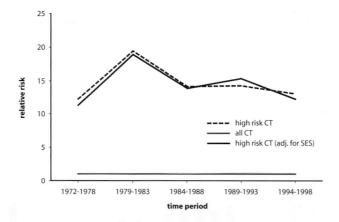

Figure 1: Age-adjusted incidence rate by place.

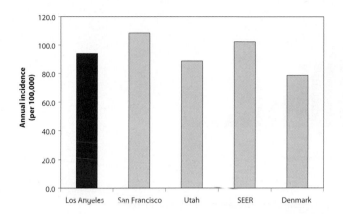

Figure 2: Age-adjusted incidence rate over the period.

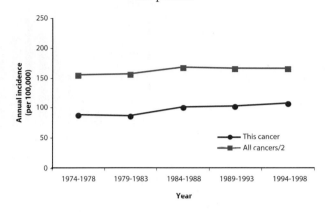

Figure 3: Age-adjusted incidence rate by age and race/ethnicity.

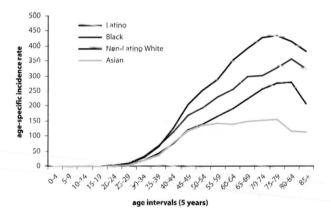

Figure 4: Age-adjusted incidence rate by social class.

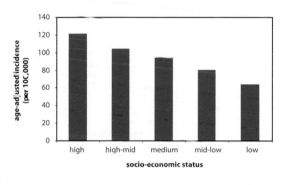

Figure 5: Distribution of the relative risk values for all census tracts.

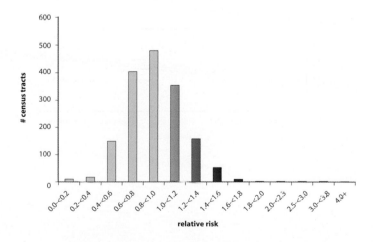

Figure 6: Census tracts by the number of cases per tract.

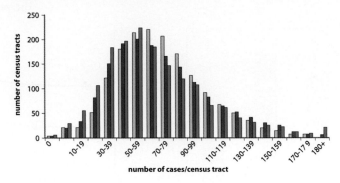

Figure 7a and b: Census tracts at high risk by the number of cases. (a) Unadjusted and (b) adjusted for social class.

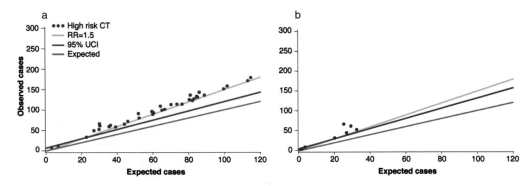

Figure 8: Risk over the period for high-risk census tracts relative to all census tracts.

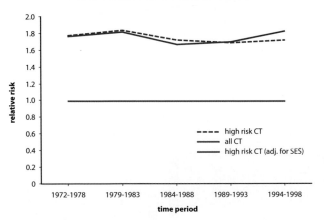

Figure 9: Map of census tracts at high risk.

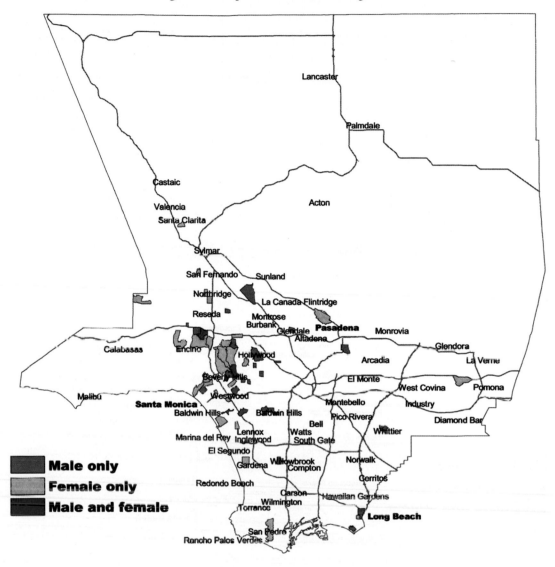

Figure 10: Male-female correlation between the relative risks for high-risk census tracts.

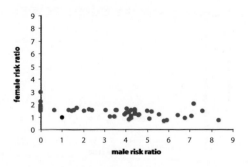

Figure 11: Map of census tracts at high risk, adjusted for social class.

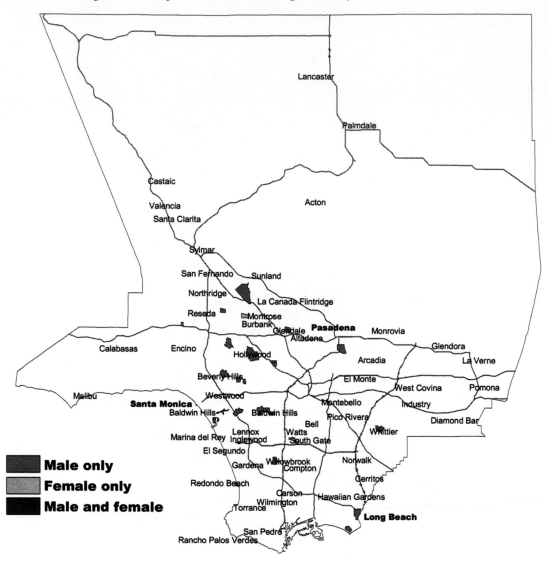

Figure 12: Male-female correlation between the relative risks for high-risk census tracts, adjusted for social class.

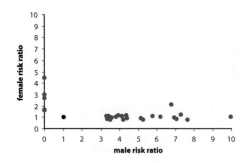

Squamous Cervix Carcinoma (Invasive)

ICDO-2 Code Anatomic Site: C 53
ICDO-2 Code Histology: 8000–8082
Age: All
Male Cases: 0
Female Cases: 12010

Background

Cervical cancer is mostly the result of an infection, caused by certain kinds of human papillomaviruses. These days most cancers are detected before they become invasive by "Pap" screening and are removed (and cured) by local biopsy. Thus an important cause of this cancer is failure to be regularly screened by Pap smear. The source of the original infection is sexual activity, and is particularly related to exposure as early as in the teenage years. Not surprisingly, the likelihood of infection and cervical abnormalities is related to the number of early sexual partners, or the past sexual experience of each of those partners.

Other viruses, such as herpes simplex virus-2, are also spread by sexual activity and are suspected of contributing to the development of cervical cancer. Smoking may enhance susceptibility to a causal virus, but the facts are as yet unclear. It is clear that immunosuppression, most notably by HIV infection and AIDS, leads to an increased likelihood of cervical cancer.

Local Pattern

This malignancy occurs more often in Los Angeles County and especially in Denmark, than in Utah. Incidence begins to increase in young adulthood, and rates are highest among Latinas, and next highest among African-Americans. Older Asian-American, especially Korean-American, women also have rates higher than those of similarly aged whites. Women of lower social class are more often affected than those of higher social class. Incidence in the county as a whole has gradually dropped over time, although that is less true among the residents of high-risk census tracts. Figure 6 shows a slight nonrandom excess of tracts with unexpectedly few or unexpectedly many cases. Although a very large number of census tracts meet the initial high-risk criteria, the number is greatly reduced after adjustment for social class. Prior to adjustment, the high-risk census tracts appeared in the low social class areas of the county, especially South Central Los Angeles, but after adjustment, no systematic distribution is evident.

Thumbnail Interpretation

The decrease in frequency of cervix carcinoma probably reflects a combination of increasingly accessible screening and a decrease in the frequency of unprotected sexual activity. The geographical distribution of high-risk census tracts reflects the poverty and poor education that promote unprotected sexual activity and less access to screening.

Figure 1: Age-adjusted incidence rate by place.

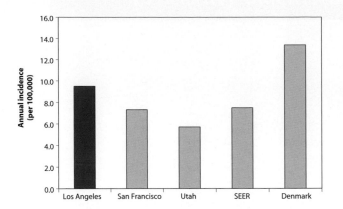

Figure 2: Age-adjusted incidence rate over the period.

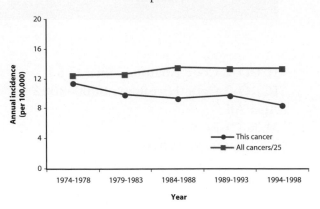

Figure 3: Age-adjusted incidence rate by age and race/ethnicity.

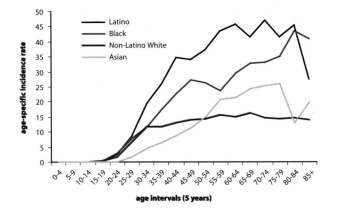

Figure 4: Age-adjusted incidence rate by social class.

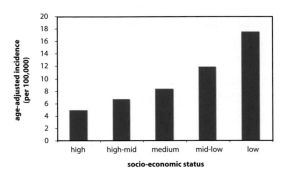

Figure 5: Distribution of the relative risk values for all census tracts.

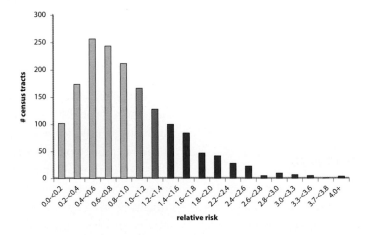

Figure 6: Census tracts by the number of cases per tract.

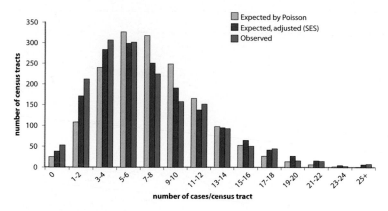

Figure 7a and b: Census tracts at high risk by the number of cases. (a) Unadjusted and (b) adjusted for social class.

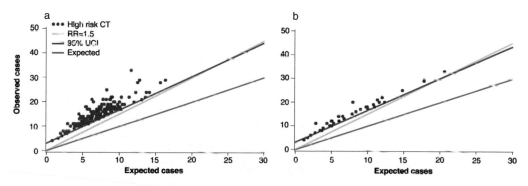

Figure 8: Risk over the period for high-risk census tracts relative to all census tracts.

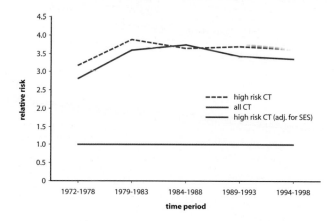

Figure 9: Map of census tracts at high risk.

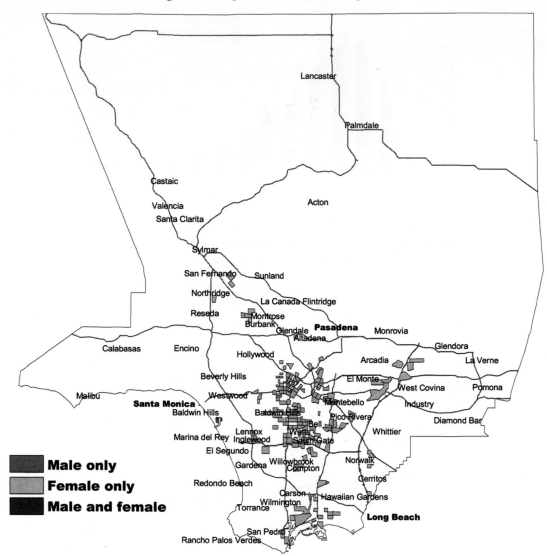

Squamous Cervix Carcinoma (Invasive)

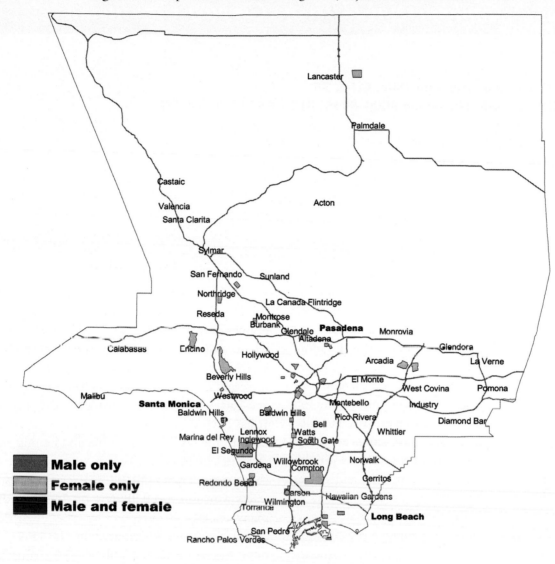

Endometrial (Uterine Corpus) Carcinoma

ICDO-2 Code Anatomic Site: C 54, 55
ICDO-2 Code Histology: 8000–8045, 8120–8560, 8570–8573
Age: All
Male Cases: 0
Female Cases: 25496

Background

Carcinoma of the endometrium (the inner lining of the body of the uterus) has long been known to be more common in the obese and those having late menopause, and to be more common in women with few or no children. When estrogen replacement therapy for menopausal symptoms came into common use, rates of endometrial cancer rose, and subsequently many studies have confirmed that prolonged use of estrogens produces growth of the endometrial cells generally, and causes carcinoma. When progestins were added to estrogens to cause the endometrial cells to mature as well as grow, forming a combination hormone replacement therapy, the frequency of endometrial cancer declined. Tamoxifen, a drug used to prevent breast cancer in high-risk women, can also cause endometrial cancer. Smoking tends to decrease the frequency of occurrence, possibly by reducing the natural production of estrogen. Women who have used oral contraceptives are less likely to develop endometrial cancer, probably because natural estrogen production is reduced. Perhaps the factor most responsible for decreasing the occurrence of endometrial cancer, however, is hysterectomy.

Local Pattern

Endometrial cancer is common in Los Angeles and other economically developed populations. It is particularly common among white women, particularly those of higher social class. The incidence of this malignancy increases in frequency just before menopause and flattens out or drops just after menopause. It has been decreasing in frequency over time, both in the county as a whole and among the residents of high-risk census tracts. Figure 6 shows a moderate nonrandom excess of census tracts with unexpectedly few or unexpectedly many cases. Geographically, the high-risk census tracts are distributed in the circle of high social class neighborhoods situated peripheral to the urban center. The number of census tracts meeting the high-risk criteria is reduced by adjustment for social class, and the small number of remaining high-risk census tracts make no evident pattern.

Thumbnail Interpretation

Women of high social class have relatively unlimited access to medical care, and thus more commonly receive hormone replacement therapy at menopause. At the beginning of the period covered, this therapy generally consisted of unopposed estrogens. In the last decade, a progestin has usually been added to the regimen, providing protection against the endometrial malignancies caused by estrogen alone. After adjustment for social class, no particular geographic pattern is evident, and no other local source of causation can be proposed.

Endometrial (Uterine Corpus) Carcinoma: Female

Figure 1: Age-adjusted incidence rate by place.

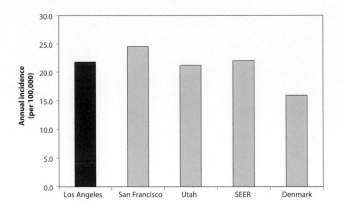

Figure 2: Age-adjusted incidence rate over the period.

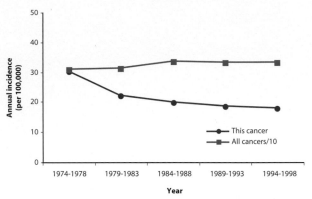

Figure 3: Age-adjusted incidence rate by age and race/ethnicity.

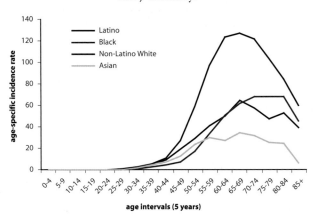

Figure 4: Age-adjusted incidence rate by social class.

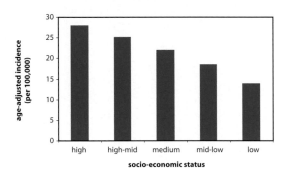

Figure 5: Distribution of the relative risk values for all census tracts.

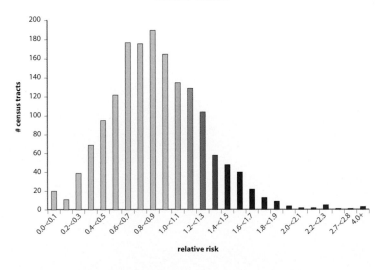

Figure 6: Census tracts by the number of cases per tract.

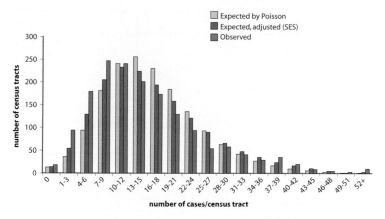

Figure 7a and b: Census tracts at high risk by the number of cases. (a) Unadjusted and (b) adjusted for social class.

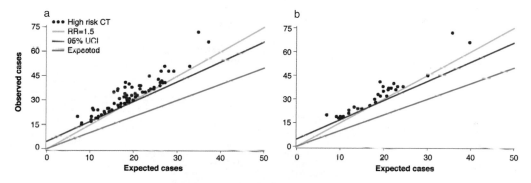

Figure 8: Risk over the period for high risk census tracts relative to all census tracts.

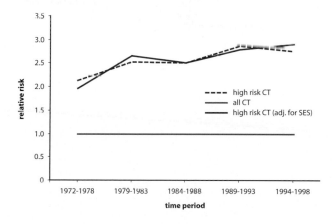

Figure 9: Map of census tracts at high risk.

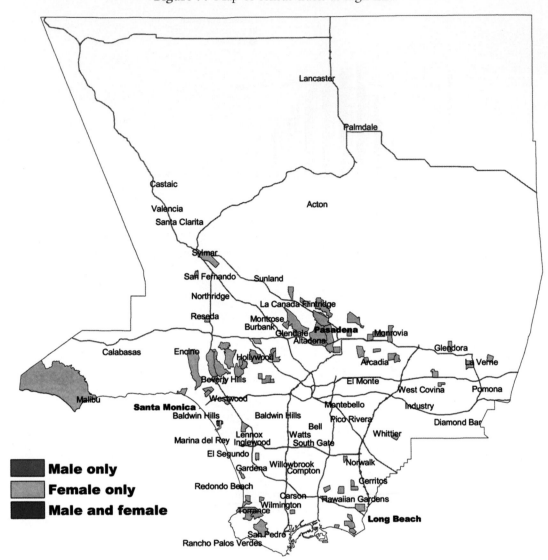

Figure 11: Map of census tracts at high risk, adjusted for social class.

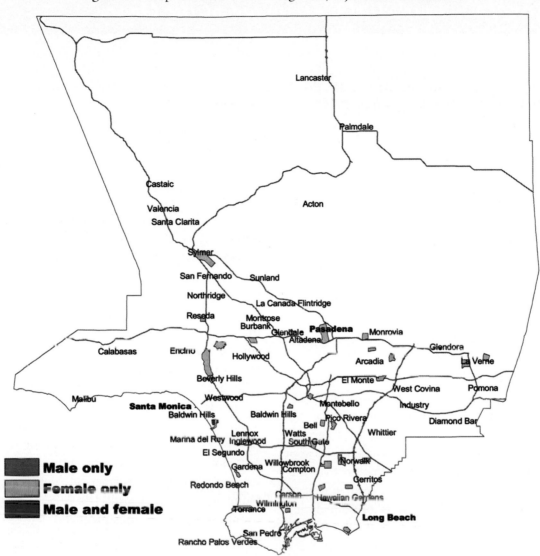

Epithelial Carcinoma of the Ovary

ICDO-2 Code Anatomic Site: C 48, 56, 76.2
ICDO-2 Code Histology: 8000–8560, 8570–8573, 9000
Age: All
Male Cases: 61
Female Cases: 16637

Background

Epithelial carcinomas are the most common malignancy of the ovary. The causes of this malignancy have not been identified, nor has even the cell of origin with certainty. It is clear that the more babies a woman has, the lower the risk, and that long-term use of oral contraceptives reduces risk. Hysterectomy and even tubal ligation also reduce risk, although the mechanism is unclear. Women who belong to one of the rare families with mutations in the genes BRCA1 or BRCA2 are at a greatly increased risk of ovarian carcinoma, as well as of breast cancer. Exposure to radiation from the atomic bomb resulted in more ovarian cancers than was expected, and it is presumed that other forms of radiation can do so as well. Epithelial ovarian cancers are thought by some to arise from the cells that line the ovary, and that same sort of cell also lines the abdominal cavity. Some cancers indistinguishable from these ovarian cancers appear in women even after the ovaries have been removed. Men do not get ovarian cancer, of course, but the cells that line the abdominal cavity are the same, and similar cancers are seen to occur among men.

Local Pattern

Epithelial ovarian cancers and the much more rare equivalent malignancies of men occur in Los Angeles with roughly the same frequency as elsewhere. Risk begins to increase in young adulthood. White women of upper social class are at slightly increased risk, whereas women of Asian origin are at decreased risk. Classification differences are the probable reason why similar cancers among men seem to be more common in Denmark. Incidence has been constant over time, both in the county as a whole and among the residents of high-risk census tracts. Figure 6 shows only a slight nonrandom excess of census tracts with unexpectedly few or unexpectedly many cases. Before adjustment for social class, the census tracts at high risk are found in the high social class regions of the county; but after adjustment no pattern is evident.

Thumbnail Interpretation

Other than the possible contribution of a small number of ovarian carcinomas in those with the rare mutations in genes such as BRCA1 and BRCA2, which are especially common among Ashkenazi Jews, there is no obvious explanation for the increased risk among upper social class women. No systematic pattern of geographical occurrence is apparent, and therefore no local source of causation can be identified.

Figure 1: Age-adjusted incidence rate by place.

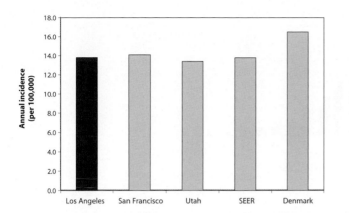

Figure 2: Age-adjusted incidence rate over the period.

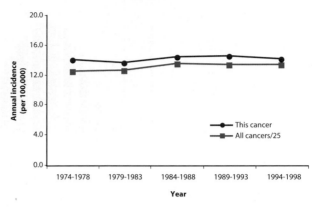

Figure 3: Age-adjusted incidence rate by age and race/ethnicity.

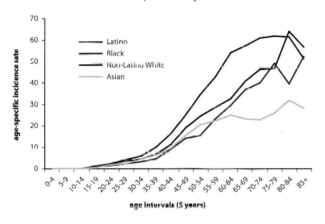

Figure 4: Age-adjusted incidence rate by social class.

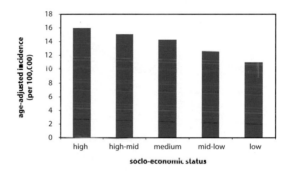

Figure 5: Distribution of the relative risk values for all census tracts.

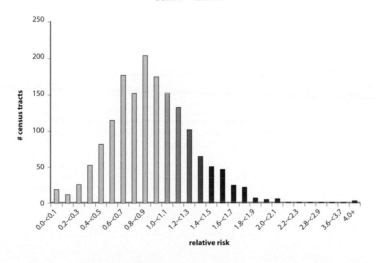

Figure 6: Census tracts by the number of cases per tract.

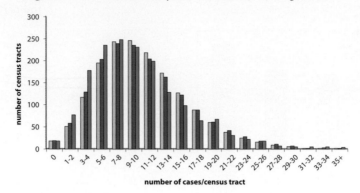

Figure 7a and b: Census tracts at high risk by the number of cases. (a) Unadjusted and (b) adjusted for social class.

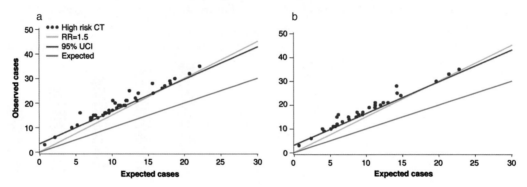

Figure 8: Risk over the period for high-risk census tracts relative to all census tracts.

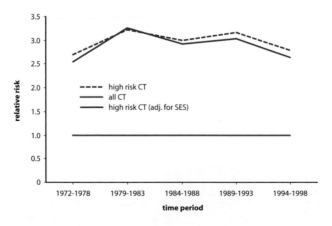

Figure 9: Map of census tracts at high risk.

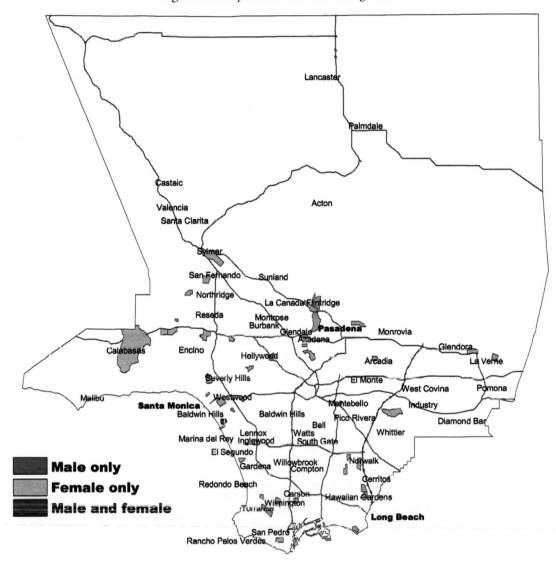

Male only
Female only
Male and female

Figure 11: Map of census tracts at high risk, adjusted for social class.

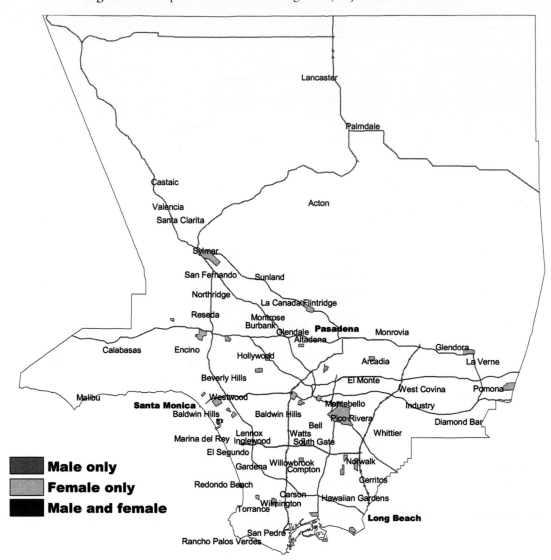

Germ Cell Malignancies

ICDO-2 Code Anatomic Site: C 0–80
ICDO-2 Code Histology: 8590, 8600–8671, 9060–9091, 9100–9110
Age: All
Male Cases: 5007
Female Cases: 1007

Background

The most important germ cell malignancies are cancers of the testis in young men, and similar malignancies which occur, more rarely, in the ovaries of women. Testicular cancers are probably caused in part by genetic factors, because they frequently run in families and occur more commonly in certain ethnic groups. In addition, because they commonly occur in men in the third or fourth decade of life, particularly in those who have developmental abnormalities of the genitalia, it is thought that unknown events occurring during gestation are partly responsible. The most important such developmental abnormality, and the best predictor of testis cancer, is failure of a testicle to properly descend into the scrotum. In such cases, the cancer does not necessarily occur in the same testicle.

Local Pattern

Testis cancers occur with equal frequency in the different parts of the United States. They are several times more common than the equivalent germ cell malignancies of the ovary, and occur almost exclusively among young adults. Asian-American and African-American men are at relatively low risk, as are those of lower social class. Figure 6 shows a slight nonrandom excess of census tracts with unexpectedly few or unexpectedly many cases. Incidence in the county as a whole as well as among residents of high-risk census tracts has been relatively constant. Census tracts at high risk for men are distributed among the high social class regions of the county before adjustment for social class. After adjustment there are few contiguous census tracts and no particular pattern is apparent. No census tract stands out on the basis of a particularly high number of excess cases.

Thumbnail Interpretation

Other than the higher risk among young white men, which is likely to be explained by a genetic factor, no inequity in risk is seen. No systematic pattern of geographical occurrence is apparent, and therefore no local source of causation can be proposed.

Figure 1: Age-adjusted incidence rate by place.

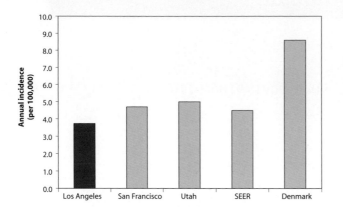

Figure 2: Age-adjusted incidence rate over the period.

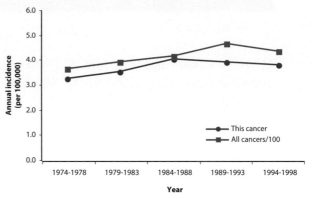

Figure 3: Age-adjusted incidence rate by age and race/ethnicity.

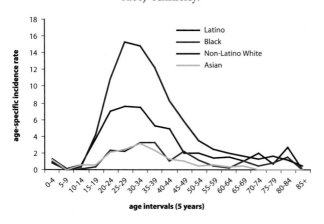

Figure 4: Age-adjusted incidence rate by social class.

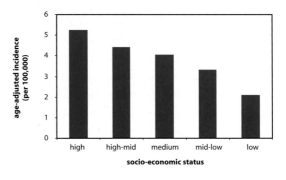

Figure 5: Distribution of the relative risk values for all census tracts.

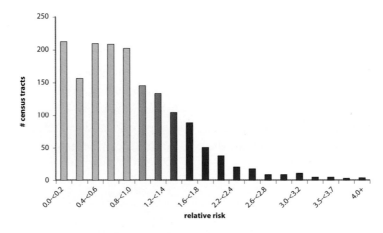

Figure 6: Census tracts by the number of cases per tract.

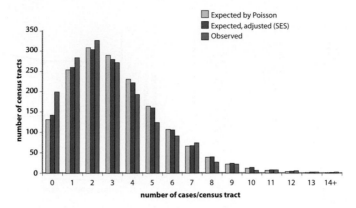

Figure 7a and b: Census tracts at high risk by the number of cases. (a) Unadjusted and (b) adjusted for social class.

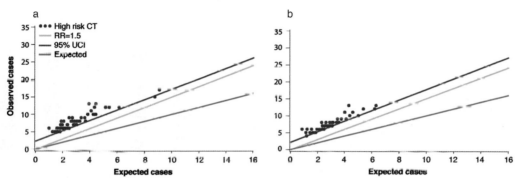

Figure 8: Risk over the period for high-risk census tracts relative to all census tracts.

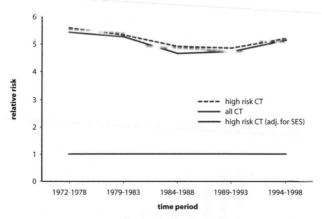

Figure 1: Age-adjusted incidence rate by place.

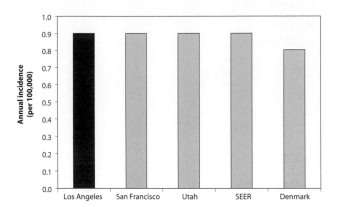

Figure 2: Age-adjusted incidence rate over the period.

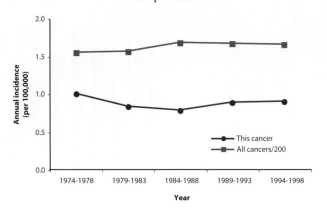

Figure 3: Age-adjusted incidence rate by age and race/ethnicity.

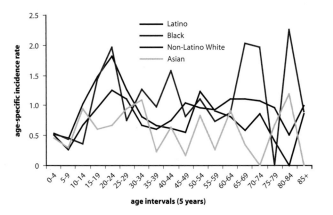

Figure 4: Age-adjusted incidence rate by social class.

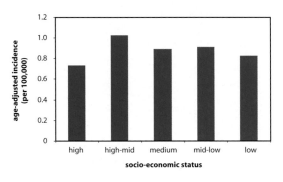

Figure 5: Distribution of the relative risk values for all census tracts.

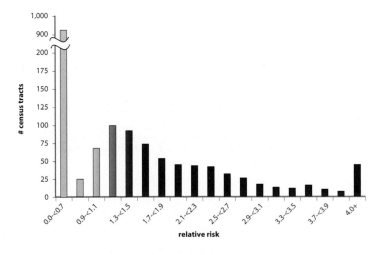

Figure 6: Census tracts by the number of cases per tract.

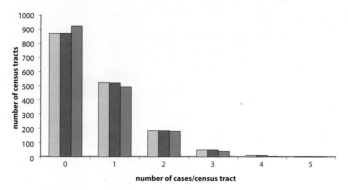

Figure 7a and b: Census tracts at high risk by the number of cases. (a) Unadjusted and (b) adjusted for social class.

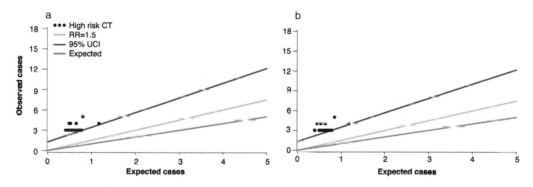

Figure 8: Risk over the period for high-risk census tracts relative to all census tracts.

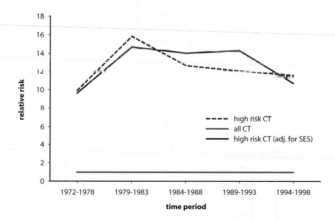

Figure 9: Map of census tracts at high risk.

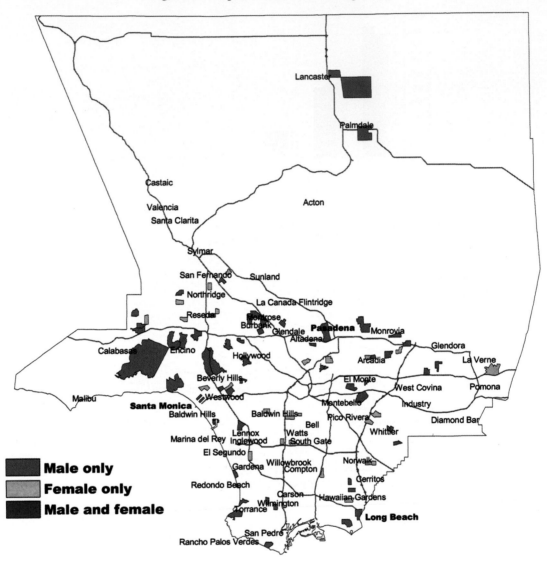

Figure 10: Male-female correlation between the relative risks for high-risk census tracts.

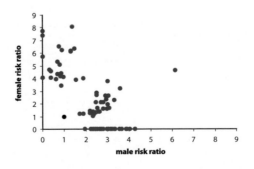

Figure 11: Map of census tracts at high risk, adjusted for social class.

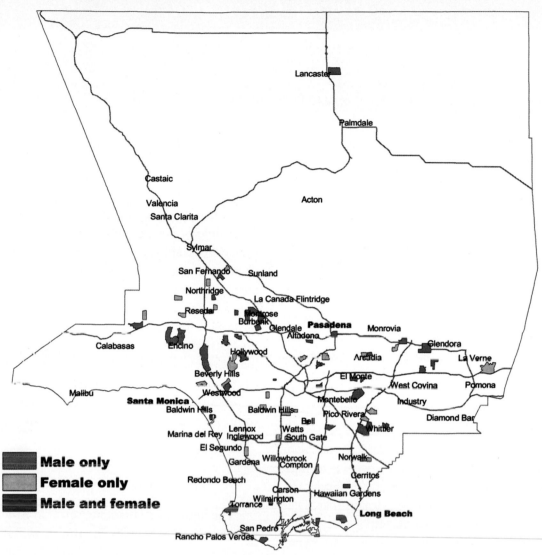

Figure 12: Male-female correlation between the relative risks for high-risk census tracts, adjusted for social class.

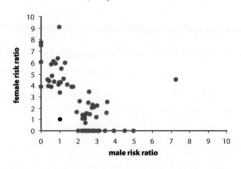

Prostate Carcinoma

ICDO-2 Code Anatomic Site: C 61
ICDO-2 Code Histology: 8000–8560, 8570–8573
Age: All
Male Cases: 88587
Female Cases: 0

Background

Prostate cancer is the most common malignancy among American men and is especially common among African-American men. The prostate functions in response to sex hormones, and prostate cancer is thought to be an abnormal response to some kind of variation in those hormones, but the exact nature and the external causes of that hormone aberration are unknown. Many prostate cancers do not spread rapidly (or at all). Since the screening test prostate-specific antigen (PSA) came into use around 1990, the disease has been diagnosed more commonly in persons of high social class with good access to medical care, although this was not true prior to that. Diet is strongly suspected of playing a role, as are genetic factors, because multiple cases occur more often in the same family than would be expected.

Local Pattern

Prostate cancer occurs slightly less frequently in Los Angeles County than in other parts of the country. Prostate cancer is much more common in African-Americans, and much less common in Asian-Americans. A substantial increase in incidence in the county as a whole occurred at about the beginning of the 1990s, although the rate subsequently decreased, and this sequence was also seen among the residents of high-risk census tracts. Figure 6 shows a moderate nonrandom excess of census tracts with unexpectedly few or unexpectedly many cases. After adjustment for social class, almost all the high-risk census tracts are located in the African-American community of South Los Angeles, where they comprise a large complex of contiguous census tracts.

Thumbnail Interpretation

The high rate of prostate cancer in African-Americans is reflected in the map. It is presumed to occur on the basis of genetic susceptibility, although environmental factors, such as diet, may also contribute. The increase in incidence in the last decade of the period undoubtedly is also related to the introduction of screening by PSA. The slight tendency for higher risk to appear among high social class communities also may be explained by higher access to screening, and probably reflects the inclusion of a greater proportion of less aggressive tumors.

Prostate Carcinoma: Male

Figure 1: Age-adjusted incidence rate by place.

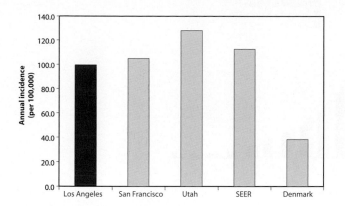

Figure 2: Age-adjusted incidence rate over the period.

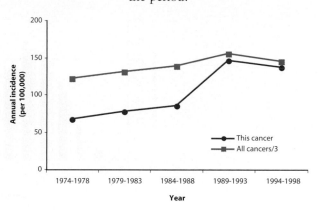

Figure 3: Age-adjusted incidence rate by age and race/ethnicity.

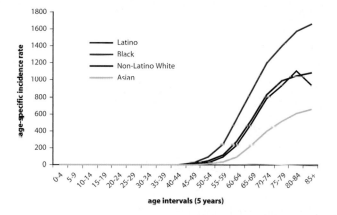

Figure 4: Age-adjusted incidence rate by social class.

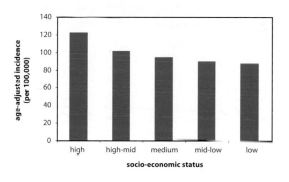

Figure 5: Distribution of the relative risk values for all census tracts.

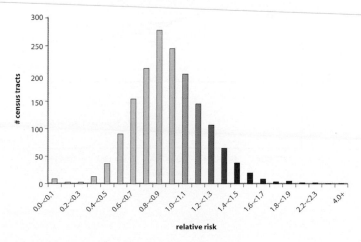

Figure 6: Census tracts by the number of cases per tract.

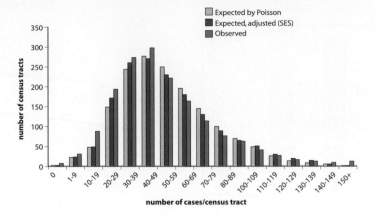

Figure 7a and b: Census tracts at high risk by the number of cases. (a) Unadjusted and (b) adjusted for social class.

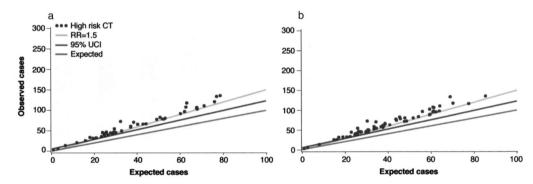

Figure 8: Risk over the period for high-risk census tracts relative to all census tracts.

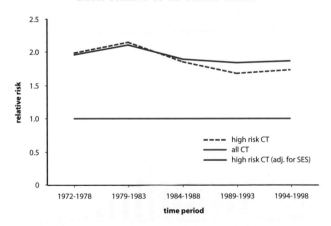

Figure 9: Map of census tracts at high risk.

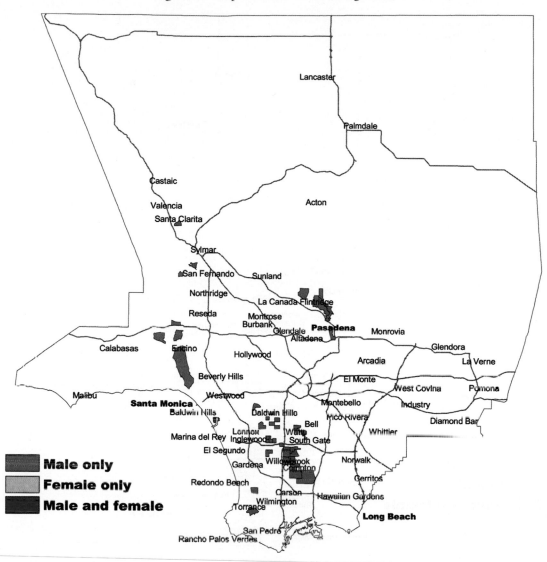

Male only
Female only
Male and female

Figure 11: Map of census tracts at high risk, adjusted for social class.

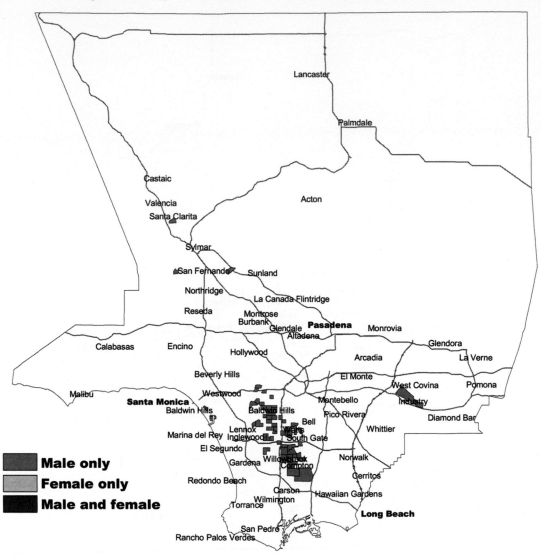

Anogenital Adenocarcinoma

ICDO-2 Code Anatomic Site: C 21, 51–53, 57, 60, 62, 63
ICDO-2 Code Histology: 8124, 8140–8560, 8570–8573
Age: All
Male Cases: 445
Female Cases: 3497

Background

Adenocarcinoma of the cervix, vulva, vagina, and anal glands may be due to the same underlying causes that result in adenocarcinoma of the endometrium. These include obesity, few or no pregnancies, and the use of estrogen at menopause. These cancers may also be caused by certain strains of human papillomavirus, suggesting that, as with cervical cancer, past sexual activity may be responsible. One form of adenocarcinoma, clear cell carcinoma of the vagina, has occurred among young women who were still in the uterus of mothers who took diethylstilbestrol (DES) to prevent spontaneous abortion during pregnancy.

Local Pattern

These malignancies are somewhat more common in Los Angeles County than in other regions of the country, and are five times more common among women as among men, especially among middle-aged adults. Middle-aged Latinas and older African-American women appear to be at especially high risk. Incidence among females has increased slightly over time in Los Angeles County as a whole, but has been relatively constant among the residents of high-risk census tracts. Figure 6 shows only a slight nonrandom excess of census tracts with unexpectedly few or unexpectedly many cases. The high-risk census tracts are scattered across Los Angeles County, without the geographical grouping, contiguous high-risk census tracts, or correlation between sex-specific high risk that would suggest systematic occurrence. No census tract stands out on the basis of a particularly high relative risk and number of excess cases.

Thumbnail Interpretation

The only suggestion of nonrandom incidence is the unexplained consistent increased risk to middle-aged Latinas and older African-American women. The suggestion of an increase in incidence over time is not explained, because rates of cervical cancers have generally decreased. Increasing accuracy of diagnosis may have been responsible.

Figure 1: Age-adjusted incidence rate by place.

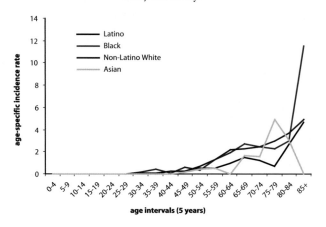

Figure 2: Age-adjusted incidence rate over the period.

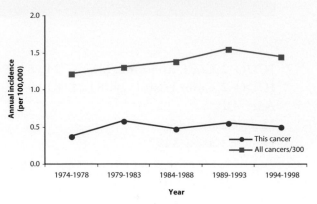

Figure 3: Age-adjusted incidence rate by age and race/ethnicity.

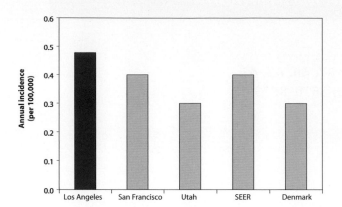

Figure 4: Age-adjusted incidence rate by social class.

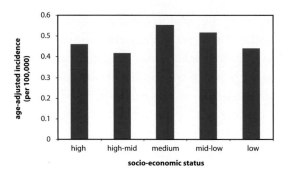

Figure 5: Distribution of the relative risk values for all census tracts.

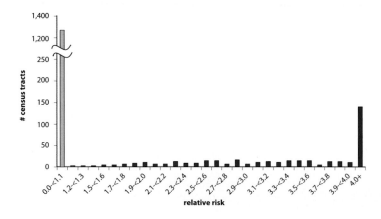

Figure 6: Census tracts by the number of cases per tract.

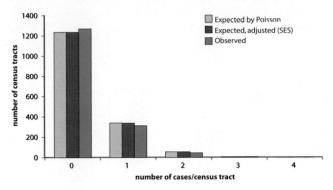

Figure 7a and b: Census tracts at high risk by the number of cases. (a) Unadjusted and (b) adjusted for social class.

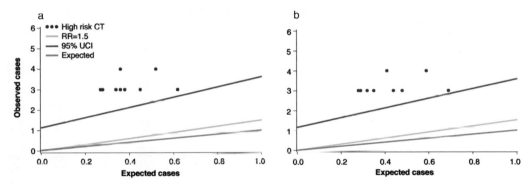

Figure 8: Risk over the period for high-risk census tracts relative to all census tracts.

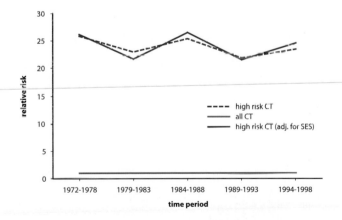

Figure 1: Age-adjusted incidence rate by place.

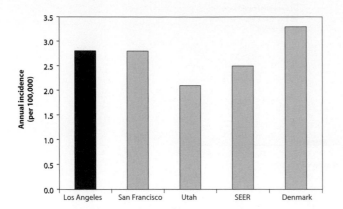

Figure 2: Age-adjusted incidence rate over the period.

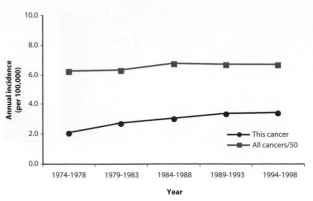

Figure 3: Age-adjusted incidence rate by age and race/ethnicity.

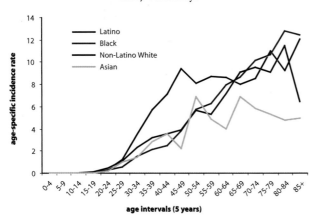

Figure 4: Age-adjusted incidence rate by social class.

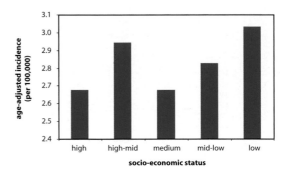

Figure 5: Distribution of the relative risk values for all census tracts.

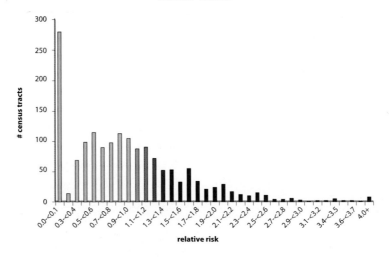

Figure 6: Census tracts by the number of cases per tract.

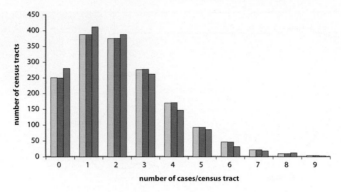

Figure 7a and b: Census tracts at high risk by the number of cases. (a) Unadjusted and (b) adjusted for social class.

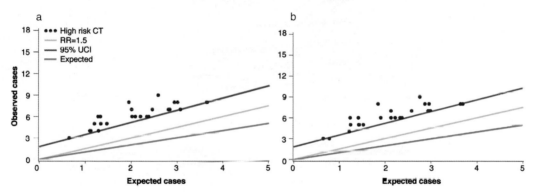

Figure 8: Risk over the period for high-risk census tracts relative to all census tracts.

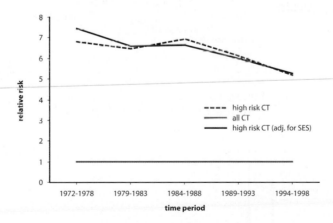

Figure 9: Map of census tracts at high risk.

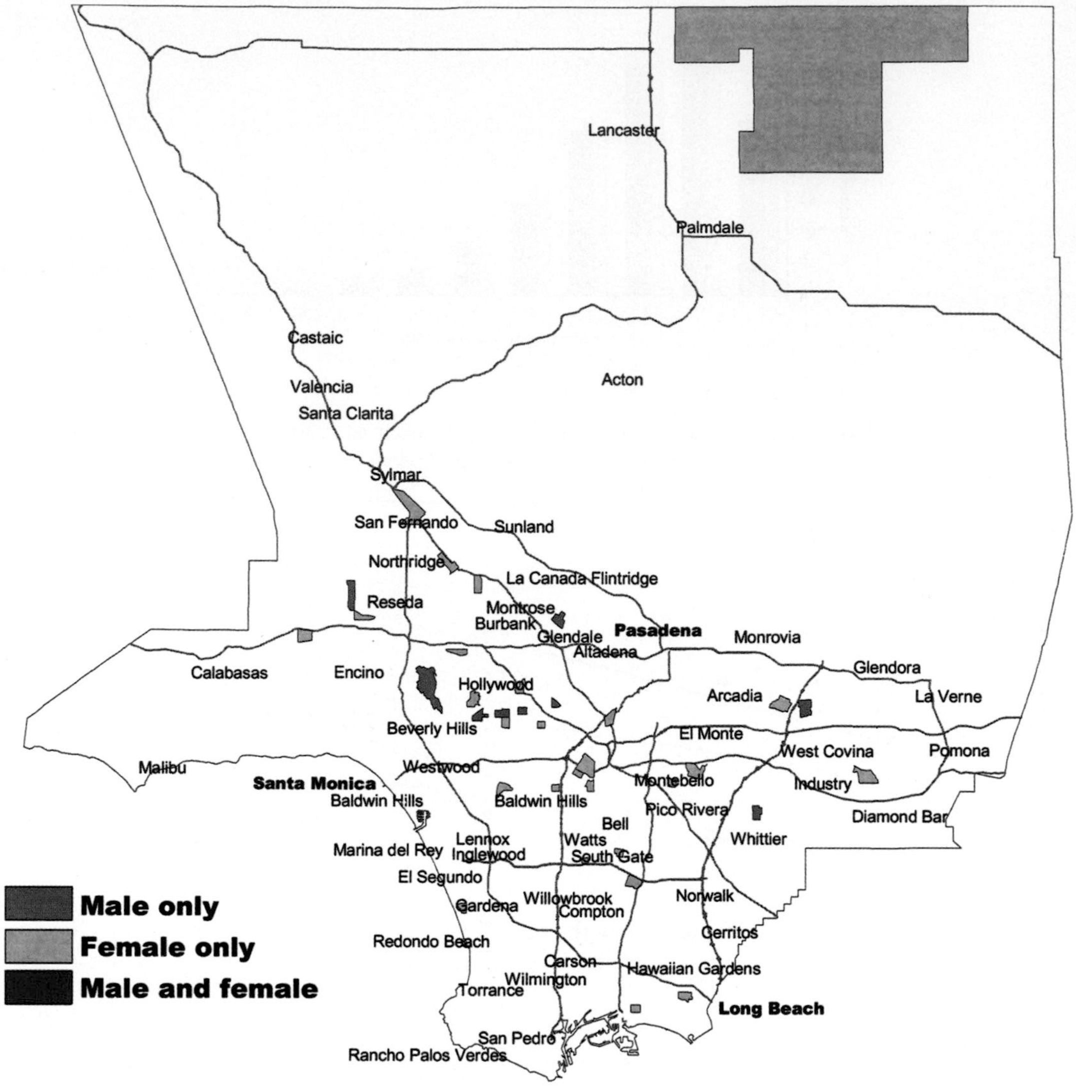

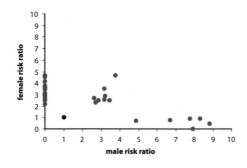

Figure 10: Male-female correlation between the relative risks for high-risk census tracts.

Figure 11: Map of census tracts at high risk, adjusted for social class.

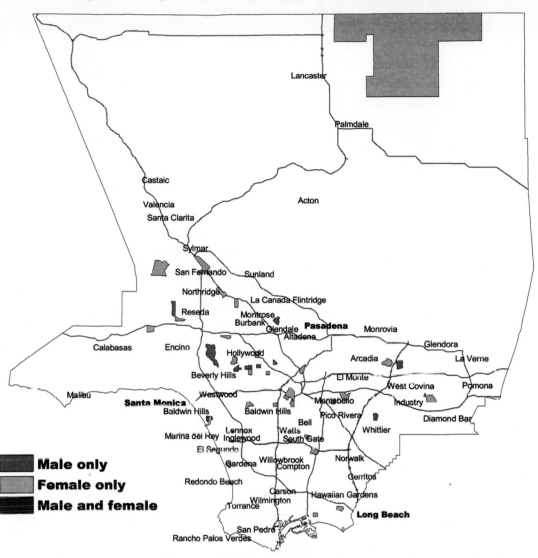

Figure 12: Male-female correlation between the relative risks for high-risk census tracts, adjusted for social class.

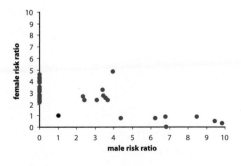

Anogenital Squamous Carcinoma

ICDO-2 Code Anatomic Site: C 21, 51, 52, 54, 55, 57, 60, 62, 63
ICDO-2 Code Histology: 8000–8082
Age: All
Male Cases: 1223
Female Cases: 3011

Background

Like squamous cervical cancer, squamous cancers of the anus and genitalia in both men and women are clearly related, in large part, to sexually transmitted infections with human papillomaviruses. Just as with squamous cervical cancer, past sexual activity on the part of both the affected person and his or her partners is presumed responsible. Human papillomaviruses are transmitted by anal as well as vaginal intercourse and do cause squamous carcinomas of the anus. Smoking and medical causes of immunosuppression are additional factors thought to increase risk.

Local Pattern

These squamous malignancies occur with the same frequency in Los Angeles County as in other regions of the country, and, except for anal carcinomas, are roughly twice as common in women as in men. Among both men and women, incidence rates begin to increase in middle age. Asian-Americans are at lower risk than those of other ethnicities, whereas whites and Latinas are at generally higher risk. Incidence in men, mostly representing anal carcinoma, is more common among those of lower social class. Incidence in the county as a whole has been constant over time, as it has among residents of high-risk census tracts. Figure 6 shows only a slight nonrandom excess of census tracts with unexpectedly few or unexpectedly many cases. After adjustment for social class, a group of contiguous male high-risk census tracts in West Hollywood, and a smaller focus in Long Beach appear.

Thumbnail Interpretation

Like cervical cancer, these cancers occur in a pattern that probably reflect levels of personal hygiene and the frequency of past sexual activity. The census tracts at high risk for women show no systematic geographical pattern. For men, however, the clusters of high-risk male census tracts in Long Beach and West Hollywood probably reflect the higher frequency of anal intercourse.

Figure 1: Age-adjusted incidence rate by place.

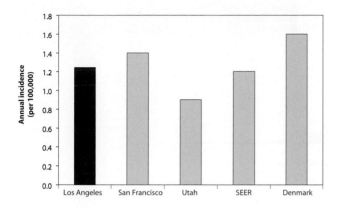

Figure 2: Age-adjusted incidence rate over the period.

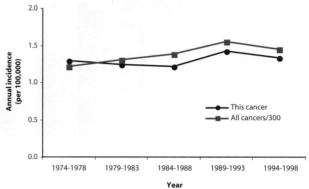

Figure 3: Age-adjusted incidence rate by age and race/ethnicity.

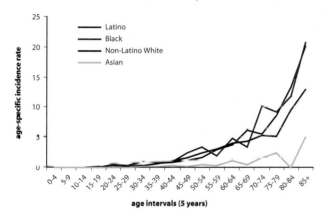

Figure 4: Age-adjusted incidence rate by social class.

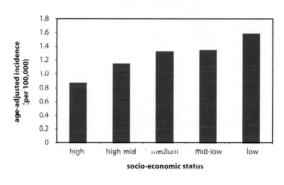

Figure 5: Distribution of the relative risk values for all census tracts.

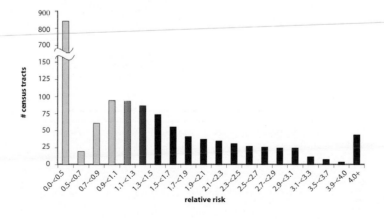

Figure 6: Census tracts by the number of cases per tract.

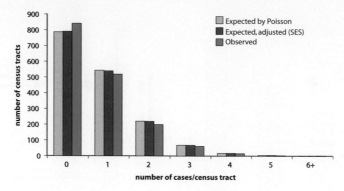

Figure 7a and b: Census tracts at high risk by the number of cases. (a) Unadjusted and (b) adjusted for social class.

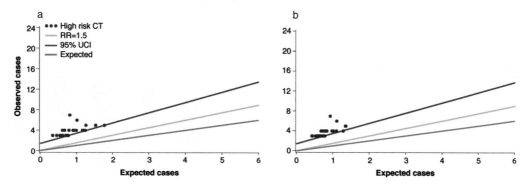

Figure 8: Risk over the period for high-risk census tracts relative to all census tracts.

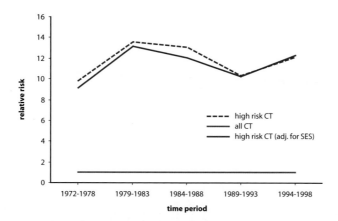

Anogenital Squamous Carcinoma: Female

Figure 1: Age-adjusted incidence rate by place.

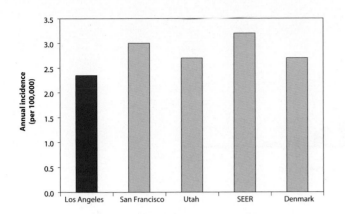

Figure 2: Age-adjusted incidence rate over the period.

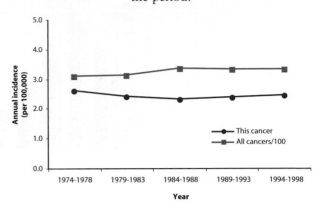

Figure 3: Age-adjusted incidence rate by age and race/ethnicity.

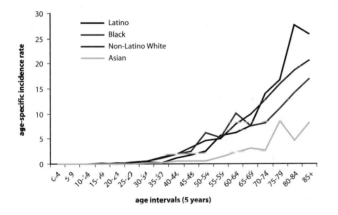

Figure 4: Age-adjusted incidence rate by social class.

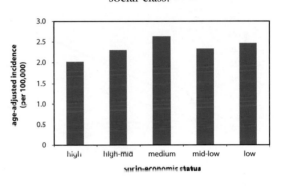

Figure 5: Distribution of the relative risk values for all census tracts.

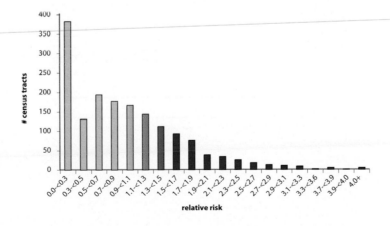

Figure 6: Census tracts by the number of cases per tract.

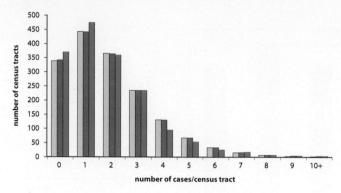

Figure 7a and b: Census tracts at high risk by the number of cases. (a) Unadjusted and (b) adjusted for social class.

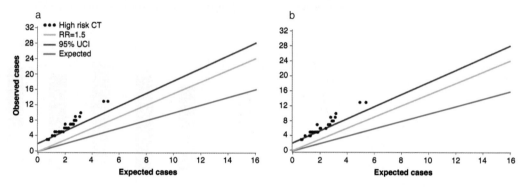

Figure 8: Risk over the period for high-risk census tracts relative to all census tracts.

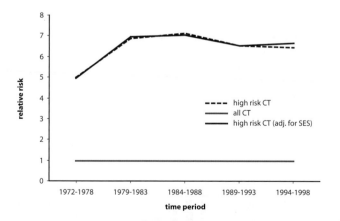

Anogenital Squamous Carcinoma

Figure 9: Map of census tracts at high risk.

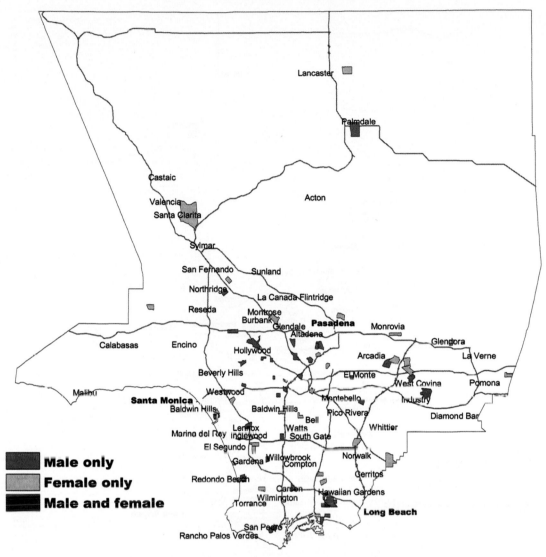

Male only
Female only
Male and female

Figure 10: Male-female correlation between the relative risks for high-risk census tracts.

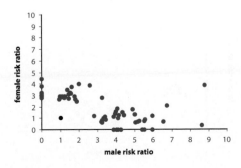

Figure 11: Map of census tracts at high risk, adjusted for social class.

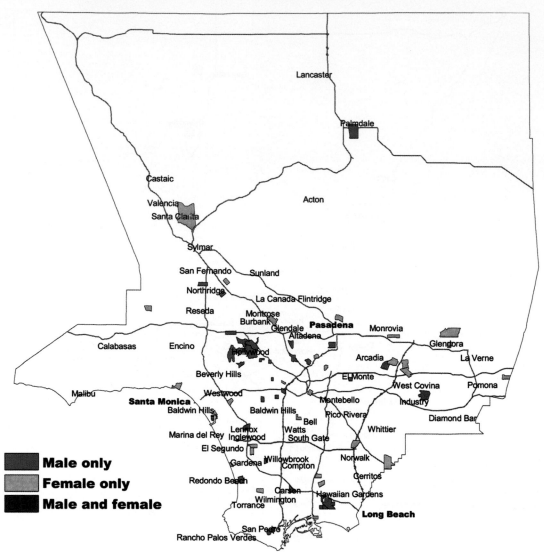

Figure 12: Male-female correlation between the relative risks for high-risk census tracts, adjusted for social class.

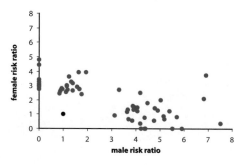

Mixed Cell Genital Tumors

ICDO-2 Code Anatomic Site: C 51–63
ICDO-2 Code Histology: 8562, 8933–8951, 8980–8982
Age: All
Male Cases: 0
Female Cases: 913

Background

These rare female tumors are developmentally related to the cells responsible for endometrial cancer and are likely to have similar causes.

Local Pattern

These malignancies occur less frequently in Los Angeles than in other parts of the country. Incidence in the county as a whole has been stable over time, although it has decreased among those residing in high-risk census tracts. The tumors occur slightly more frequently among African-Americans, but there is no relationship to social class. Figure 6 shows only a slight nonrandom excess of census tracts with unexpectedly few or unexpectedly many cases. High-risk census tracts are scattered throughout the county before and after adjustment for social class with no contiguous census tracts. No census tract stands out on the basis of a particularly high number of excess cases.

Thumbnail Interpretation

No high social class pattern of occurrence, as seen in endometrial cancer, is apparent. No systematic pattern of geographical occurrence is apparent, and therefore no local source of causation can be proposed.

Figure 1: Age-adjusted incidence rate by place.

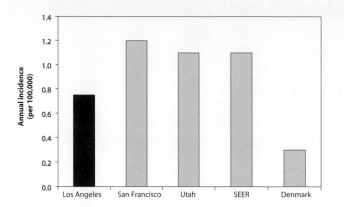

Figure 2: Age-adjusted incidence rate over the period.

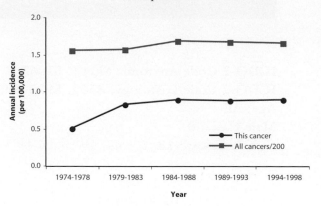

Figure 3: Age-adjusted incidence rate by age and race/ethnicity.

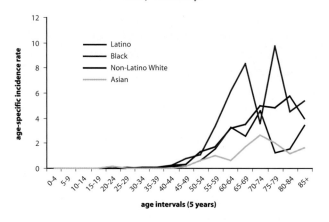

Figure 4: Age-adjusted incidence rate by social class.

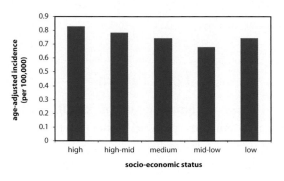

Figure 5: Distribution of the relative risk values for all census tracts.

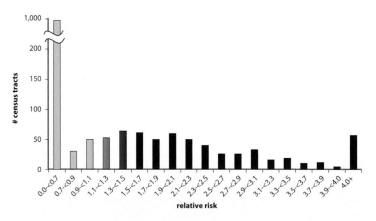

Figure 6: Census tracts by the number of cases per tract.

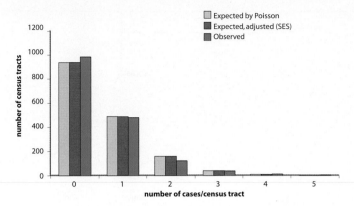

Figure 7a and b: Census tracts at high risk by the number of cases. (a) Unadjusted and (b) adjusted for social class.

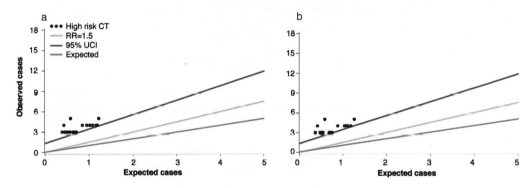

Figure 8: Risk over the period for high-risk census tracts relative to all census tracts.

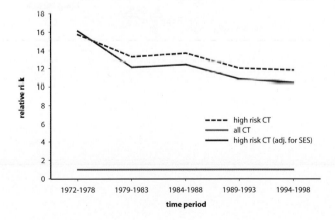

Figure 9: Map of census tracts at high risk.

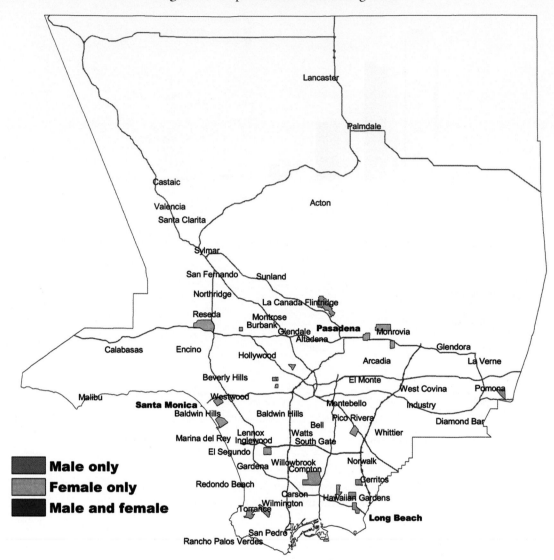

Figure 11: Map of census tracts at high risk, adjusted for social class.

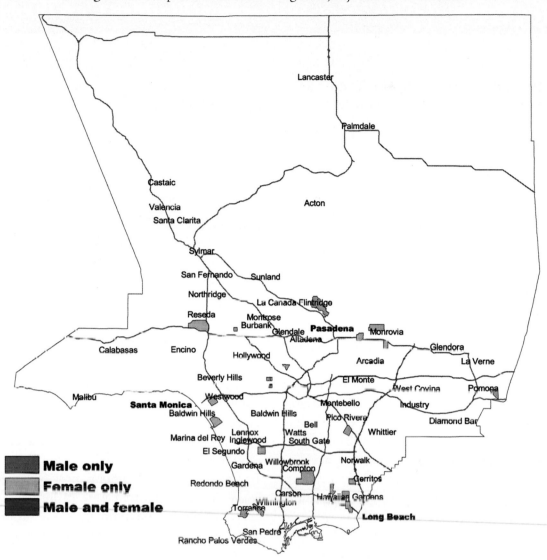

Transitional Cell (Non-Squamous) Bladder Carcinoma

ICDO-2 Code Anatomic Site: C 65–68
ICDO-2 Code Histology: 8000–8045, 8120–8560, 8570–8573
Age: All
Male Cases: 18659
Female Cases: 7509

Background

Transitional cell carcinoma is by far the most important type of cancer in the lower urinary tract and can occur anatomically anywhere in the urinary collection system, from the pelvis of the kidney down to the bladder and the urethra, where the urine exits the body. The most important current cause is cigarette smoking, and several genetic factors have been identified that slow the excretion of the carcinogens from tobacco smoke, thus prolonging contact with the bladder and increasing risk. A variety of occupational exposures also have been linked to bladder cancer, including aromatic amines such as benzidine, which are no longer used. Metal workers, rubber workers, smelter workers, and persons in a number of other occupations have been observed to be at high risk, and several different workplace exposures might be responsible. In addition to the aromatic amines, these include polyaromatic hydrocarbons (such as carbon black or soot), cutting oils, and metal dusts. Curiously, certain analgesics (common pain medicines) appear to produce cancer in the upper collecting system but not the bladder. Barbiturates, certain cancer drugs, and radiation have been thought responsible for some of these cancers under certain conditions. Permanent hair dyes have been found associated with bladder cancer in women. Arsenic exposure in the workplace, in medicines, and even in the water supply have been known to cause bladder cancer.

Local Pattern

This common cancer occurs less frequently in Los Angeles than in any of the other communities shown. It is nearly three times as common among men as women, appears with increasing frequency in middle age, and is especially common among white men. It occurs less commonly among those of very low social class, and is decreasing in incidence in Los Angeles among both men and women. Figure 6 shows only a slight nonrandom excess of census tracts with unexpectedly few or unexpectedly many cases. The high-risk areas appear to have maintained a roughly constant incidence. For women, the high-risk areas appear scattered around the county, whereas for men, they appear largely on the east side of Los Angeles County. There are a few small aggregates of high-risk census tracts, notably on the coast centered in Redondo Beach and in the eastern San Gabriel foothills along the 210 freeway. The former is reduced and the latter enhanced by adjustment for social class.

Thumbnail Interpretation

The pattern neither suggests the pattern of heavy smoking nor occupational exposure, and indeed no obvious explanation is available. The reasons for the few small clusters of high-risk census tracts are not known. None of the malignancies believed related to smoking showed a similar pattern, whether those occurring in the gastrointestinal, the respiratory, or the genitourinary organ systems.

Figure 1: Age-adjusted incidence rate by place.

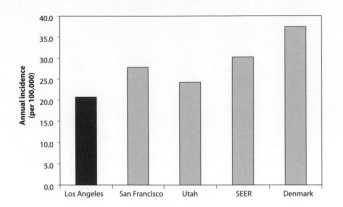

Figure 2: Age-adjusted incidence rate over the period.

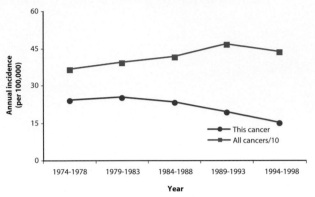

Figure 3: Age-adjusted incidence rate by age and race/ethnicity.

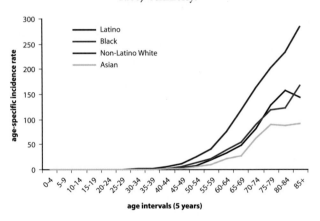

Figure 4: Age-adjusted incidence rate by social class.

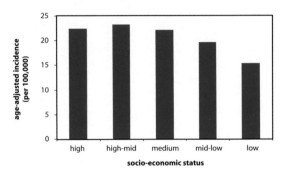

Figure 5: Distribution of the relative risk values for all census tracts.

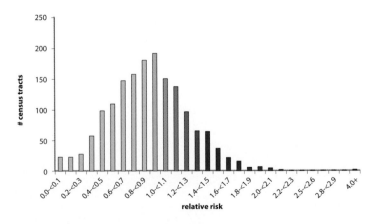

Figure 6: Census tracts by the number of cases per tract.

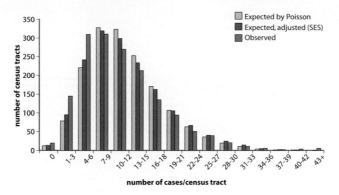

Figure 7a and b: Census tracts at high risk by the number of cases. (a) Unadjusted and (b) adjusted for social class.

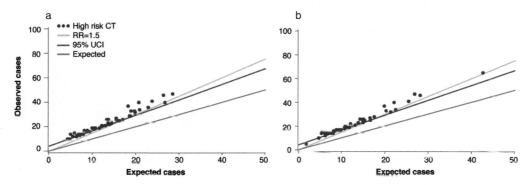

Figure 8: Risk over the period for high-risk census tracts relative to all census tracts.

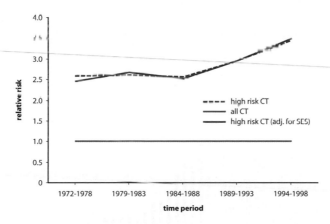

Transitional Cell (Non-Squamous) Bladder Carcinoma: Female

Figure 1: Age-adjusted incidence rate by place.

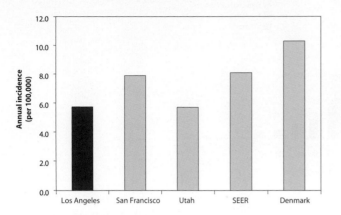

Figure 2: Age-adjusted incidence rate over the period.

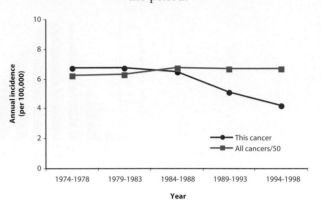

Figure 3: Age-adjusted incidence rate by age and race/ethnicity.

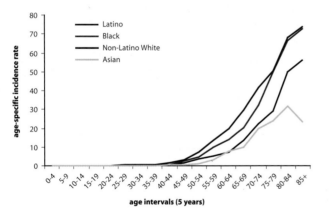

Figure 4: Age-adjusted incidence rate by social class.

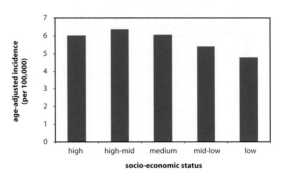

Figure 5: Distribution of the relative risk values for all census tracts.

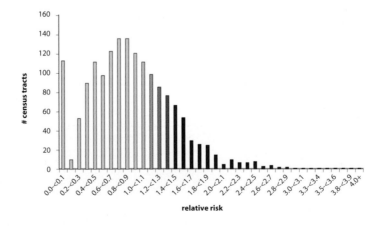

Figure 6: Census tracts by the number of cases per tract.

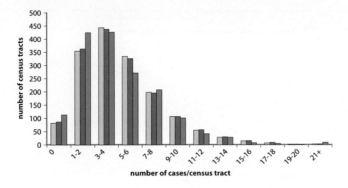

Figure 7a and b: Census tracts at high risk by the number of cases. (a) Unadjusted and (b) adjusted for social class.

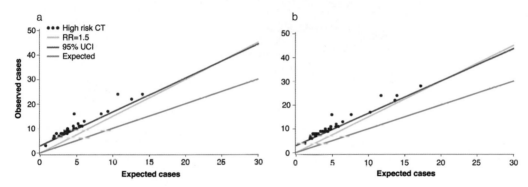

Figure 8: Risk over the period for high-risk census tracts relative to all census tracts.

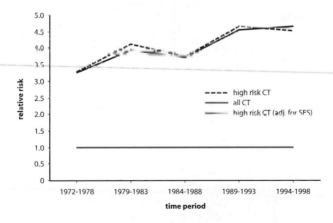

Figure 9: Map of census tracts at high risk.

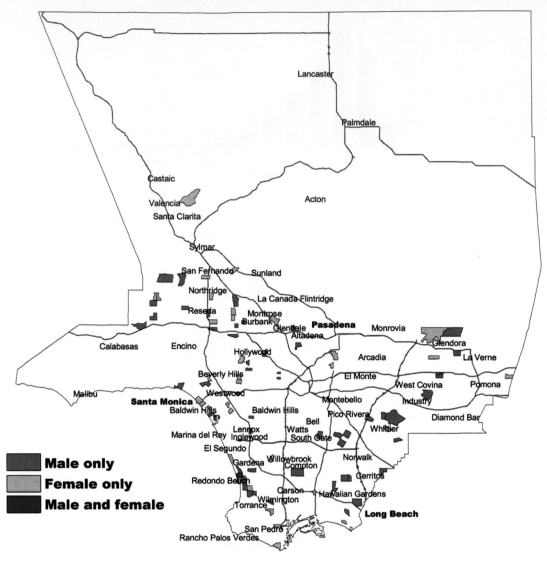

Figure 10: Male-female correlation between the relative risks for high-risk census tracts.

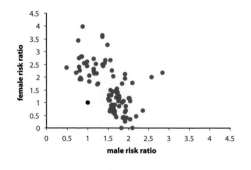

Figure 11: Map of census tracts at high risk, adjusted for social class.

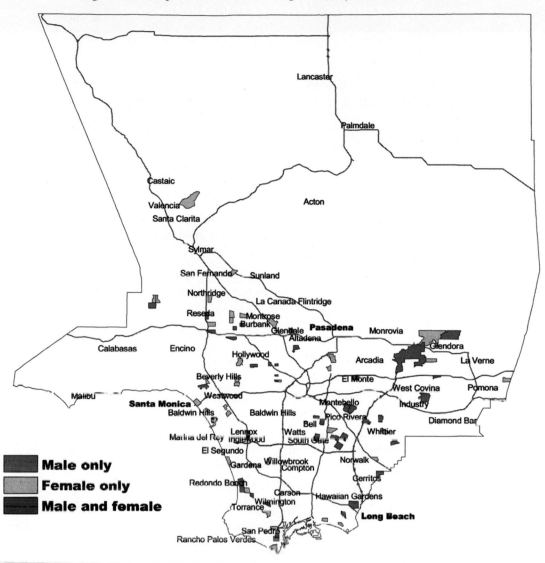

Figure 12: Male-female correlation between the relative risks for high-risk census tracts, adjusted for social class.

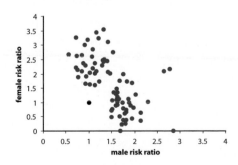

Squamous Bladder Carcinoma

ICDO-2 Code Anatomic Site: C 65–68
ICDO-2 Code Histology: 8050–8082
Age: All
Male Cases: 841
Female Cases: 529

Background

The most common cause worldwide of squamous bladder cancer is the liver fluke (*Schistosoma hematobium*). That parasite, common in east Africa, migrates to the tissues around the bladder and causes scarring, which presumably is a cause of the malignancy. In the United States, this form of bladder cancer is much less common than transitional cell carcinoma. It occurs in heavy smokers and also in persons who have been scarred from repeated bacterial bladder infections, such as older persons with long-standing indwelling urinary catheters.

Local Pattern

This malignancy is slightly less common in Los Angeles County than in other areas, except for Utah, where the rate is even lower. It occurs twice as commonly among men as among women, and is less common among Asians than it is among the members of other racial/ethnic groups. It is decreasing in frequency over time in Los Angeles County as a whole, although the rate has been relatively constant among the residents of census tracts at high risk. Incidence is unrelated to social class. Figure 6 shows only a slight nonrandom excess of census tracts with unexpectedly few or unexpectedly many cases. There is an aggregation of high-risk census tracts in Beverly Hills that is unchanged after adjustment for social class.

Thumbnail Interpretation

The higher risk in men, the higher risk in Los Angeles County than in Utah, and the decrease in incidence over time are all consistent with causation by smoking. However, the pattern in Los Angeles County does not follow that characteristic of any other smoking-related malignancy, and the reason for the increased risk in Beverly Hills and adjacent census tracts is not evident. The cases in the high-risk census tracts have tended to occur among older Jewish men and women, but were diagnosed in the same facilities responsible for the diagnosis of the more common transitional cell bladder cancer.

Figure 1: Age-adjusted incidence rate by place.

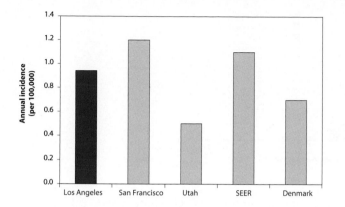

Figure 2: Age-adjusted incidence rate over the period.

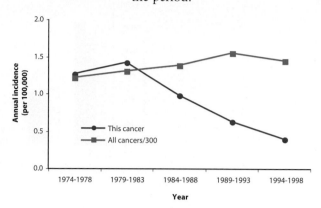

Figure 3: Age-adjusted incidence rate by age and race/ethnicity.

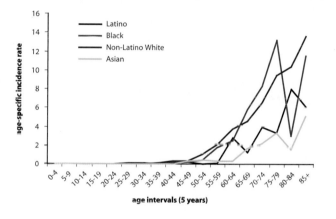

Figure 4: Age-adjusted incidence rate by social class.

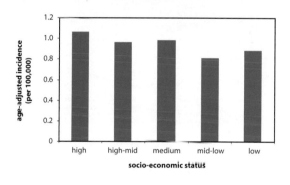

Figure 5: Distribution of the relative risk values for all census tracts.

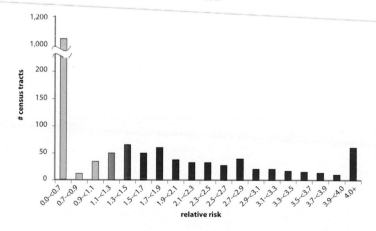

Figure 6: Census tracts by the number of cases per tract.

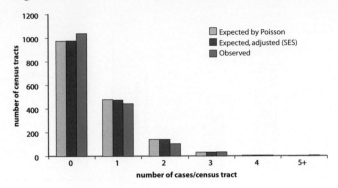

Figure 7a and b: Census tracts at high risk by the number of cases. (a) Unadjusted and (b) adjusted for social class.

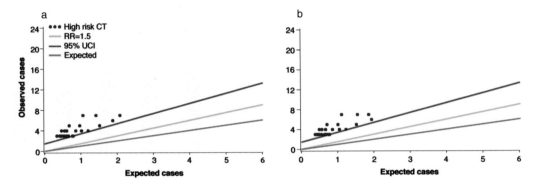

Figure 8: Risk over the period for high-risk census tracts relative to all census tracts.

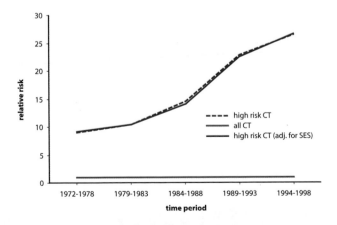

Figure 1: Age-adjusted incidence rate by place.

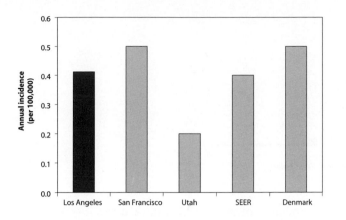

Figure 2: Age-adjusted incidence rate over the period.

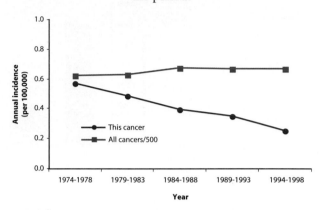

Figure 3: Age-adjusted incidence rate by age and race/ethnicity.

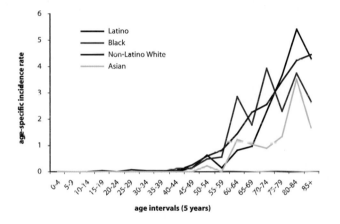

Figure 4: Age-adjusted incidence rate by social class.

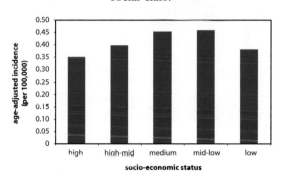

Figure 5: Distribution of the relative risk values for all census tracts.

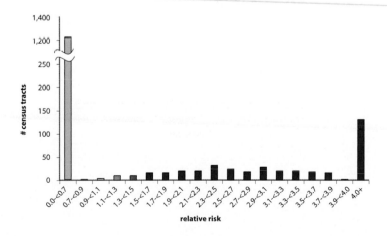

Figure 6: Census tracts by the number of cases per tract.

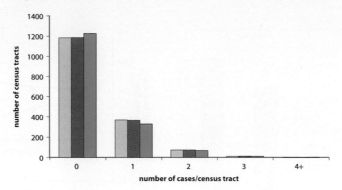

Figure 7a and b: Census tracts at high risk by the number of cases. (a) Unadjusted and (b) adjusted for social class.

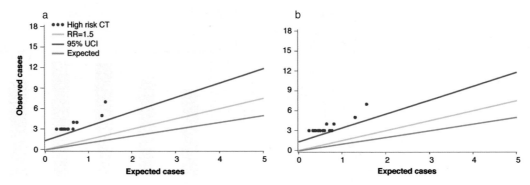

Figure 8: Risk over the period for high-risk census tracts relative to all census tracts.

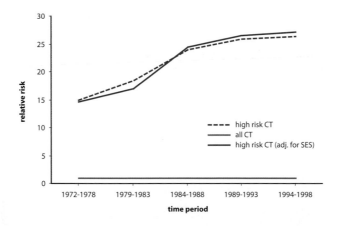

Figure 9: Map of census tracts at high risk.

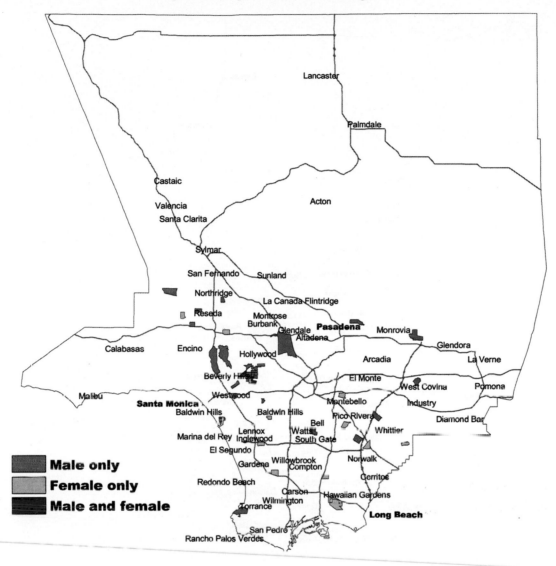

Figure 10· Male-female correlation between the relative risks for high-risk census tracts.

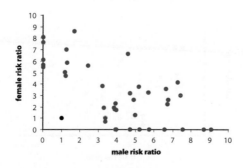

Figure 11: Map of census tracts at high risk, adjusted for social class.

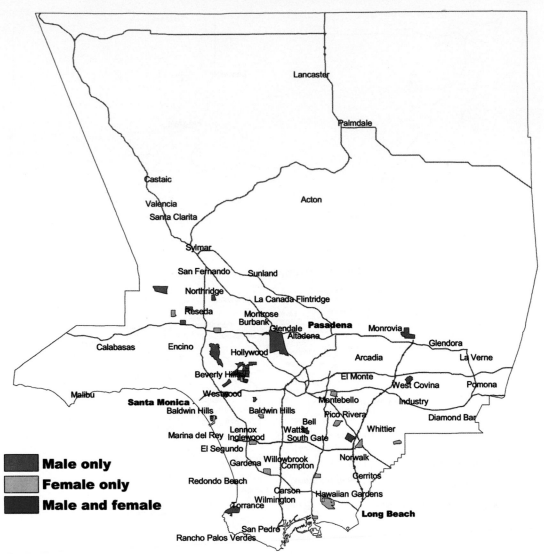

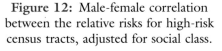

Figure 12: Male-female correlation between the relative risks for high-risk census tracts, adjusted for social class.

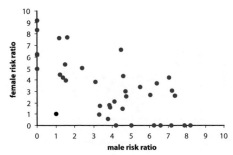

Renal Cell Carcinoma

ICDO-2 Code Anatomic Site: C 64
ICDO-2 Code Histology: 8000–8560, 8570–8573
Age: All
Male Cases: 8000
Female Cases: 4538

Background

Obesity and hypertension are well-established risk factors for renal (kidney) carcinoma, although the mechanisms are unclear. Smoking, diet pills, and analgesics have also been strongly suspected of playing a causal role. A small proportion of cases occur in persons who have one of several rare genetic syndromes.

Local Pattern

Renal cell carcinoma occurs at about the same frequency in Los Angeles as it does in other developed country populations. It begins to increase in frequency in middle age, and is twice as common among men as among women. The malignancy is less common among Asian-Americans generally, and more common among African American women, than in the members of other racial/ethnic groups. Incidence has gradually increased both in the county as a whole and among residents of high-risk census tracts. Renal cell carcinoma occurs equally in the members of all social classes. Figure 6 shows only a slight nonrandom excess of census tracts with unexpectedly few or unexpectedly many cases. The census tracts at high risk are widely scattered in a pattern that is unchanged by adjustment for social class. No census tract stands out on the basis of a particularly high number of excess cases.

Thumbnail Interpretation

The pattern of occurrence by race/ethnicity, sex, age, and even calendar time are consistent with the effects of obesity. No systematic pattern of geographical occurrence is apparent, and therefore no other local source of causation can be proposed.

Figure 1: Age-adjusted incidence rate by place.

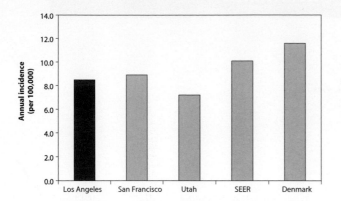

Figure 2: Age-adjusted incidence rate over the period.

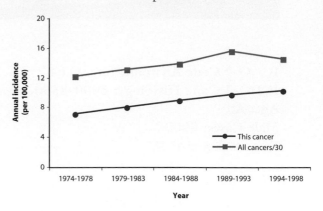

Figure 3: Age-adjusted incidence rate by age and race/ethnicity.

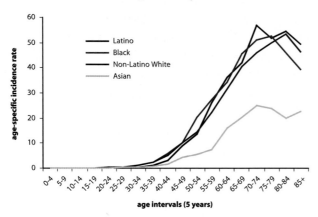

Figure 4: Age-adjusted incidence rate by social class.

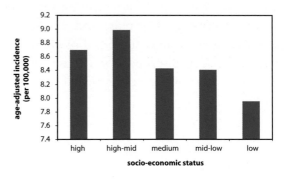

Figure 5: Distribution of the relative risk values for all census tracts.

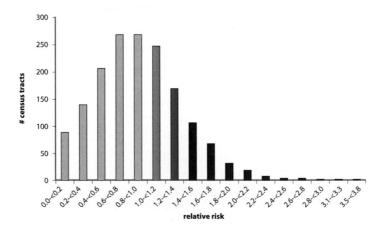

Figure 6: Census tracts by the number of cases per tract.

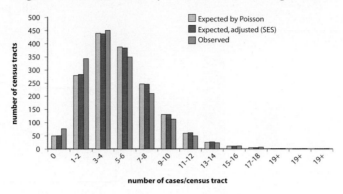

Figure 7a and b: Census tracts at high risk by the number of cases. (a) Unadjusted and (b) adjusted for social class.

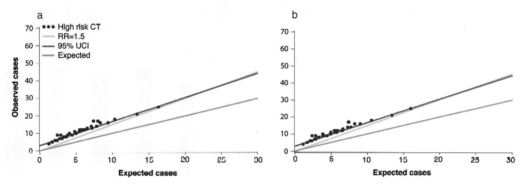

Figure 8: Risk over the period for high-risk census tracts relative to all census tracts.

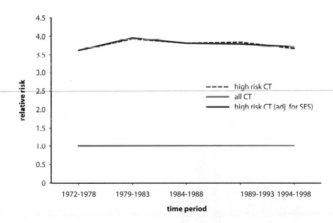

Renal Cell Carcinoma: Female

Figure 1: Age-adjusted incidence rate by place.

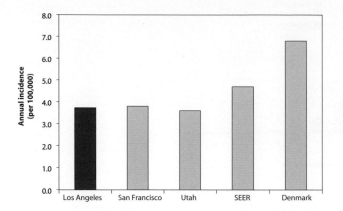

Figure 2: Age-adjusted incidence rate over the period.

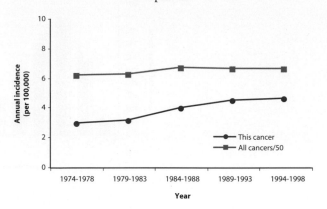

Figure 3: Age-adjusted incidence rate by age and race/ethnicity.

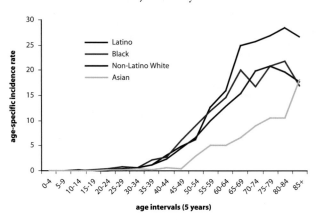

Figure 4: Age-adjusted incidence rate by social class.

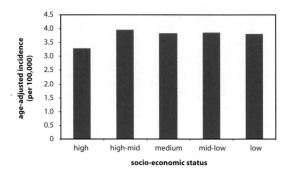

Figure 5: Distribution of the relative risk values for all census tracts.

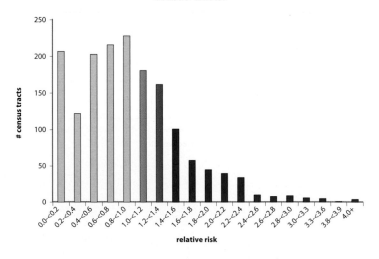

Figure 6: Census tracts by the number of cases per tract.

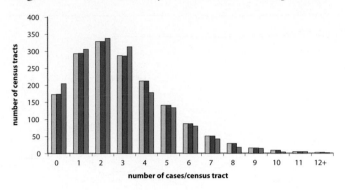

Figure 7a and b: Census tracts at high risk by the number of cases. (a) Unadjusted and (b) adjusted for social class.

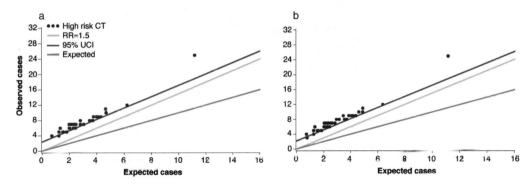

Figure 8: Risk over the period for high-risk census tracts relative to all census tracts.

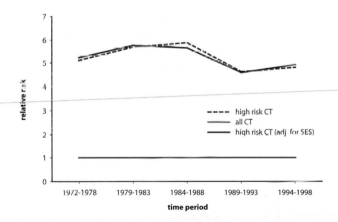

Figure 9: Map of census tracts at high risk.

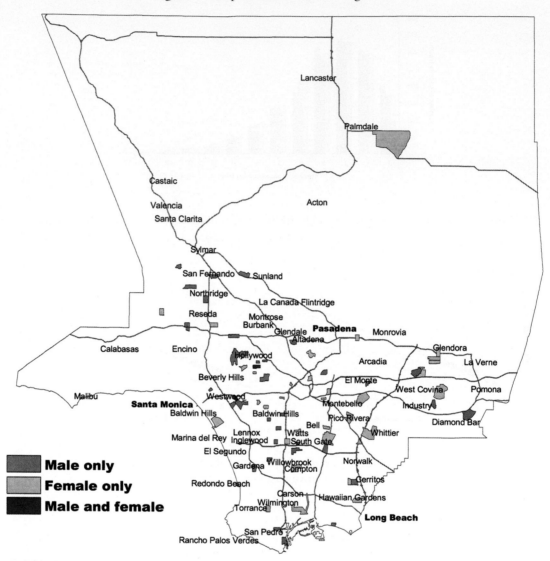

Figure 10: Male-female correlation between the relative risks for high-risk census tracts.

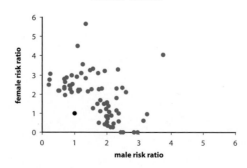

Figure 11: Map of census tracts at high risk, adjusted for social class.

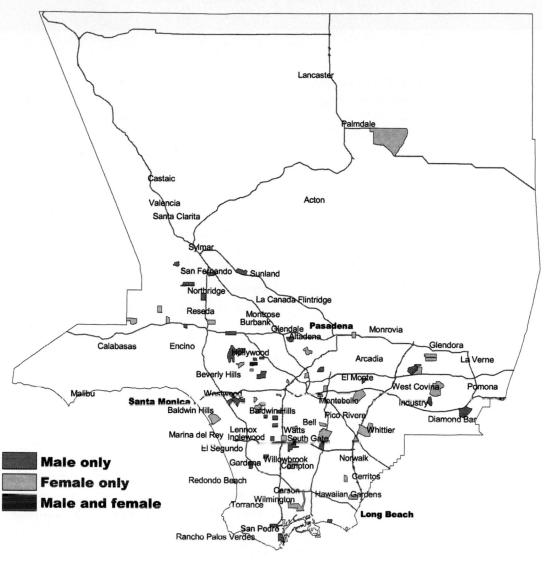

Male only
Female only
Male and female

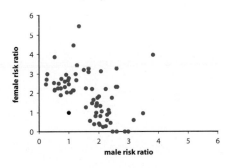

Figure 12: Male-female correlation between the relative risks for high-risk census tracts, adjusted for social class.

Wilms Tumor (Nephroblastoma)

ICDO-2 Code Anatomic Site: C 0–80
ICDO-2 Code Histology: 8960–8963
Age: All
Male Cases: 193
Female Cases: 258

Background

A small percentage of these rare childhood malignancies are inherited, and most of the others seem to result from changes in the genetic message that become apparent over the course of physical development, probably beginning in the fetal period. Wilms tumors occur more commonly in the presence of any of a number of congenital anomalies. The underlying environmental causes, if they exist, are unknown.

Local Pattern

These rare malignancies, mostly occurring in childhood, are about as common in Los Angeles County as elsewhere and occur with equal frequency in both sexes and among the members of all social classes, although they occur less often among Asian-Americans. Incidence in Los Angeles County has been relatively constant over time. Figure 6 shows no nonrandom excess of census tracts with unexpectedly few or unexpectedly many cases. The high-risk criteria were fulfilled for only three widely separated census tracts.

Thumbnail Interpretation

With the exception of some racial difference, this malignancy has occurred with roughly equal frequency among all the children of Los Angeles County. That pattern is consistent with causation by genetic or other factors that act during development. No systematic pattern of geographical occurrence is apparent, and therefore no local source of causation can be proposed.

Figure 1: Age-adjusted incidence rate by place.

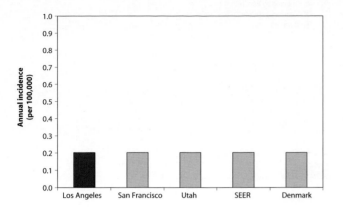

Figure 2: Age-adjusted incidence rate over the period.

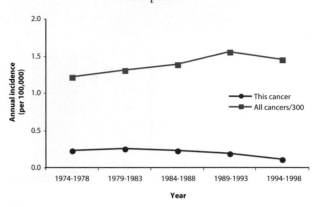

Figure 3: Age-adjusted incidence rate by age and race/ethnicity.

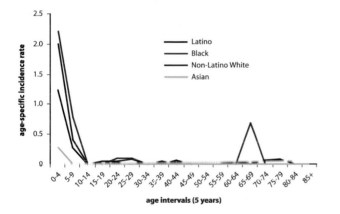

Figure 4: Age-adjusted incidence rate by social class.

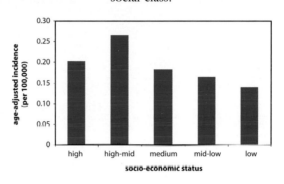

Figure 5: Distribution of the relative risk values for all census tracts.

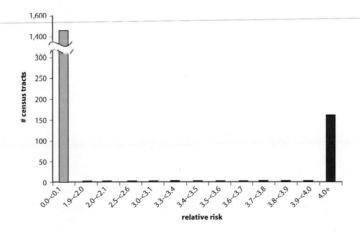

Wilms Tumor (Nephroblastoma): Male

Figure 6: Census tracts by the number of cases per tract.

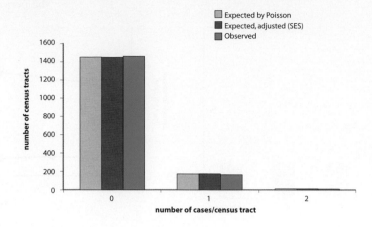

Figure 7a and b: Census tracts at high risk by the number of cases. (a) Unadjusted and (b) adjusted for social class.

There were no high risk census tracts.

Figure 8: Risk over the period for high-risk census tracts relative to all census tracts.

There were no high risk census tracts.

Wilms Tumor (Nephroblastoma): Female

Figure 1: Age-adjusted incidence rate by place.

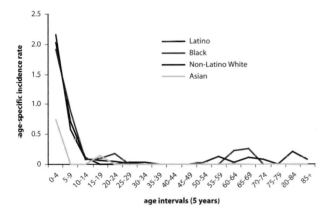

Figure 2: Age-adjusted incidence rate over the period.

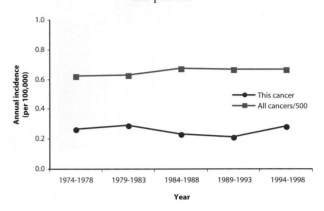

Figure 3: Age-adjusted incidence rate by age and race/ethnicity.

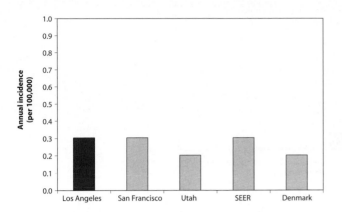

Figure 4: Age adjusted incidence rate by social class.

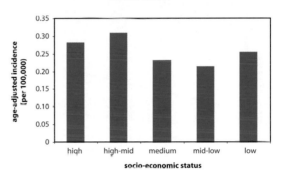

Figure 5: Distribution of the relative risk values for all census tracts.

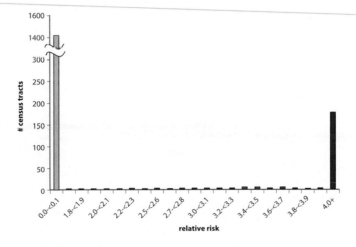

Figure 6: Census tracts by the number of cases per tract.

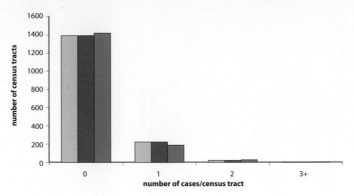

Figure 7a and b: Census tracts at high risk by the number of cases. (a) Unadjusted and (b) adjusted for social class.

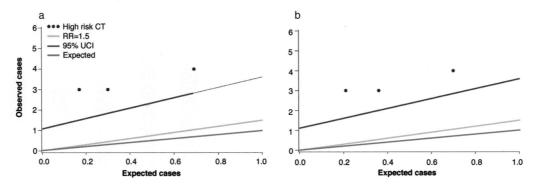

Figure 8: Risk over the period for high-risk census tracts relative to all census tracts.

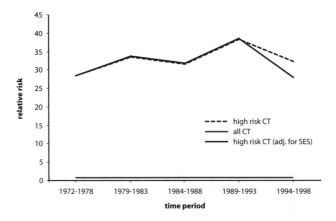

Figure 9: Map of census tracts at high risk.

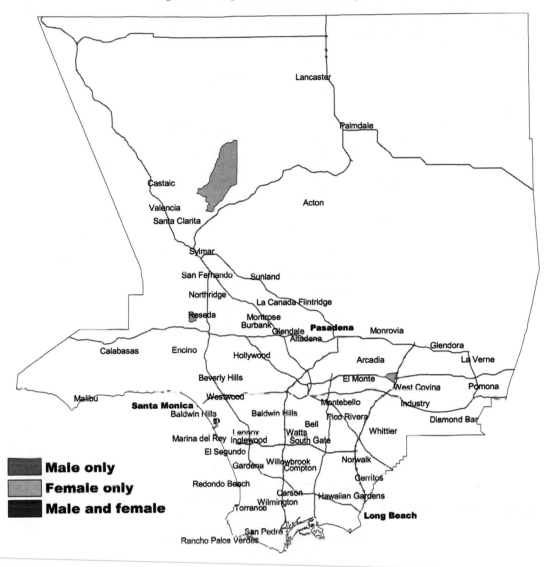

Figure 10: Male-female correlation between the relative risks for high-risk census tracts.

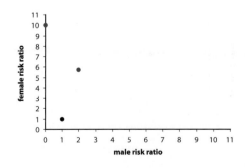

Figure 11: Map of census tracts at high risk, adjusted for social class.

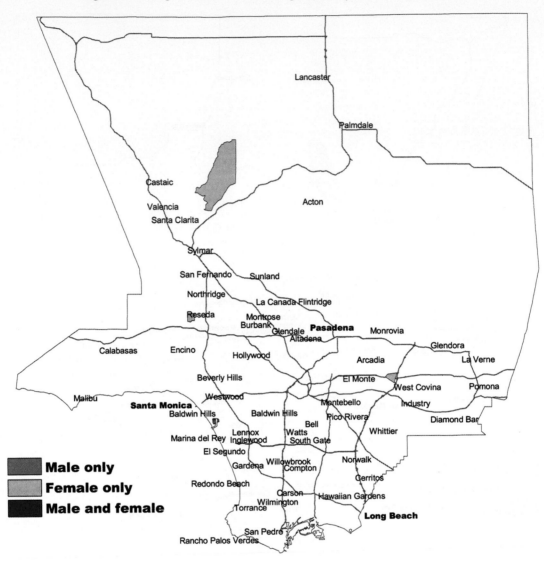

Figure 12: Male-female correlation between the relative risks for high-risk census tracts, adjusted for social class.

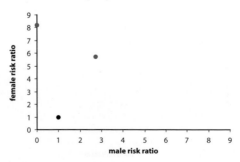

Retinoblastoma

ICDO-2 Code Anatomic Site: C 0–80
ICDO-2 Code Histology: 9510–9523
Age: All
Male Cases: 164
Female Cases: 148

Background

Retinoblastoma is a malignancy of the eye, usually occurring in young children. About half of these malignancies occur in persons who have inherited a gene causing a specific susceptibility. The causes of the others are unknown, although the mechanism of action is by means of a mutation in the same gene. This may be produced by an environmental exposure such as radiation, or it may occur for unknown reasons.

Local Pattern

The incidence of retinoblastoma in Los Angeles County is identical to that in other populations and identical in the two sexes. Incidence is unrelated to either racial/ethnic identity or social class, and the occurrence in Los Angeles County as a whole has been constant over time. Figure 6 shows no nonrandom excess of census tracts with unexpectedly few or unexpectedly many cases. Only one census tract fulfilled the high-risk criteria.

Thumbnail Interpretation

No systematic pattern of occurrence is apparent, and therefore no local source of causation can be proposed.

Figure 1: Age-adjusted incidence rate by place.

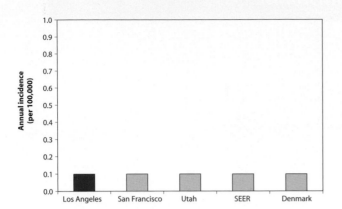

Figure 2: Age-adjusted incidence rate over the period.

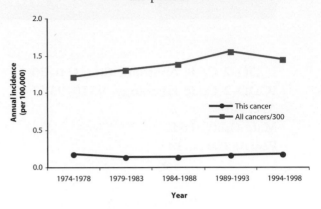

Figure 3: Age-adjusted incidence rate by age and race/ethnicity.

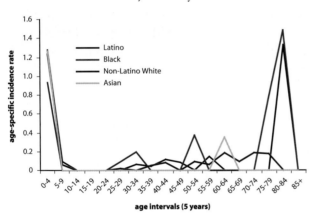

Figure 4: Age-adjusted incidence rate by social class.

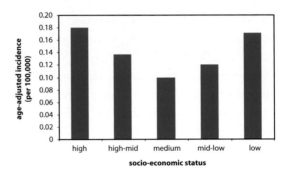

Figure 5: Distribution of the relative risk values for all census tracts.

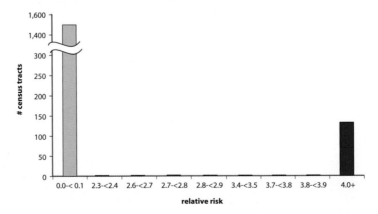

Figure 6: Census tracts by the number of cases per tract.

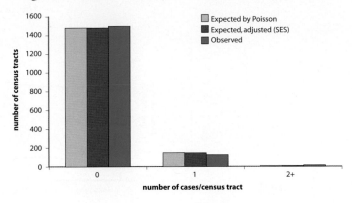

Figure 7a and b: Census tracts at high risk by the number of cases. (a) Unadjusted and (b) adjusted for social class.

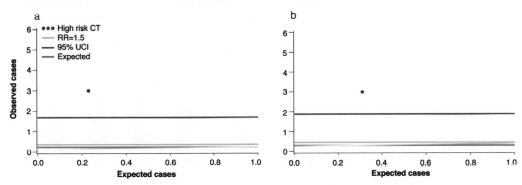

Figure 8: Risk over the period for high-risk census tracts relative to all census tracts.

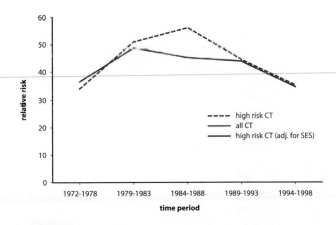

Figure 1: Age-adjusted incidence rate by place.

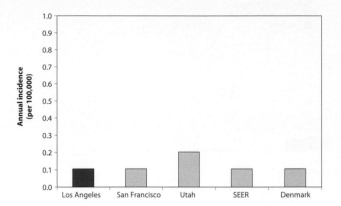

Figure 2: Age-adjusted incidence rate over the period.

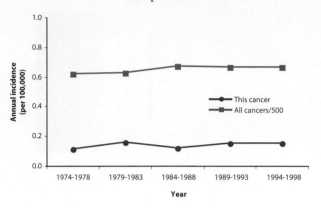

Figure 3: Age-adjusted incidence rate by age and race/ethnicity.

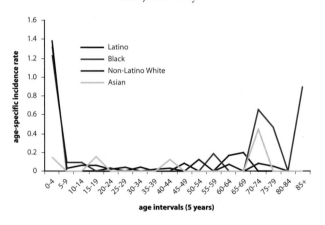

Figure 4: Age-adjusted incidence rate by social class.

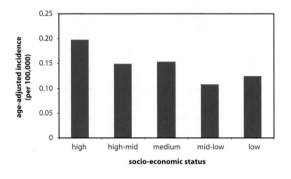

Figure 5: Distribution of the relative risk values for all census tracts.

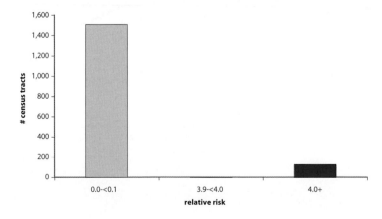

Figure 6: Census tracts by the number of cases per tract.

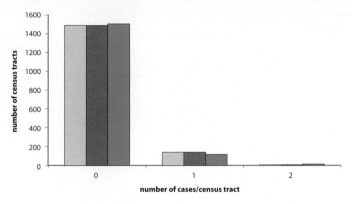

Figure 7a and b: Census tracts at high risk by the number of cases. (a) Unadjusted and (b) adjusted for social class.

There were no high risk census tracts.

Figure 8: Risk over the period for high-risk census tracts relative to all census tracts.

There were no high risk census tracts.

Figure 9: Map of census tracts at high risk.

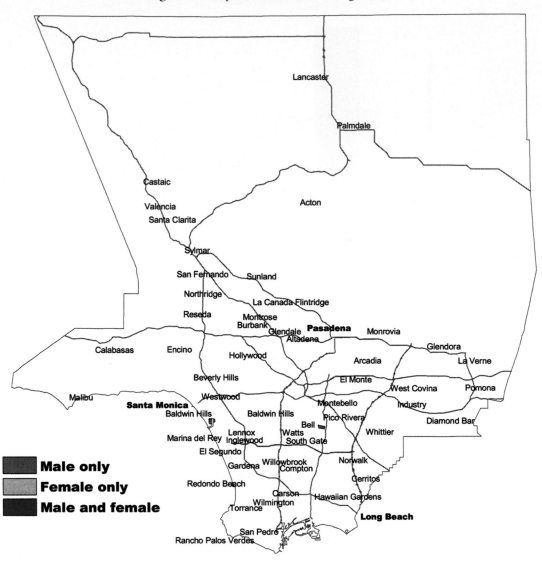

Figure 10: Male-female correlation between the relative risks for high-risk census tracts.

There were no high risk census tracts.

Figure 11: Map of census tracts at high risk, adjusted for social class.

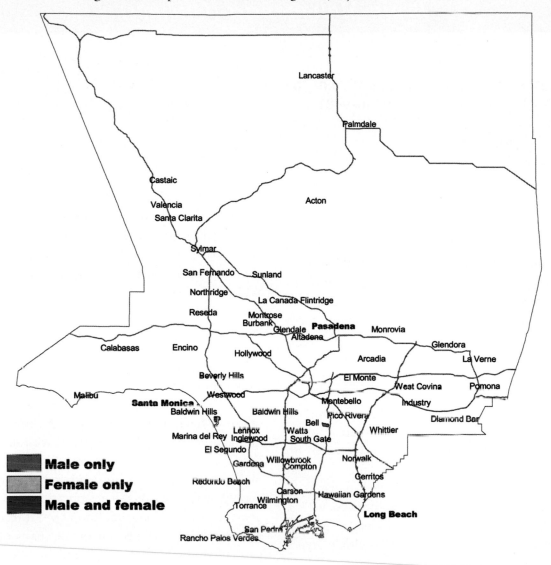

Male only
Female only
Male and female

Figure 12: Male-female correlation between the relative risks for high-risk census tracts, adjusted for social class.

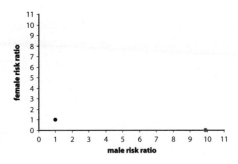

Brain Malignancies (Gliomas)

ICDO-2 Code Anatomic Site: C 70.0, 70.9, 71
ICDO-2 Code Histology: 9380–9481
Age: All
Male Cases: 5463
Female Cases: 4292

Background

The only clearly recognized cause of brain malignancies is ionizing radiation. While children exposed to the atomic bomb went on to develop many more brain tumors than were expected, radiation otherwise accounts for a very small proportion of the total. Although no specific causes have been firmly established, workers in several occupations have been considered by some investigators, but not others, to be at higher risk. Among these are workers in the petrochemical, electrical, rubber, agricultural, and medical industries. However, no specific workplace exposures are known to lead to brain tumors. A very small proportion of brain malignancies appear in persons with a family history of brain tumors or genetic syndromes associated with multiple types of malignancy.

in other localities. They are prominent among childhood malignancies, although they may occur at any age, and are somewhat more common among males than among females. Whites are at especially high and Asian-Americans at especially low risk in comparison with other racial/ethnic groups, and persons of upper social class are at slightly increased risk. These tumors have appeared with relatively constant frequency over time in the county as a whole, as well as among residents of census tracts at high risk. Figure 6 shows no nonrandom excess of census tracts with unexpectedly few or unexpectedly many cases. The high-risk census tracts are scattered over the county with relatively few appearing in the geographical regions of lower social class. No census tract stands out on the basis of a particularly high relative risk and number of excess cases.

Local Pattern

Gliomas occur with roughly the same frequency in Los Angeles County as they do

Thumbnail Interpretation

No systematic pattern of occurrence is apparent, and therefore no local source of causation can be proposed.

Brain Malignancies (Gliomas): Male

Figure 1: Age-adjusted incidence rate by place.

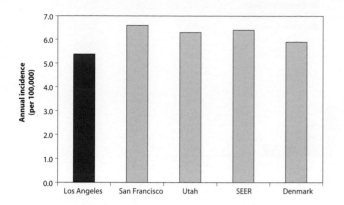

Figure 2: Age-adjusted incidence rate over the period.

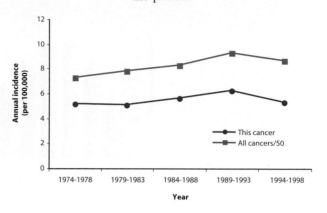

Figure 3: Age-adjusted incidence rate by age and race/ethnicity.

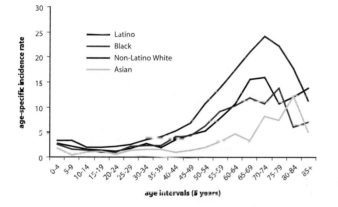

Figure 4: Age-adjusted incidence rate by social class.

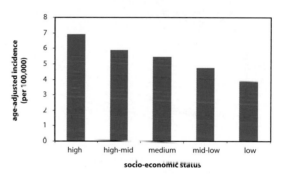

Figure 5: Distribution of the relative risk values for all census tracts.

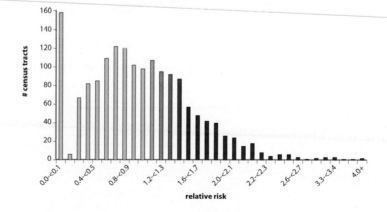

Figure 6: Census tracts by the number of cases per tract.

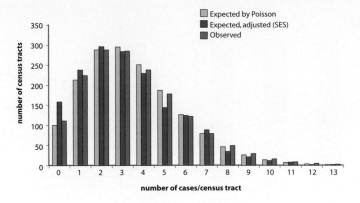

Figure 7a and b: Census tracts at high risk by the number of cases. (a) Unadjusted and (b) adjusted for social class.

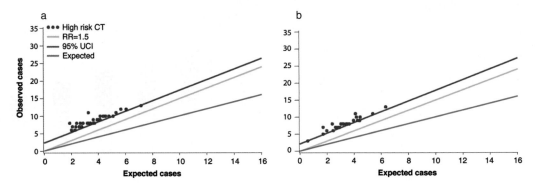

Figure 8: Risk over the period for high-risk census tracts relative to all census tracts.

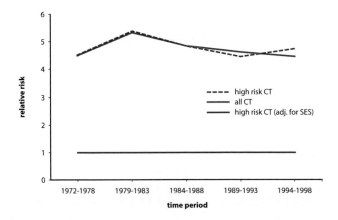

Brain Malignancies (Gliomas): Female

Figure 1: Age-adjusted incidence rate by place.

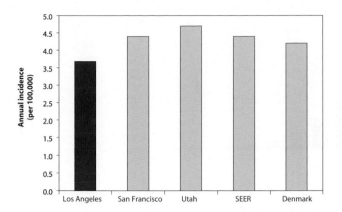

Figure 2: Age-adjusted incidence rate over the period.

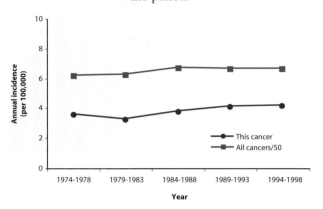

Figure 3: Age-adjusted incidence rate by age and race/ethnicity.

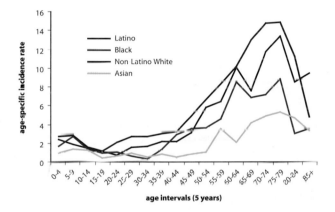

Figure 4: Age-adjusted incidence rate by social class.

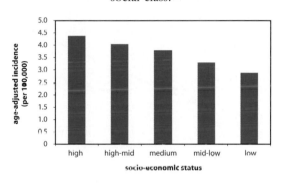

Figure 5: Distribution of the relative risk values for all census tracts.

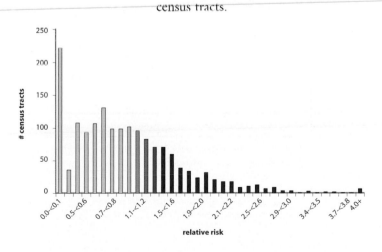

Figure 6: Census tracts by the number of cases per tract.

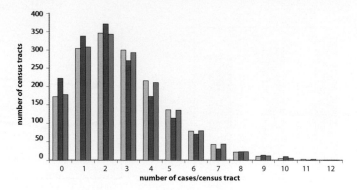

Figure 7a and b: Census tracts at high risk by the number of cases. (a) Unadjusted and (b) adjusted for social class.

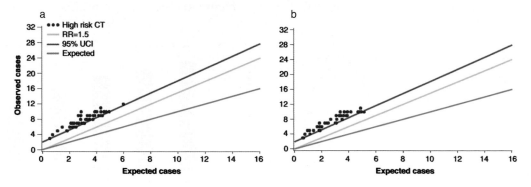

Figure 8: Risk over the period for high-risk census tracts relative to all census tracts.

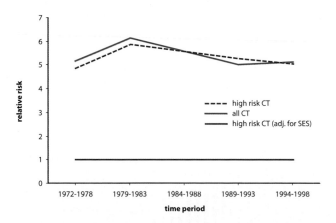

Brain Malignancies (Gliomas)

Figure 9: Map of census tracts at high risk.

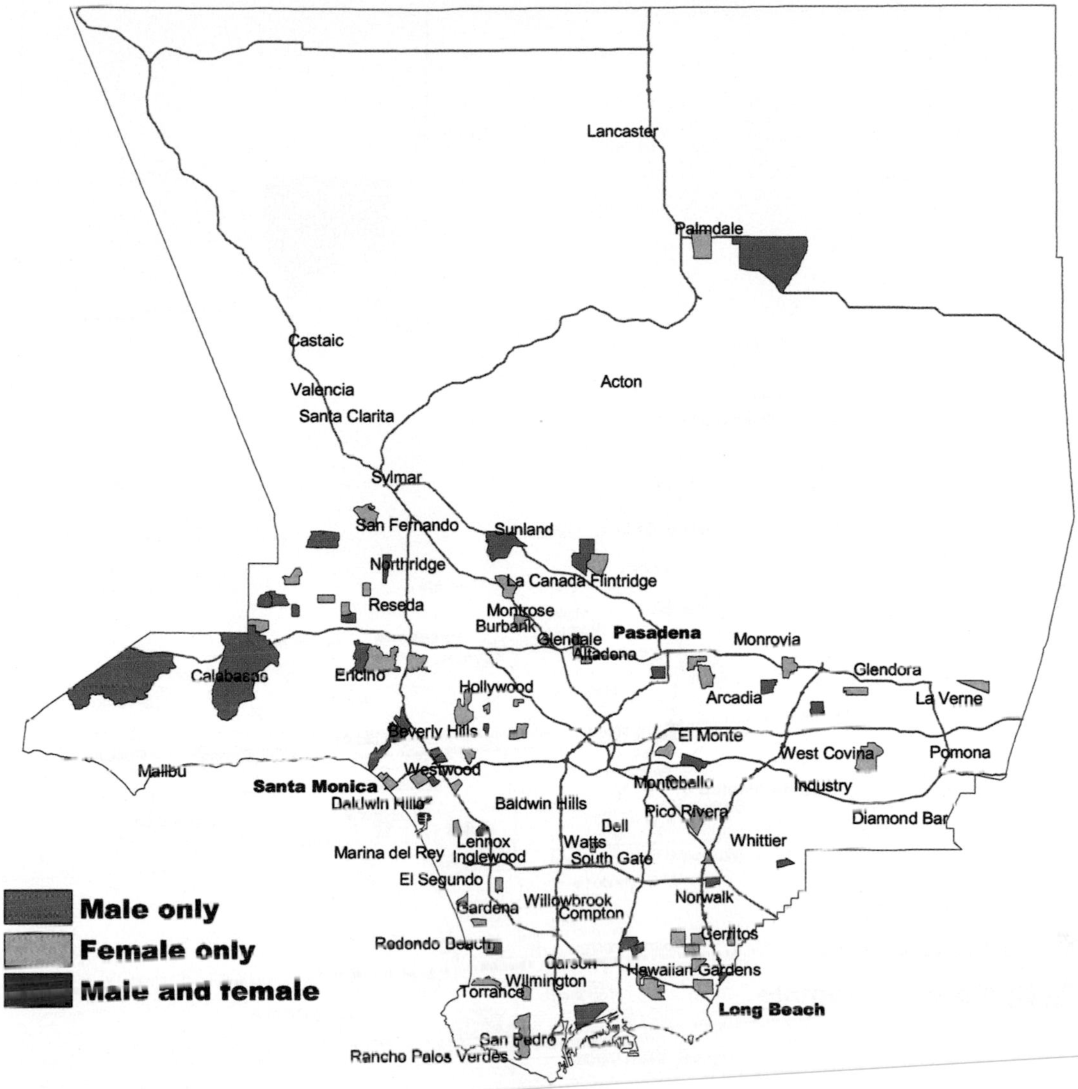

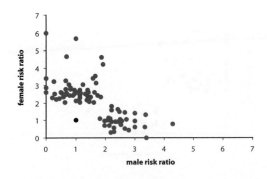

Figure 10: Male-female correlation between the relative risks for high-risk census tracts.

Figure 11: Map of census tracts at high risk, adjusted for social class.

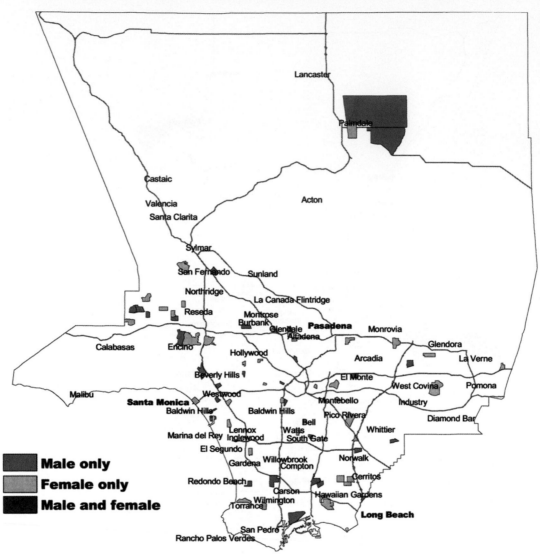

Figure 12: Male-female correlation between the relative risks for high-risk census tracts, adjusted for social class.

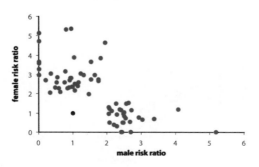

Spinal Cord Malignancies

ICDO-2 Code Anatomic Site: C 70.1, 72
ICDO-2 Code Histology: 9380–9481
Age: All
Male Cases: 180
Female Cases: 160

Background

These cancers are presumed to have the same spectrum of causes as brain malignancies.

Local Pattern

These rare tumors occur with equal frequency in Los Angeles County and elsewhere, among persons of either sex, at all ages, and among people of all racial/ethnicity groups and all social classes. Incidence does not change dramatically with age. The rate of occurrence has been rather constant over the time period. Figure 6 shows no nonrandom excess of census tracts with unexpectedly few or unexpectedly many cases, and the high-risk criteria are fulfilled for no census tracts.

Thumbnail Interpretation

No systematic pattern of occurrence is apparent, and therefore no local source of causation can be proposed.

Spinal Cord Malignancies: Male

Figure 1: Age-adjusted incidence rate by place.

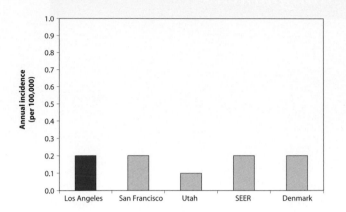

Figure 2: Age-adjusted incidence rate over the period.

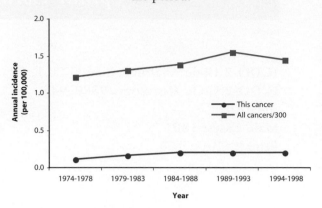

Figure 3: Age-adjusted incidence rate by age and race/ethnicity.

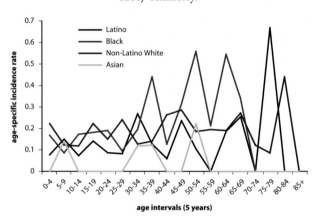

Figure 4: Age-adjusted incidence rate by social class.

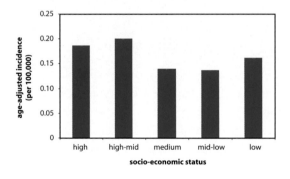

Figure 5: Distribution of the relative risk values for all census tracts.

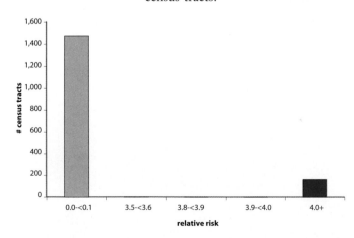

Figure 6: Census tracts by the number of cases per tract.

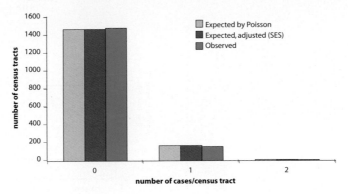

Figure 7a and b: Census tracts at high risk by the number of cases. (a) Unadjusted and (b) adjusted for social class.

There were no high risk census tracts.

Figure 8: Risk over the period for high-risk census tracts relative to all census tracts.

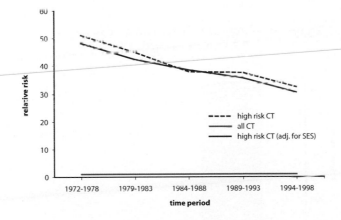

Figure 1: Age-adjusted incidence rate by place.

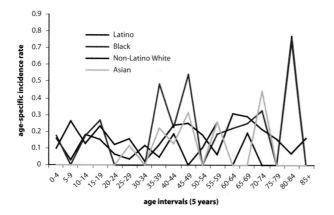

Figure 2: Age-adjusted incidence rate over the period.

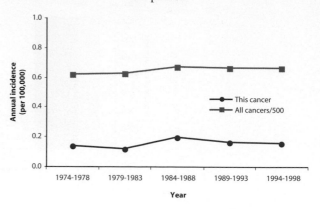

Figure 3: Age-adjusted incidence rate by age and race/ethnicity.

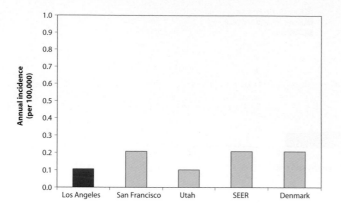

Figure 4: Age-adjusted incidence rate by social class.

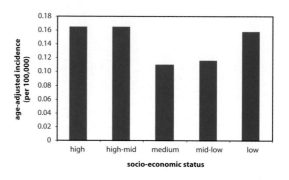

Figure 5: Distribution of the relative risk values for all census tracts.

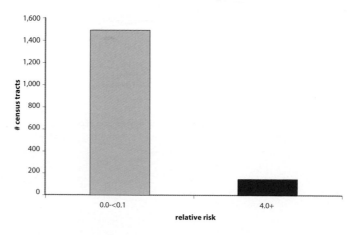

Figure 6: Census tracts by the number of cases per tract.

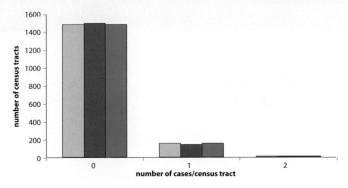

Figure 7a and b: Census tracts at high risk by the number of cases. (a) Unadjusted and (b) adjusted for social class.

There were no high risk census tracts.

Figure 8: Risk over the period for high-risk census tracts relative to all census tracts.

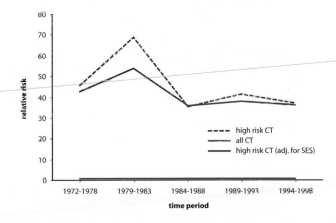

Neurolemoma (Nerve Sheath Malignancy)

ICDO-2 Code Anatomic Site: C 0–80
ICDO-2 Code Histology: 9540–9570
Age: All
Male Cases: 213
Female Cases: 178

Background

Neurolemomas are rare malignancies that resemble sarcomas in many respects. Like sarcomas, neurolemomas occur in the members of families with inherited genetic syndromes, especially the several genetic abnormalities collectively called neurofibromatosis. Like other tumors of the nervous system, neurolemomas can result from by ionizing radiation, probably including the radiation used for diagnostic or dental purposes. One group of neurolemomas, those of the acoustic nerve, have been associated with chronic exposure to very loud sounds, either occupational or recreational. Otherwise, the causes of these rare malignancies are unknown.

Local Pattern

Nerve sheath malignancies occur with rough equality in men and women and among the residents of Los Angeles and other regions. They begin to occur in childhood and appear with equal frequency among the members of all racial-ethnic groups and all social classes. Incidence has been relatively constant over time. Figure 6 shows no nonrandom excess of census tracts with unexpectedly few or unexpectedly many cases. Only two census tracts met the high-risk criteria.

Thumbnail Interpretation

No systematic pattern of occurrence is apparent, and therefore no local source of causation can be proposed.

Neuroblastoma

ICDO-2 Code Anatomic Site: C 0–80
ICDO-2 Code Histology: 9490–9523
Age: All
Male Cases: 264
Female Cases: 220

Background

Neuroblastoma is usually a malignancy of children but does sometimes affect adults. A very small proportion of these malignancies occur in families with previous cases. Otherwise there are no known specific causes. Because a large proportion of these malignancies occur in infants or toddlers, much speculation has centered on exposures to the mother while the child was *in utero*. Those that have been mentioned have included drugs, including hormones, and exposures at the parental workplace.

Local Pattern

Neuroblastomas have occurred less commonly in Los Angeles County than elsewhere, but they have appeared in the county with a relatively constant rate over time. Incidence is similar in boys and girls, and there is no particular relationship to social class. Figure 6 shows no nonrandom excess of census tracts with unexpectedly few or unexpectedly many cases. Only a single census tract met the high-risk criteria.

Thumbnail Interpretation

No systematic pattern of occurrence is apparent, and therefore no local source of causation can be proposed

Neuroblastoma: Male

Figure 1: Age-adjusted incidence rate by place.

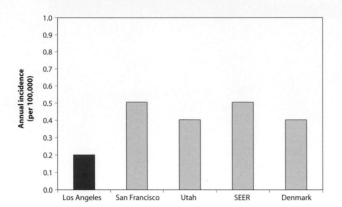

Figure 2: Age-adjusted incidence rate over the period.

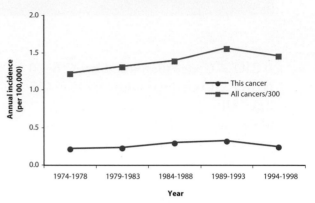

Figure 3: Age-adjusted incidence rate by age and race/ethnicity.

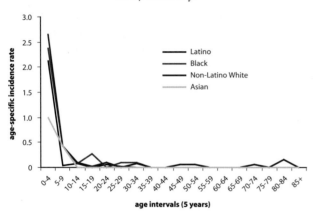

Figure 4: Age-adjusted incidence rate by social class.

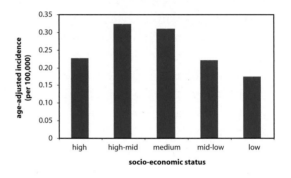

Figure 5: Distribution of the relative risk values for all census tracts.

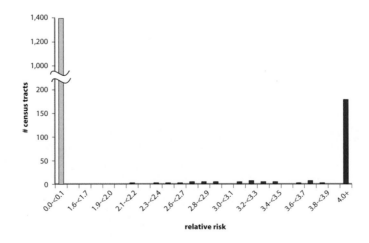

Figure 6: Census tracts by the number of cases per tract.

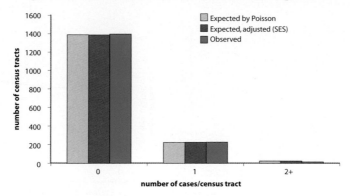

Figure 7a and b: Census tracts at high risk by the number of cases. (a) Unadjusted and (b) adjusted for social class.

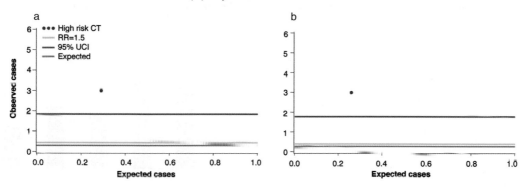

Figure 8: Risk over the period for high-risk census tracts relative to all census tracts.

There were no high risk census tracts.

Figure 1: Age-adjusted incidence rate by place.

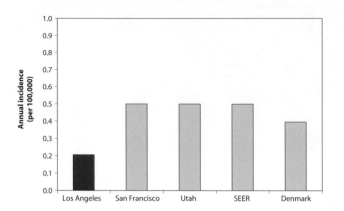

Figure 2: Age-adjusted incidence rate over the period.

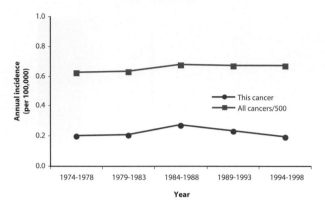

Figure 3: Age-adjusted incidence rate by age and race/ethnicity.

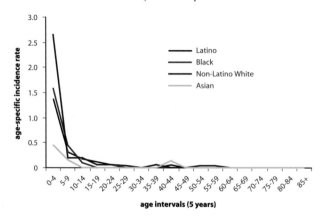

Figure 4: Age-adjusted incidence rate by social class.

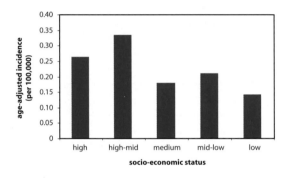

Figure 5: Distribution of the relative risk values for all census tracts.

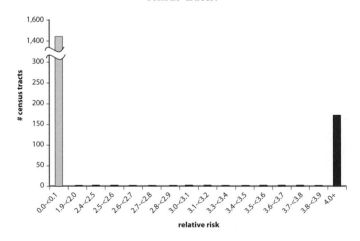

Figure 6: Census tracts by the number of cases per tract.

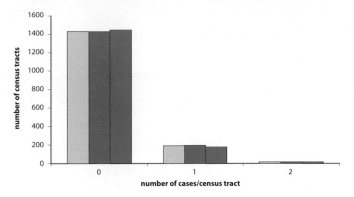

Figure 7a and b: Census tracts at high risk by the number of cases. (a) Unadjusted and (b) adjusted for social class.

There were no high risk census tracts.

Figure 8: Risk over the period for high-risk census tracts relative to all census tracts.

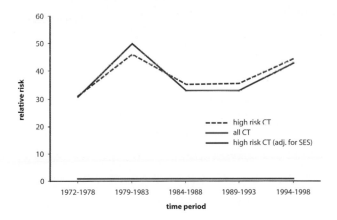

Figure 9: Map of census tracts at high risk.

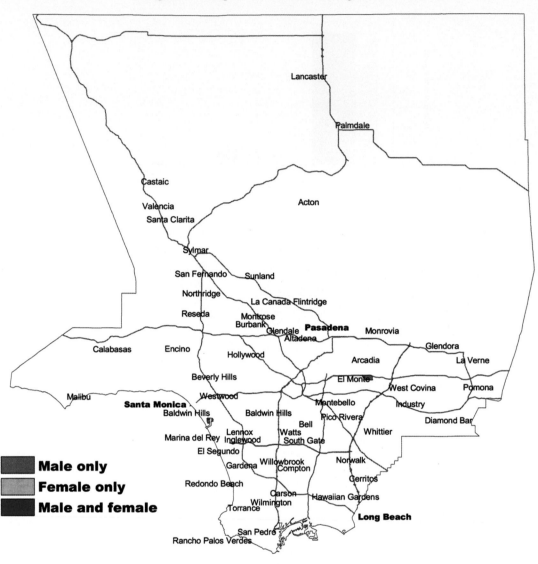

Figure 10: Male-female correlation between the relative risks for high-risk census tracts.

There were no high risk census tracts.

Figure 11: Map of census tracts at high risk, adjusted for social class.

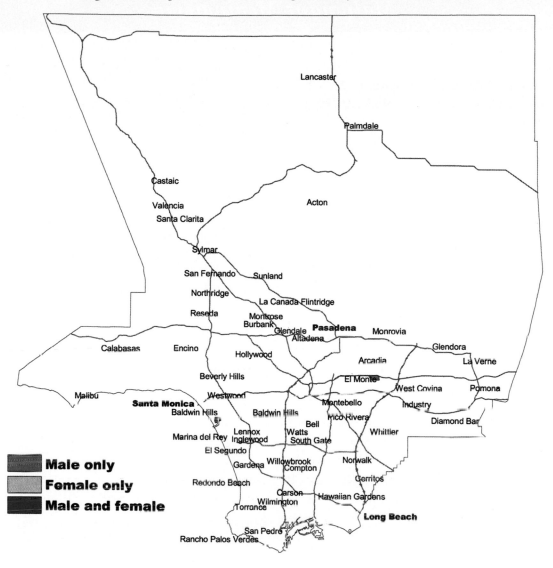

Male only
Female only
Male and female

Figure 12: Male-female correlation
between the relative risks for high risk
census tracts, adjusted for social class.

There were no high risk census tracts.

Malignant Meningioma

ICDO-2 Code Anatomic Site: C 70–72
ICDO-2 Code Histology: 9530–9539
Age: All
Male Cases: 104
Female Cases: 106

Background

These malignancies occur in the tissues surrounding the brain and spinal cord, and are related to neurolemomas. Ionizing radiation is the only exposure believed to cause them, and even that is probably responsible for a very small proportion of cases.

Local Pattern

Unlike benign meningiomas (which are also potentially life-threatening because growth occurs within the confined space of the skull), malignant meningioma is very rare. It occurs at about the same rate in Los Angeles County as in other populations. Incidence is also similar in the two genders, in all racial/ethnic groups, and in all social classes. Incidence has not varied over time, and is unrelated to social class. Figure 6 shows no nonrandom excess of census tracts with unexpectedly few or unexpectedly many cases. No high-risk census tracts met the high-risk criteria.

Thumbnail Interpretation

No systematic pattern of occurrence is apparent, and therefore no local source of causation can be proposed.

Figure 1: Age-adjusted incidence rate by place.

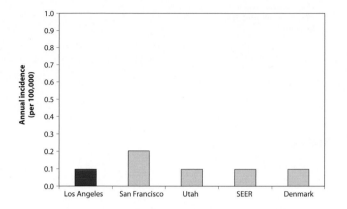

Figure 2: Age-adjusted incidence rate over the period.

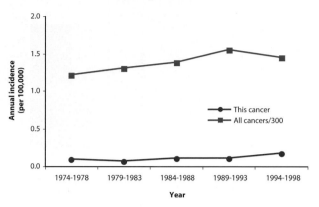

Figure 3: Age-adjusted incidence rate by age and race/ethnicity.

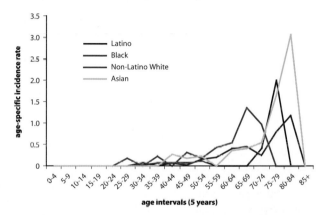

Figure 4: Age-adjusted incidence rate by social class.

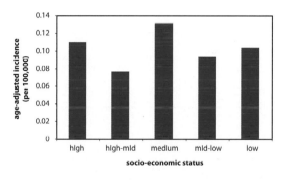

Figure 5: Distribution of the relative risk values for all census tracts.

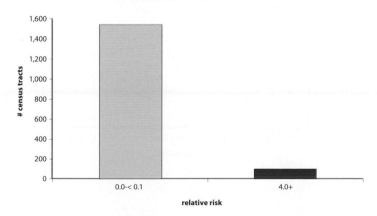

Figure 6: Census tracts by the number of cases per tract.

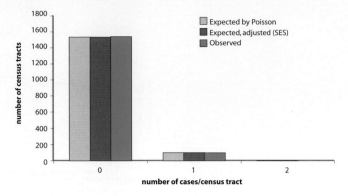

Figure 7a and b: Census tracts at high risk by the number of cases. (a) Unadjusted and (b) adjusted for social class.

There were no high risk census tracts.

Figure 8: Risk over the period for high-risk census tracts relative to all census tracts.

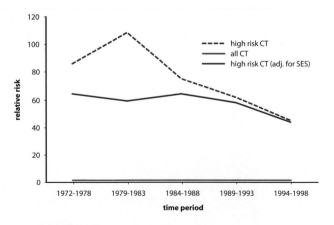

Figure 1: Age-adjusted incidence rate by place.

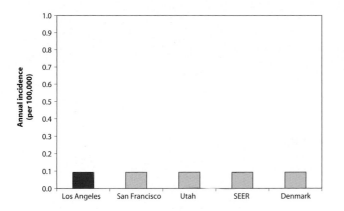

Figure 2: Age-adjusted incidence rate over the period.

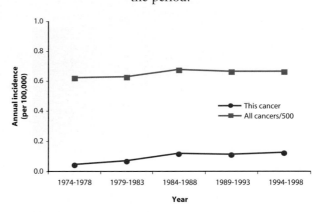

Figure 3: Age-adjusted incidence rate by age and race/ethnicity.

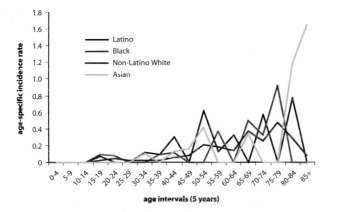

Figure 4: Age-adjusted incidence rate by social class.

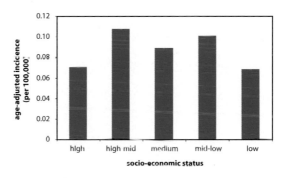

Figure 5: Distribution of the relative risk values for all census tracts.

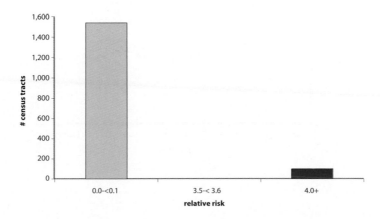

Figure 6: Census tracts by the number of cases per tract.

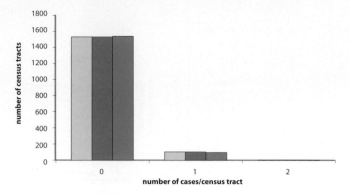

Figure 7a and b: Census tracts at high risk by the number of cases. (a) Unadjusted and (b) adjusted for social class.

There were no high risk census tracts.

Figure 8: Risk over the period for high-risk census tracts relative to all census tracts.

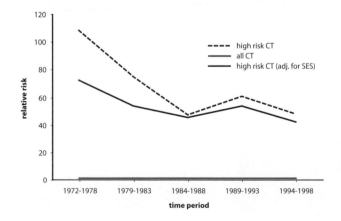

Papillary Carcinoma of the Thyroid

ICDO-2 Code Anatomic Site: C 73
ICDO-2 Code Histology: 8050–8053, 8360, 8340, 8450, 8452
Age: All
Male Cases: 1953
Female Cases: 6026

Background

Papillary carcinoma is the most common form of thyroid cancer, and has been reported generally to be increasing in frequency over time. However, these malignancies tend to grow very slowly, and may never produce symptoms beyond the presence of a nodule in the thyroid. Ionizing radiation (nuclear disasters, diagnostic x-rays, radioactive iodine) is the only certain cause of this malignancy. Speculation based on knowledge of thyroid physiology, the female predominance, and the worldwide pattern of occurrence has focused on benign thyroid disease, iodine consumption, other dietary components, and hormonal aberrations as other possible causes. Evidence for each of these is inconsistent. Most studies have not distinguished between the various cellular subtypes.

Local Pattern

This malignancy is more common in Los Angeles County and other parts of the United States than it is in Denmark. Incidence begins to increase early in adulthood, especially in women, who are at higher risk than men. Whites are at higher risk than African-Americans or Latinos. Incidence is higher among those of higher social class. Incidence is particularly high among Filipinas, although Asian-Americans in general are at low risk of papillary thyroid cancer. In recent decades, incidence in Los Angeles County has increased slightly, both in the county as a whole and among the residents of high-risk census tracts. Figure 6 shows a moderate nonrandom excess of census tracts with unexpectedly few or unexpectedly many cases. A large complex of high-risk census tracts, including several showing high risk for both males and females, extends from Santa Monica on the west and Encino on the north through Beverly Hills, extending east into Los Angeles.

Thumbnail Interpretation

No explanation is available for either the variations in risk by gender, race/ethnicity, or social class, nor is there one for the increase in risk over time in Los Angeles County or elsewhere. There is also no clear explanation for the striking geographical distribution within Los Angeles County. Because of the wide geographical area involved, no exposure emanating from a single specific point provides a plausible explanation. However, the cases that have occurred in the high-risk census tracts include an unexpectedly high proportion of young, unmarried California natives of both sexes with rather small tumors. That suggests that the factor responsible acts at a relatively early age that it takes place locally and that personal behavior of some kind plays a role. None of the factors that have aroused causal speculation would clearly explain this distribution. The localities at high risk are among those with very high quality medical care, and there is a possibility that persons living there have been screened for thyroid nodules with more than usual efficiency and it is possible that among those tumors diagnosed in those areas are some that would never have enlarged to the point of recognition. Because radiation is the most well-documented determinant, the search for the explanation should be sure to rule out the possibility that cumulative exposure to x-rays might play a role.

Figure 1: Age-adjusted incidence rate by place.

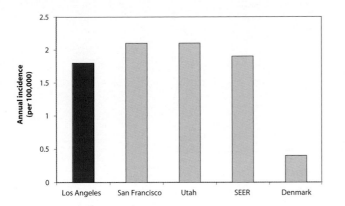

Figure 2: Age-adjusted incidence rate over the period.

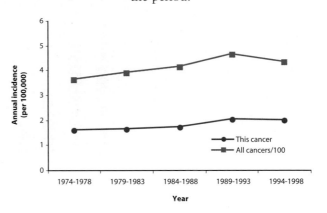

Figure 3: Age-adjusted incidence rate by age and race/ethnicity.

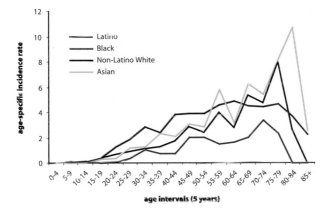

Figure 4: Age-adjusted incidence rate by social class.

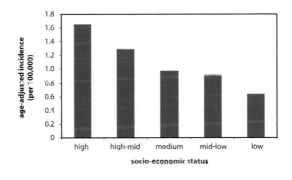

Figure 5: Distribution of the relative risk values for all census tracts.

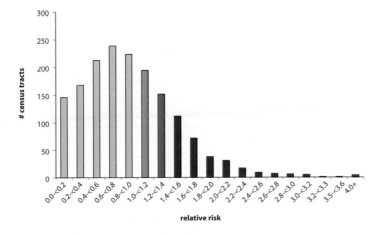

Figure 6: Census tracts by the number of cases per tract.

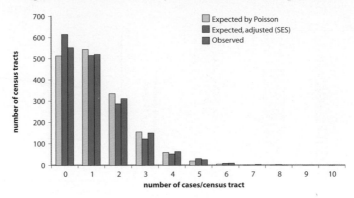

Figure 7a and b: Census tracts at high risk by the number of cases. (a) Unadjusted and (b) adjusted for social class.

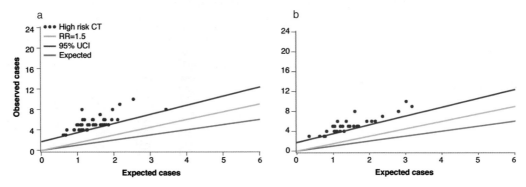

Figure 8: Risk over the period for high-risk census tracts relative to all census tracts.

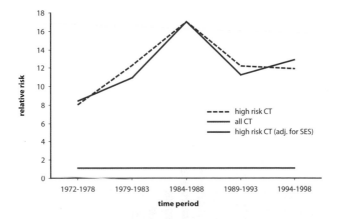

Figure 1: Age-adjusted incidence rate by place.

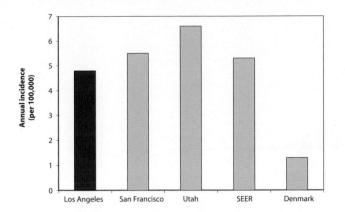

Figure 2: Age-adjusted incidence rate over the period.

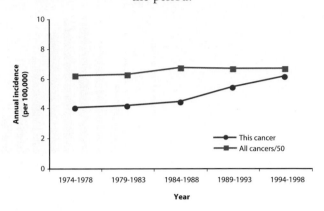

Figure 3: Age-adjusted incidence rate by age and race/ethnicity.

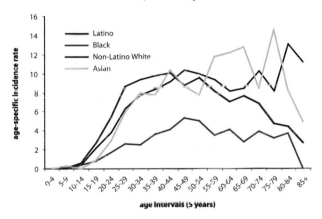

Figure 4: Age-adjusted incidence rate by social class.

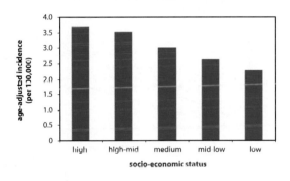

Figure 5: Distribution of the relative risk values for all census tracts.

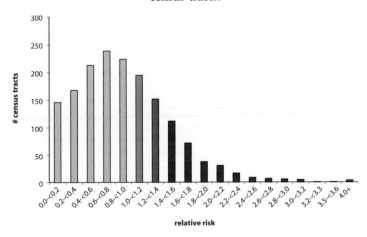

Papillary Carcinoma of the Thyroid: Female

Figure 6: Census tracts by the number of cases per tract.

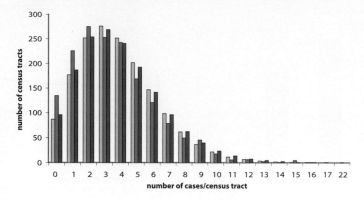

Figure 7a and b: Census tracts at high risk by the number of cases. (a) Unadjusted and (b) adjusted for social class.

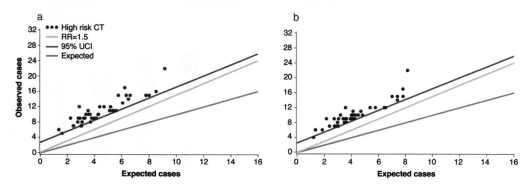

Figure 8: Risk over the period for high-risk census tracts relative to all census tracts.

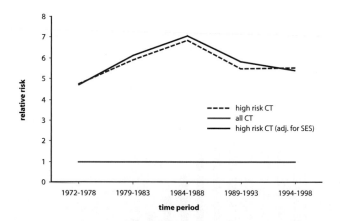

Figure 9: Map of census tracts at high risk.

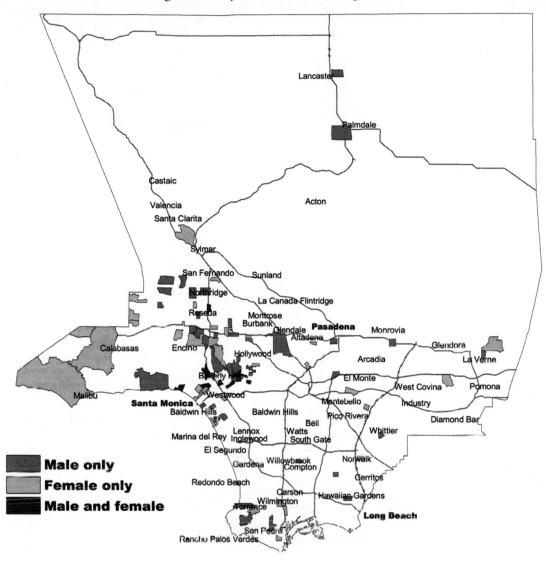

Figure 10: Male-female correlation between the relative risks for high-risk census tracts.

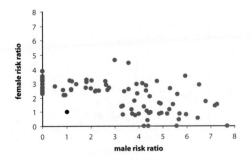

Figure 11: Map of census tracts at high risk, adjusted for social class.

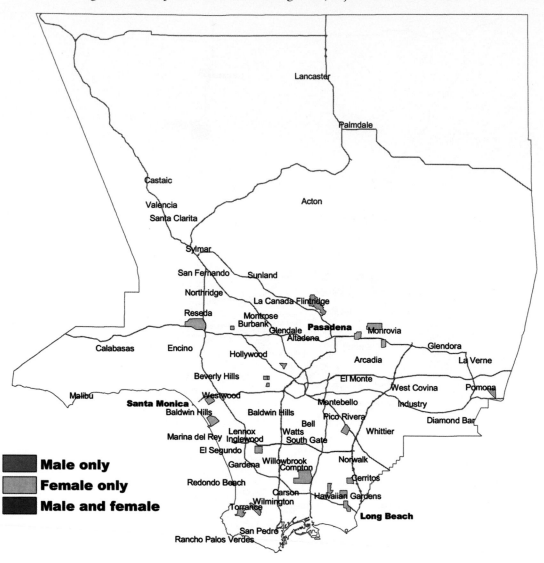

Figure 12: Male-female correlation between the relative risks for high-risk census tracts, adjusted for social class.

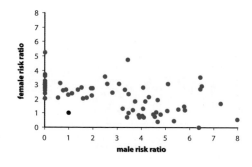

Follicular Carcinoma of the Thyroid

ICDO-2 Code Anatomic Site: C 73
ICDO-2 Code Histology: 8330–8332
Age: All
Male Cases: 345
Female Cases: 904

Background

Ionizing radiation is the only certain cause of thyroid cancer, although iodine deficiency, hormonal aberrations, and benign thyroid abnormalities have also been linked to thyroid cancer in some circumstances, but not others. Most studies have not distinguished between the various cellular subtypes.

Local Pattern

Follicular thyroid cancer occurs twice as often in women as men, and at about the same rate in Los Angeles County as in other regions of the country. Incidence begins to increase in very early adulthood, indeed among female teenagers, and all racial/ethnic groups appear to be at about equal risk. Among men, but not women, the malignancy has occurred more commonly among those of higher social class. Although male incidence in Los Angeles County as a whole decreased in frequency in the initial part of the time period, it sub-

sequently stabilized. Among women in Los Angeles County as a whole, risk has decreased over time, but this is less apparent among those residing in high-risk census tracts. Figure 6 shows only a slight nonrandom excess of census tracts with unexpectedly few or unexpectedly many cases. Before and after adjustment for social class, census tracts at high-risk appear scattered throughout the county with only a single pair of contiguous census tracts. No census tract stands out on the basis of a particularly high number of excess cases.

Thumbnail Interpretation

Certain aspects of the pattern of occurrence of follicular thyroid cancer are similar to the pattern of papillary thyroid cancer, but the striking geographical distribution is absent. No explanation for the trend over age and sex or over social class is available. No systematic pattern of occurrence is apparent, and therefore no local source of causation can be proposed.

Figure 1: Age-adjusted incidence rate by place.

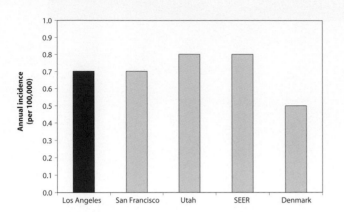

Figure 2: Age-adjusted incidence rate over the period.

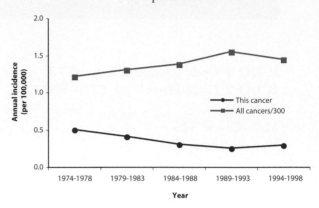

Figure 3: Age-adjusted incidence rate by age and race/ethnicity.

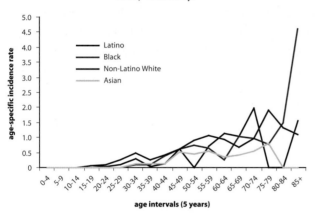

Figure 4: Age-adjusted incidence rate by social class.

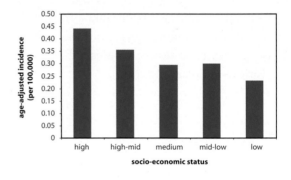

Figure 5: Distribution of the relative risk values for all census tracts.

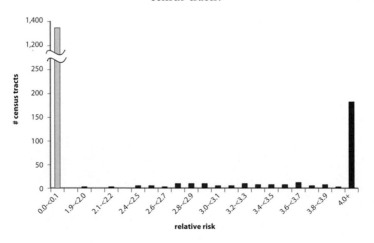

Figure 6: Census tracts by the number of cases per tract.

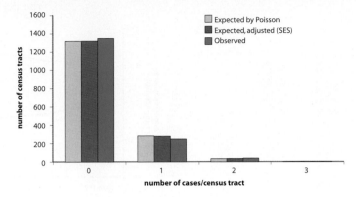

Figure 7a and b: Census tracts at high risk by the number of cases. (a) Unadjusted and (b) adjusted for social class.

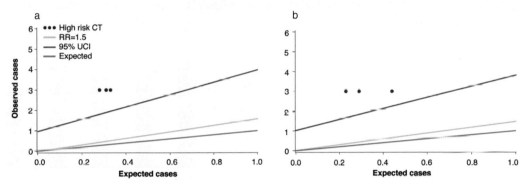

Figure 8: Risk over the period for high-risk census tracts relative to all census tracts.

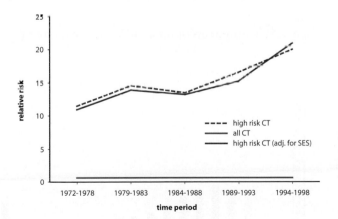

Follicular Carcinoma of the Thyroid: Female

Figure 1: Age-adjusted incidence rate by place.

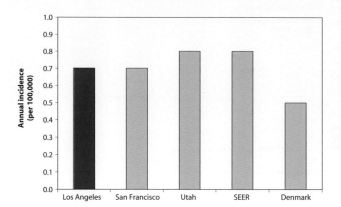

Figure 2: Age-adjusted incidence rate over the period.

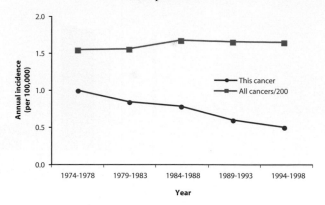

Figure 3: Age-adjusted incidence rate by age and race/ethnicity.

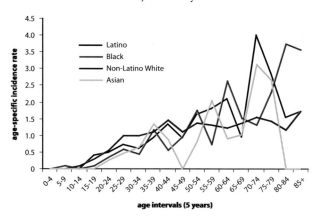

Figure 4: Age-adjusted incidence rate by social class.

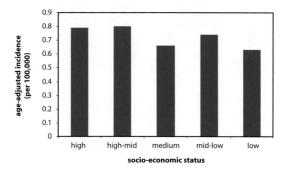

Figure 5: Distribution of the relative risk values for all census tracts.

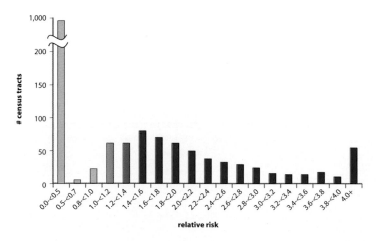

Figure 6: Census tracts by the number of cases per tract.

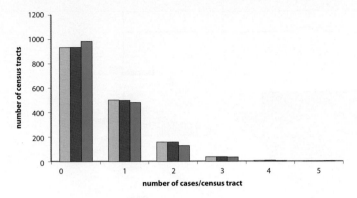

Figure 7a and b: Census tracts at high risk by the number of cases. (a) Unadjusted and (b) adjusted for social class.

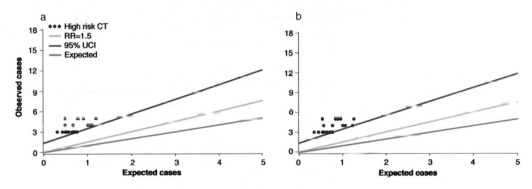

Figure 8: Risk over the period for high-risk census tracts relative to all census tracts.

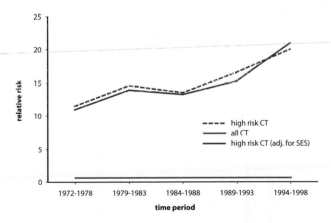

Figure 9: Map of census tracts at high risk.

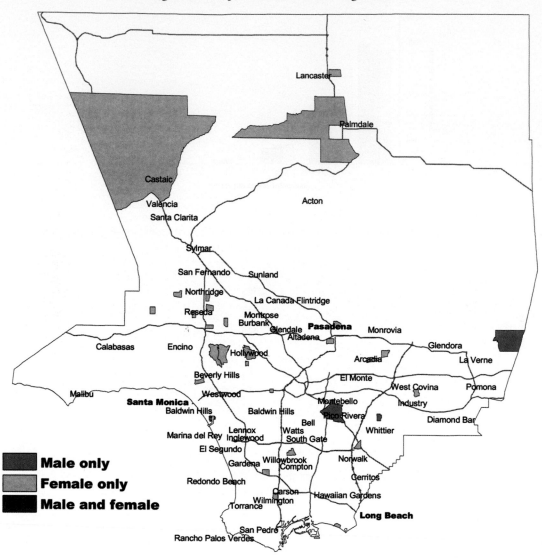

Figure 10: Male-female correlation between the relative risks for high-risk census tracts.

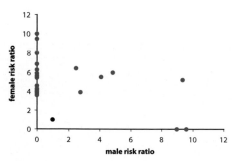

Figure 11: Map of census tracts at high risk, adjusted for social class.

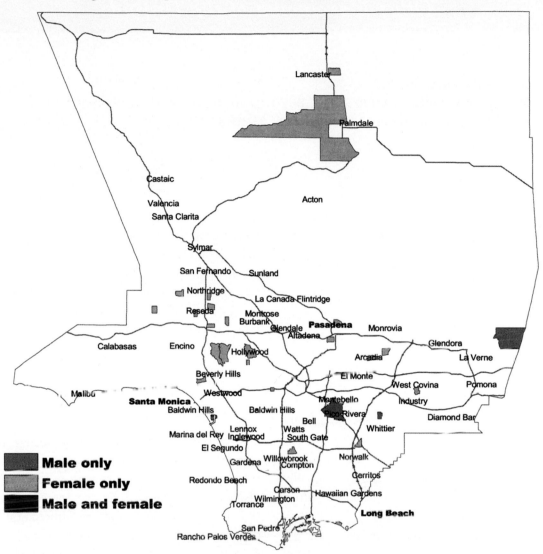

Figure 12: Male-female correlation between the relative risks for high-risk census tracts, adjusted for social class.

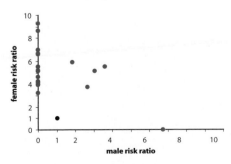

Other Carcinoma of the Thyroid

ICDO-2 Code Anatomic Site: C 73
ICDO-2 Code Histology: 8000–8045, 8070–8130, 8140–8251, 8261–8281,
8300–8324, 8350, 8380–8442, 8460–8506, 8520–8573
Age: All
Male Cases: 194
Female Cases: 456

Background

This category of thyroid cancer includes a variety of other histological types as well as the thyroid malignancies that were not specifically categorized. These include both relatively aggressive and relatively indolent varieties.

Local Pattern

As a group these malignancies are equally common across the United States and in Denmark. They occur among the members of different racial/ethnic groups equally, although they have occurred more often among the residents of higher social class neighborhoods. Unlike papillary thyroid cancer, they are diagnosed more often later in life. In the county as a whole, incidence has decreased somewhat over time. Figure 6 shows only a slight nonrandom excess of census tracts with unexpectedly few or unexpectedly many cases. The high-risk tract criteria are only met for females, and those few census tracts are scattered with no apparent pattern. No census tract stands out on the basis of a particularly high relative risk and number of excess cases.

Thumbnail Interpretation

Except for the apparent increased frequency among the affluent, which could be the result of very slow-growing tumors being discovered by more frequent screenings, or the erroneous inclusion of unrecognized papillary thyroid cancers among these cases reported to be of other histologies, no systematic pattern of occurrence is apparent, and therefore no local source of causation can be proposed.

Figure 1: Age-adjusted incidence rate by place.

Figure 2: Age-adjusted incidence rate over the period.

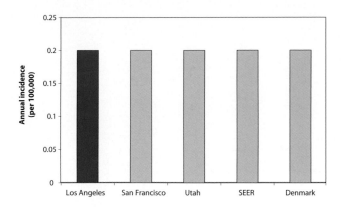

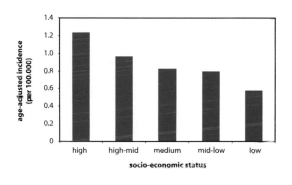

Figure 3: Age-adjusted incidence rate by age and race/ethnicity.

Figure 4: Age-adjusted incidence rate by social class.

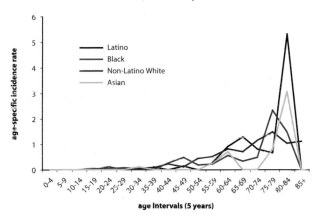

Figure 5: Distribution of the relative risk values for all census tracts.

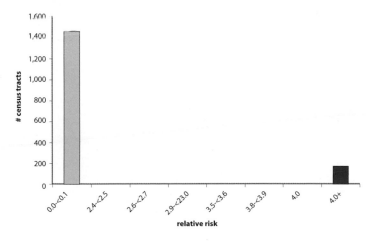

Other Carcinoma of the Thyroid: Male

Figure 6: Census tracts by the number of cases per tract.

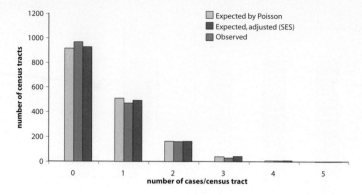

Figure 7a and b: Census tracts at high risk by the number of cases. (a) Unadjusted and (b) adjusted for social class.

There were no high risk census tracts.

Figure 8: Risk over the period for high-risk census tracts relative to all census tracts.

There were no high risk census tracts.

Figure 1: Age-adjusted incidence rate by place.

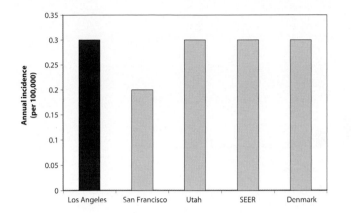

Figure 2: Age-adjusted incidence rate over the period.

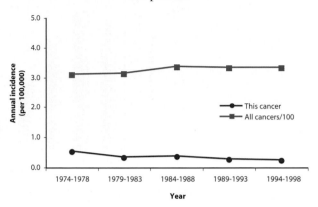

Figure 3: Age-adjusted incidence rate by age and race/ethnicity.

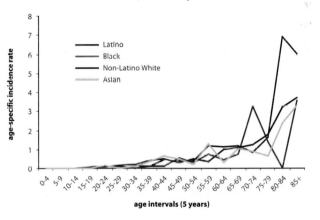

Figure 4: Age-adjusted incidence rate by social class.

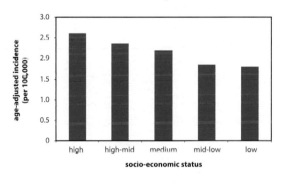

Figure 5: Distribution of the relative risk values for all census tracts.

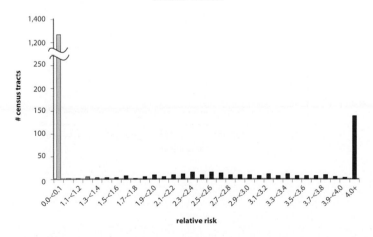

Figure 6: Census tracts by the number of cases per tract.

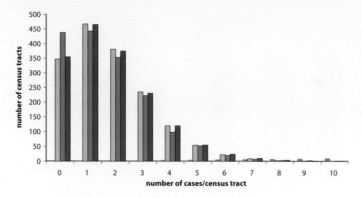

Figure 7a and b: Census tracts at high risk by the number of cases. (a) Unadjusted and (b) adjusted for social class.

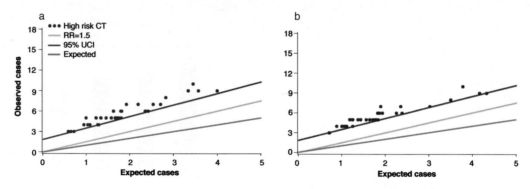

Figure 8: Risk over the period for high-risk census tracts relative to all census tracts.

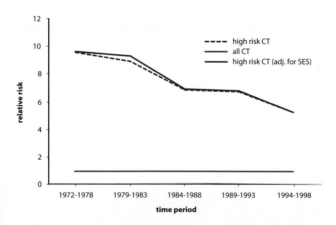

Figure 9: Map of census tracts at high risk.

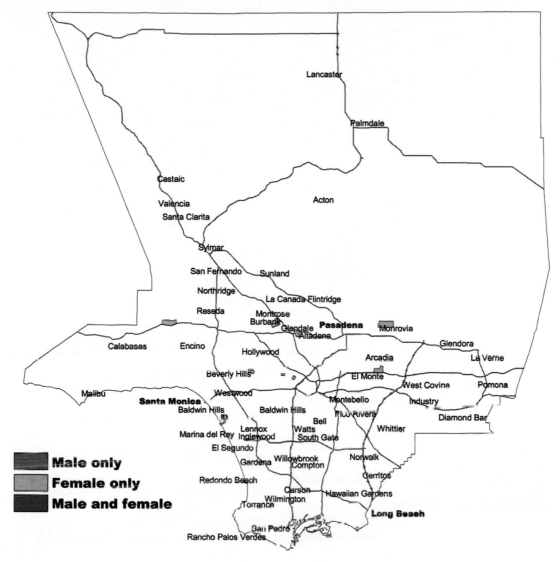

Figure 10: Male-female correlation between the relative risks for high-risk census tracts.

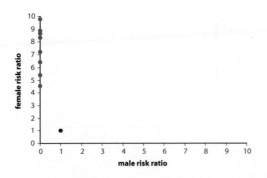

Figure 11: Map of census tracts at high risk, adjusted for social class.

Lancaster
Palmdale
Castaic
Acton
Valencia
Santa Clarita
Sylmar
San Fernando
Sunland
Northridge
La Canada Flintridge
Reseda
Montrose
Burbank
Glendale
Pasadena
Monrovia
Altadena
Glendora
Calabasas
Encino
Hollywood
Arcadia
La Verne
Beverly Hills
El Monte
West Covina
Pomona
Malibu
Westwood
Santa Monica
Montebello
Industry
Baldwin Hills
Baldwin Hills
Pico Rivera
Bell
Diamond Bar
Lennox
Watts
Whittier
Marina del Rey
Inglewood
South Gate
El Segundo
Male only
Female only
Male and female
Gardena
Willowbrook
Compton
Norwalk
Cerritos
Redondo Beach
Carson
Hawaiian Gardens
Wilmington
Torrance
Long Beach
San Pedro
Rancho Palos Verdes

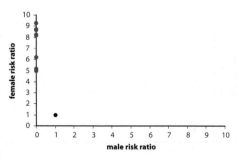

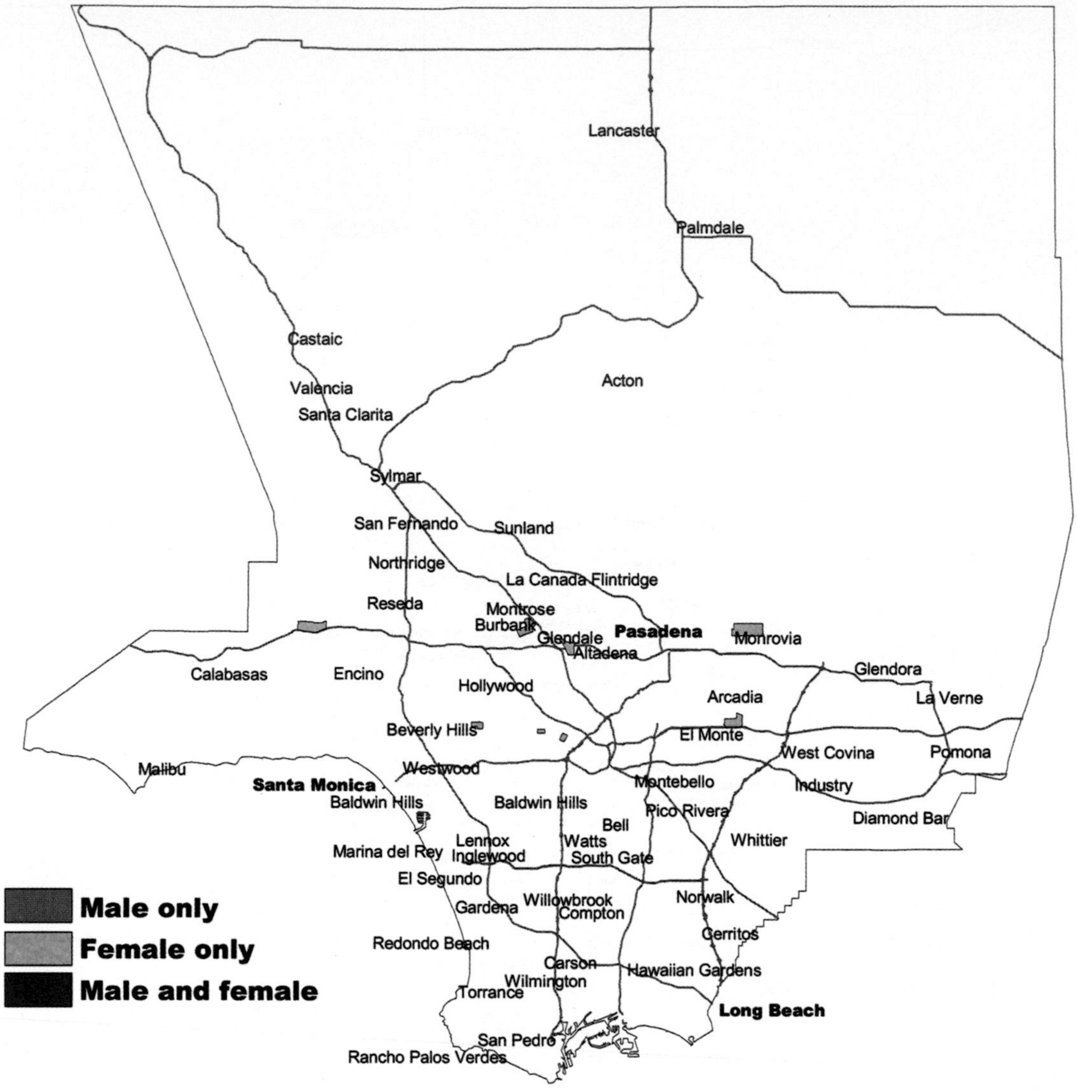

Figure 12: Male-female correlation between the relative risks for high-risk census tracts, adjusted for social class.

female risk ratio
male risk ratio
0
1
2
3
4
5
6
7
8
9
10

Pituitary Carcinoma

ICDO-2 Code Anatomic Site: C 75.1, 75.2
ICDO-2 Code Histology: 8000–8560, 8570–8573
Age: All
Male Cases: 31
Female Cases: 25

Background

The causes of this very rare malignancy are unknown.

Local Pattern

This tumor occurs with equal frequency in the sexes as well as in the various regions of the country, including Los Angeles County. It affects all races and incidence has been constant over time. There is no consistent relationship between risk and social class. Figure 6 shows no nonrandom excess of census tracts with unexpectedly few or unexpectedly many cases. No census tracts fulfilled the high-risk criteria.

Thumbnail Interpretation

No systematic pattern of occurrence is apparent, and therefore no local source of causation can be proposed.

Figure 1: Age-adjusted incidence rate by place.

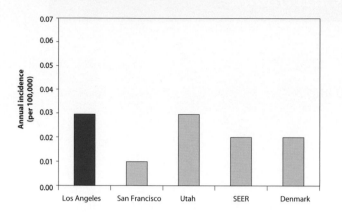

Figure 2: Age-adjusted incidence rate over the period.

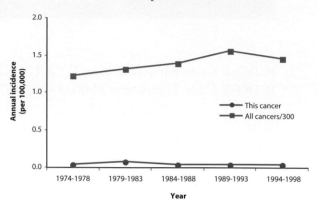

Figure 3: Age-adjusted incidence rate by age and race/ethnicity.

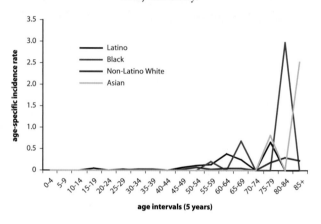

Figure 4: Age-adjusted incidence rate by social class.

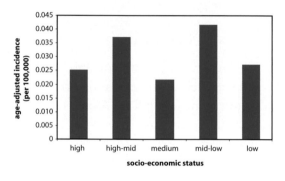

Figure 5: Distribution of the relative risk values for all census tracts.

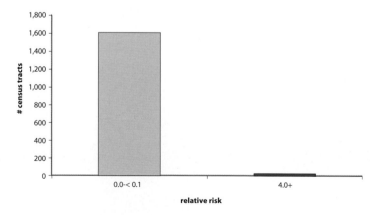

Figure 6: Census tracts by the number of cases per tract.

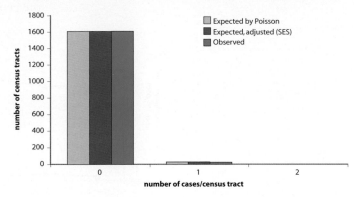

Figure 7a and b: Census tracts at high risk by the number of cases. (a) Unadjusted and (b) adjusted for social class.

There were no high risk census tracts.

Figure 8: Risk over the period for high-risk census tracts relative to all census tracts.

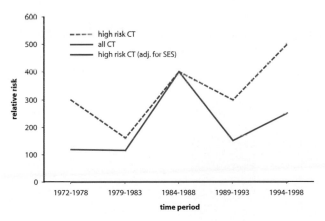

Figure 1: Age-adjusted incidence rate by place.

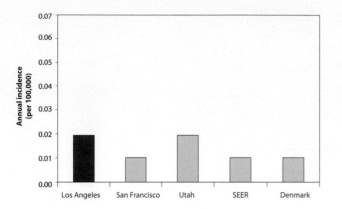

Figure 2: Age-adjusted incidence rate over the period.

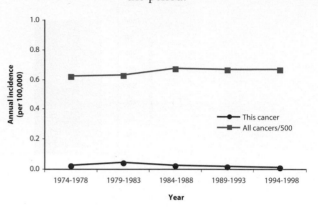

Figure 3: Age-adjusted incidence rate by age and race/ethnicity.

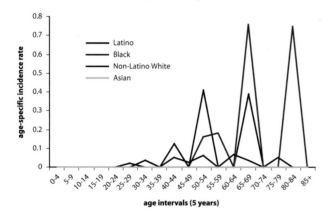

Figure 4: Age-adjusted incidence rate by social class.

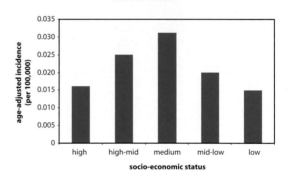

Figure 5: Distribution of the relative risk values for all census tracts.

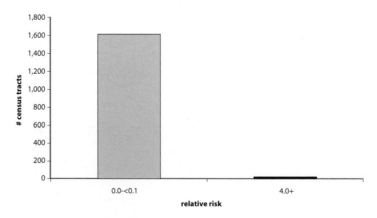

Figure 6: Census tracts by the number of cases per tract.

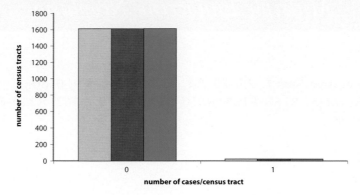

Figure 7a and b: Census tracts at high risk by the number of cases. (a) Unadjusted and (b) adjusted for social class.

There were no high risk census tracts.

Figure 8: Risk over the period for high-risk census tracts relative to all census tracts.

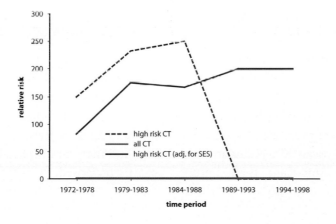

Multiple Endocrine Neoplasia

ICDO-2 Code Anatomic Site: C 25, 73, 74, 75.0, 75.3–75.9, 0–80
ICDO-2 Code Histology: 8150–8155, 8510–8512, 8290, 9362, 8240–8247,
8360–8375, 8680–8713
Age: All
Male Cases: 2319
Female Cases: 2241

Background

These malignancies occur in the members of diverse families afflicted with any of several related heritable (genetic) syndromes. Individuals may have one or more neoplasms occurring in patterns specific to each heritable syndrome. There are no known environmental causes, although any one of these tumors might occur in the absence of any recognized genetic background. The particular neoplasm or neoplasms likely to occur in any individual person cannot be predicted in advance.

Local Pattern

As a group, these tumors are diagnosed less often in Los Angeles than in other parts of the country. Incidence begins to rise in middle age, the tumors occurring slightly more often among men than among women. They appear more often among African-Americans and less often among Asian-Americans than among whites and Latinos. These tumors are being diagnosed more frequently over time in the county as a whole, although that is not the case among the residents of high-risk census tracts. Figure 6 shows a slight nonrandom excess of census tracts with unexpectedly few or unexpectedly many cases. Those census tracts that are at high risk are scattered throughout the county before and after adjustment for social class, with few census tracts contiguous to each other or showing high risk for both sexes. No census tract stands out on the basis of a particularly high number of excess cases.

Thumbnail Interpretation

The reasons for the higher incidence among males and the increasing trend are unknown, although one alternative is increasing scrutiny and more exhaustive screening. No systematic pattern of occurrence is apparent, and therefore no local source of causation can be proposed.

Figure 1: Age-adjusted incidence rate by place.

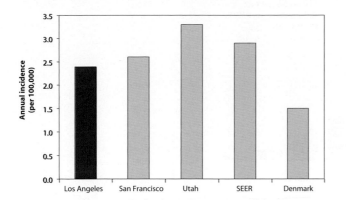

Figure 2: Age-adjusted incidence rate over the period.

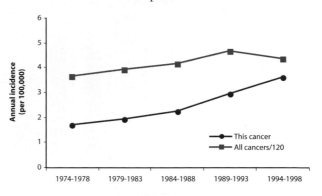

Figure 3: Age-adjusted incidence rate by age and race/ethnicity.

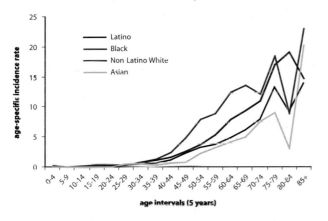

Figure 4: Age-adjusted incidence rate by social class.

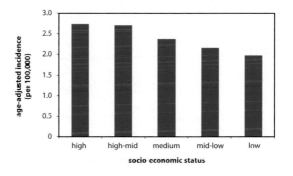

Figure 5: Distribution of the relative risk values for all census tracts

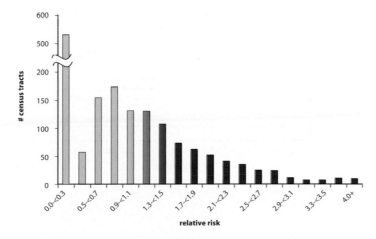

Figure 6: Census tracts by the number of cases per tract.

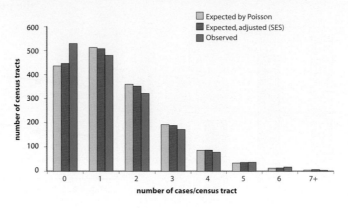

Figure 7a and b: Census tracts at high risk by the number of cases. (a) Unadjusted and (b) adjusted for social class.

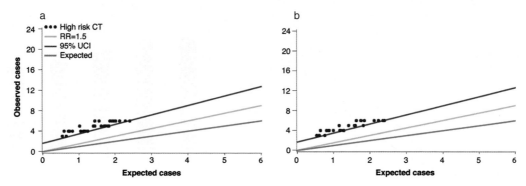

Figure 8: Risk over the period for high-risk census tracts relative to all census tracts.

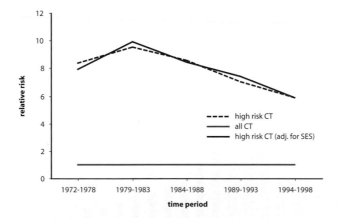

Multiple Endocrine Neoplasia: Female

Figure 1: Age-adjusted incidence rate by place.

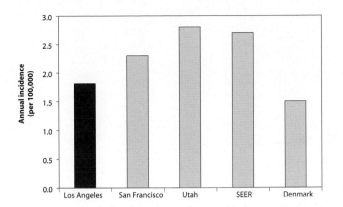

Figure 2: Age-adjusted incidence rate over the period.

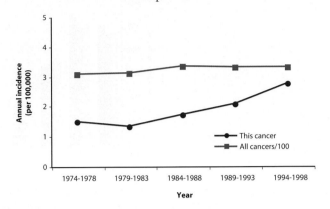

Figure 3: Age-adjusted incidence rate by age and race/ethnicity.

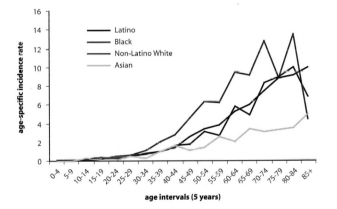

Figure 4: Age-adjusted incidence rate by social class.

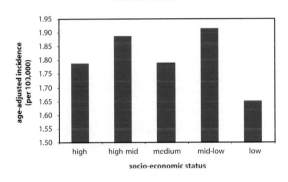

Figure 5: Distribution of the relative risk values for all census tracts.

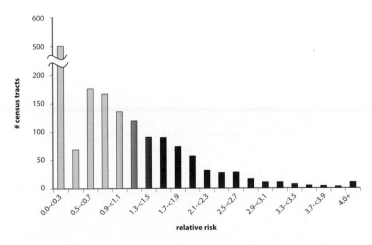

Figure 6: Census tracts by the number of cases per tract.

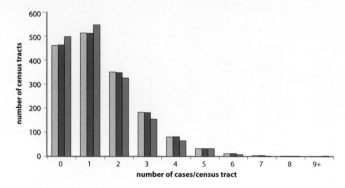

Figure 7a and b: Census tracts at high risk by the number of cases. (a) Unadjusted and (b) adjusted for social class.

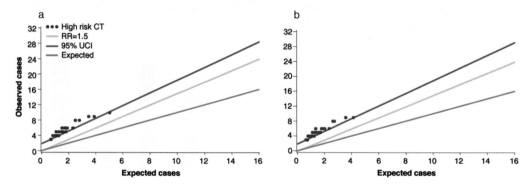

Figure 8: Risk over the period for high-risk census tracts relative to all census tracts.

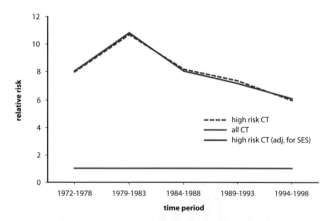

Figure 9: Map of census tracts at high risk.

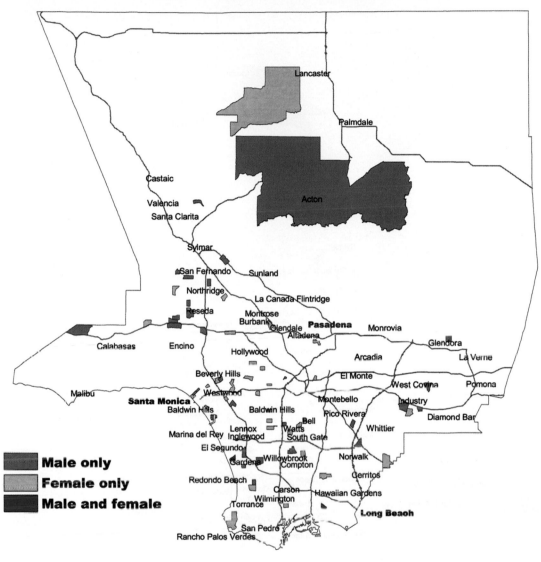

Figure 10: Male-female correlation between the relative risks for high-risk census tracts.

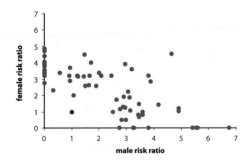

Figure 11: Map of census tracts at high risk, adjusted for social class.

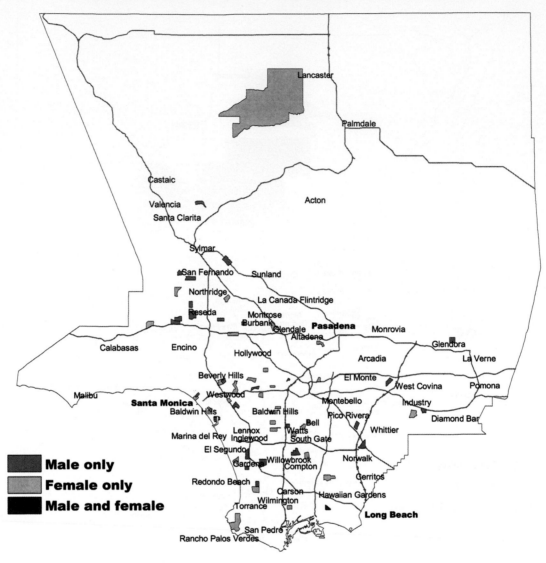

Figure 12: Male-female correlation between the relative risks for high-risk census tracts, adjusted for social class.

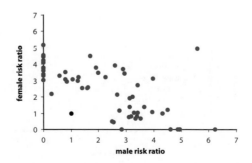

Carcinoma of the Thymus

ICDO-2 Code Anatomic Site: C 37
ICDO-2 Code Histology: 8000–8560, 8570–8573
Age: All
Male Cases: 155
Female Cases: 108

Background

The causes of this very rare malignancy are unknown.

Local Pattern

Thymus carcinoma is more common among men than among women and occurs with equal frequency in Los Angeles County and in other regions of the country. Asian-Americans are at slightly higher risk than the members of other groups. Rates of occurrence begin to increase in middle age. They have been constant over the period in the county as a whole. Figure 6 shows no nonrandom excess of census tracts with unexpectedly few or unexpectedly many cases. A single census tract met the high-risk criteria.

Thumbnail Interpretation

No reason for the excess among women is known. No systematic pattern of occurrence is apparent, and therefore no local source of causation can be proposed.

Figure 1: Age-adjusted incidence rate by place.

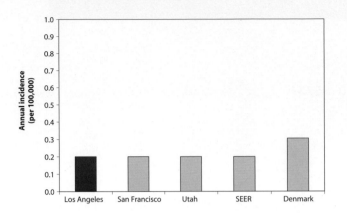

Figure 2: Age-adjusted incidence rate over the period.

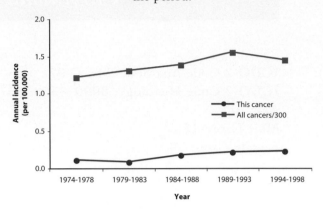

Figure 3: Age-adjusted incidence rate by age and race/ethnicity.

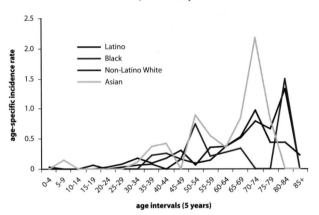

Figure 4: Age-adjusted incidence rate by social class.

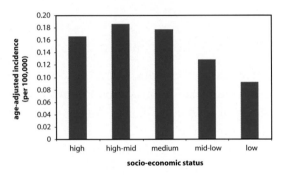

Figure 5: Distribution of the relative risk values for all census tracts.

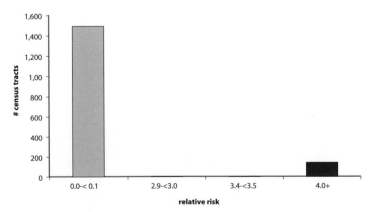

Figure 6: Census tracts by the number of cases per tract.

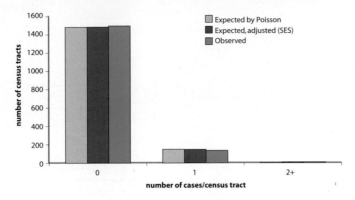

Figure 7a and b: Census tracts at high risk by the number of cases. (a) Unadjusted and (b) adjusted for social class.

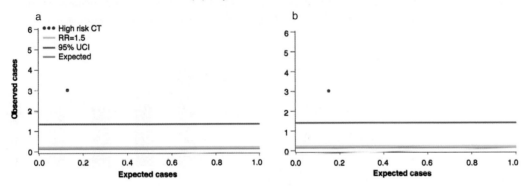

Figure 8: Risk over the period for high-risk census tracts relative to all census tracts.

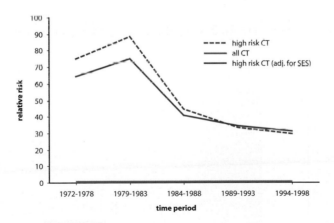

Carcinoma of the Thymus: Female

Figure 1: Age-adjusted incidence rate by place.

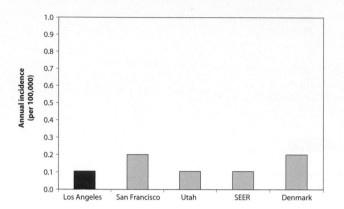

Figure 2: Age-adjusted incidence rate over the period.

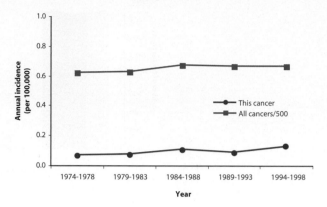

Figure 3: Age-adjusted incidence rate by age and race/ethnicity.

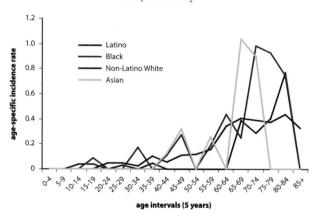

Figure 4: Age-adjusted incidence rate by social class.

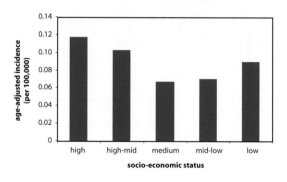

Figure 5: Distribution of the relative risk values for all census tracts.

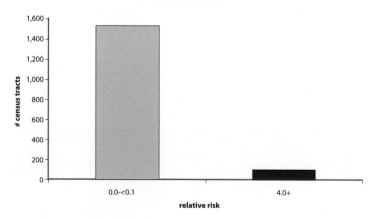

Figure 6: Census tracts by the number of cases per tract.

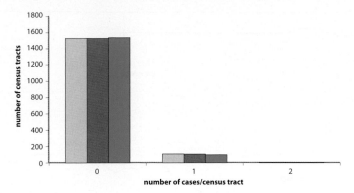

Figure 7a and b: Census tracts at high risk by the number of cases. (a) Unadjusted and (b) adjusted for social class.

There were no high risk census tracts.

Figure 8: Risk over the period for high-risk census tracts relative to all census tracts.

There were no high risk census tracts.

Figure 9: Map of census tracts at high risk.

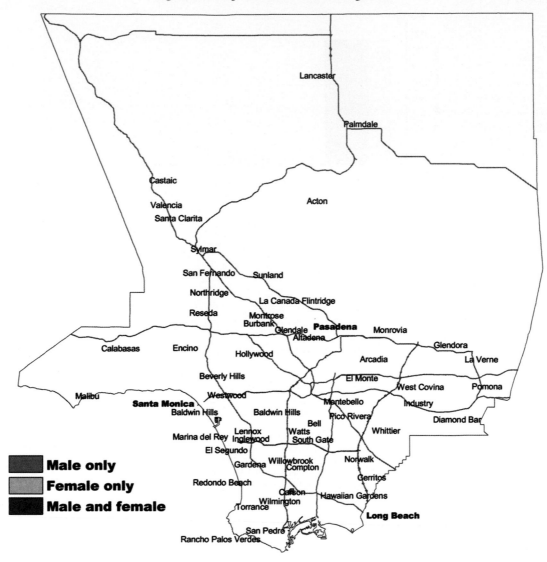

Male only
Female only
Male and female

Figure 10: Male-female correlation between the relative risks for high-risk census tracts.

There were no high risk census tracts.

Figure 11: Map of census tracts at high risk, adjusted for social class.

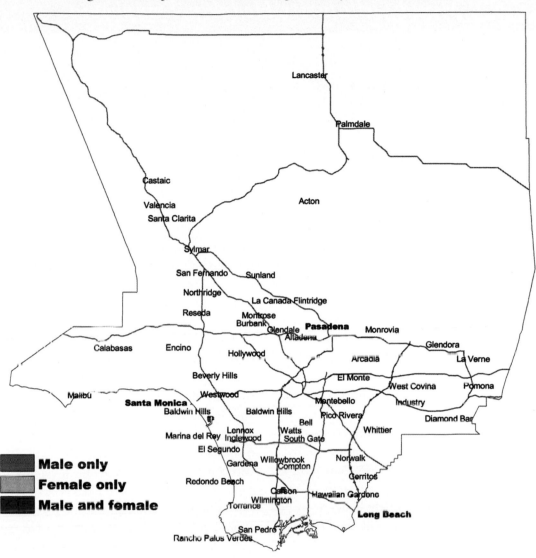

Figure 12: Male-female correlation between the relative risks for high-risk census tracts, adjusted for social class.

There were no high risk census tracts.

Mixed Cellularity Hodgkin Lymphoma

ICDO-2 Code Anatomic Site: C 0–80
ICDO-2 Code Histology: 9652
Age: All
Male Cases: 958
Female Cases: 510

Background

Hodgkin lymphoma is comprised of several malignancies having a common characteristic—an identical malignant cell of origin present in small numbers within a tumor mass largely comprised of other cell types. Each of these malignancies shows a characteristic microscopic appearance and occurs in a characteristic pattern in the population. Mixed cell Hodgkin lymphoma is the most common type in children and older persons, occurs especially in males, and is more likely to affect persons born in developing countries. Evidence of one particular virus, Epstein-Barr, is often seen within the malignant cells of this tumor, and that virus is widely believed to be the causal agent. This virus will have infected most persons, even in the United States, before they reach adulthood, and some speculate that persons who are infected at a particularly late age are likely to develop the malignancy. Incidence is also increased in the presence of AIDS. Hodgkin lymphoma in general has been linked to woodworking occupations, but this association has not been found consistently or linked to one cell type.

Local Pattern

Mixed cellularity Hodgkin lymphoma is more common in Los Angeles than in other regions, more common among males than among females, and somewhat more common among Latinos. Relative to the other forms of Hodgkin lymphoma, the mixed cellularity form occurs more commonly among children and older people. Incidence has decreased over the period in the county as a whole, but not among the residents of high-risk census tracts. There is no clear relation to social class. Figure 6 shows no nonrandom excess of census tracts with unexpectedly few or unexpectedly many cases. Census tracts at high risk are scattered throughout the county, and the pattern is not altered by social class adjustment. No census tract stands out on the basis of a particularly high number of excess cases.

Thumbnail Interpretation

The reasons for the higher risk in Latinos, children, and older persons are unknown as is the reason for the decrease in incidence over time. No systematic pattern of geographical occurrence is apparent, and therefore no local source of causation can be proposed.

Figure 1: Age-adjusted incidence rate by place.

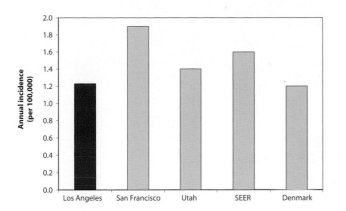

Figure 2: Age-adjusted incidence rate over the period.

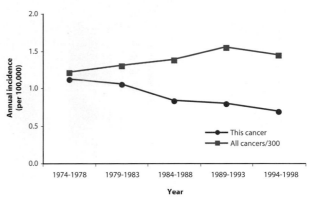

Figure 3: Age-adjusted incidence rate by age and race/ethnicity.

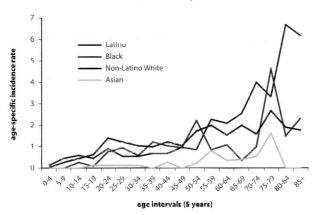

Figure 4: Age-adjusted incidence rate by social class.

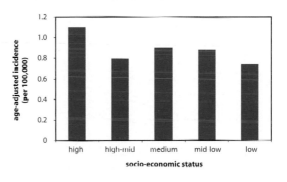

Figure 5: Distribution of the relative risk values for all census tracts.

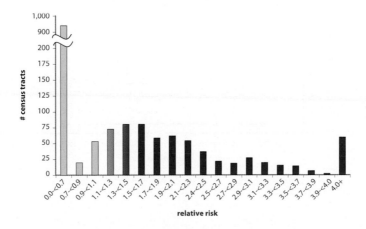

Figure 6: Census tracts by the number of cases per tract.

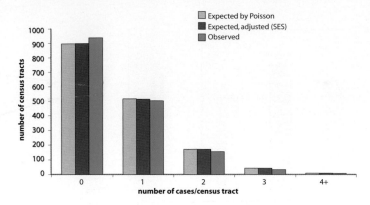

Figure 7a and b: Census tracts at high risk by the number of cases. (a) Unadjusted and (b) adjusted for social class.

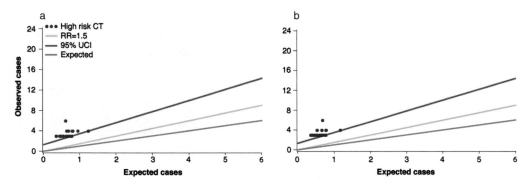

Figure 8: Risk over the period for high-risk census tracts relative to all census tracts.

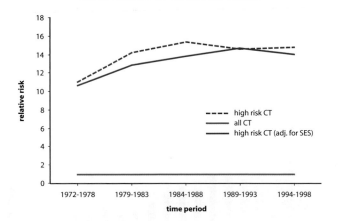

Figure 1: Age-adjusted incidence rate by place.

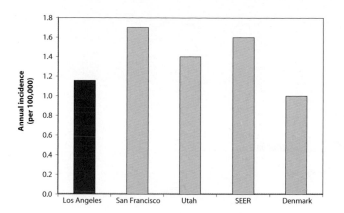

Figure 2: Age-adjusted incidence rate over the period.

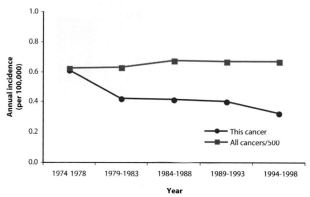

Figure 3: Age-adjusted incidence rate by age and race/ethnicity.

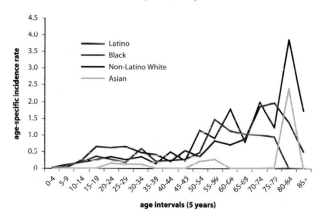

Figure 4: Age-adjusted incidence rate by social class.

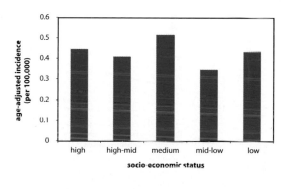

Figure 5: Distribution of the relative risk values for all census tracts.

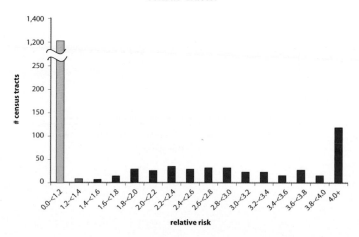

Figure 6: Census tracts by the number of cases per tract.

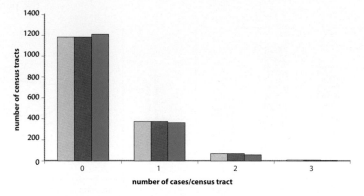

Figure 7a and b: Census tracts at high risk by the number of cases. (a) Unadjusted and (b) adjusted for social class.

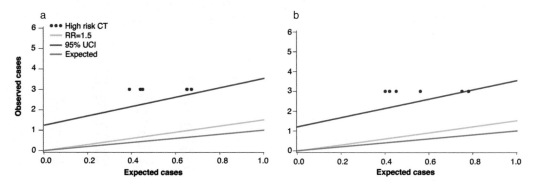

Figure 8: Risk over the period for high-risk census tracts relative to all census tracts.

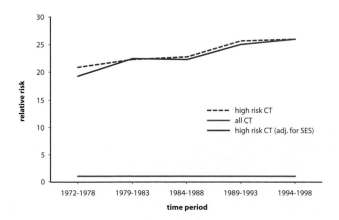

Figure 9: Map of census tracts at high risk.

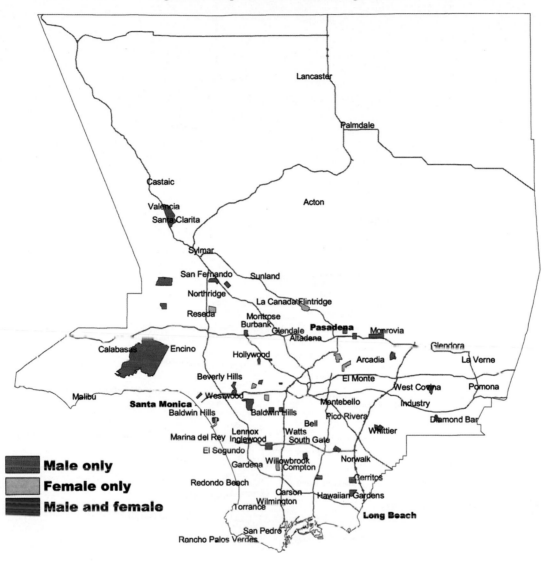

Figure 10: Male-female correlation between the relative risks for high-risk census tracts.

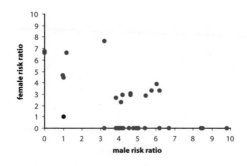

Figure 11: Map of census tracts at high risk, adjusted for social class.

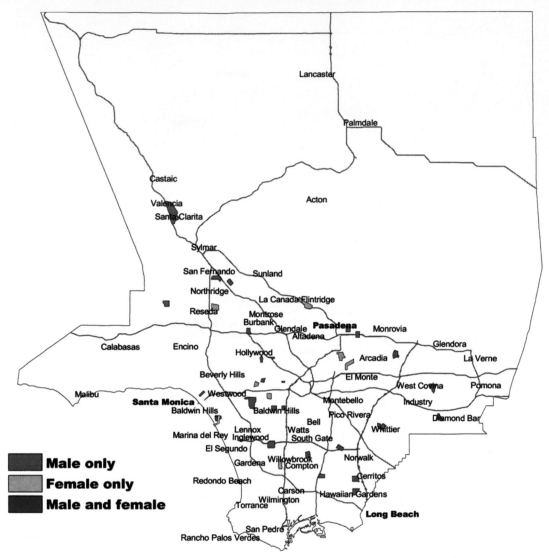

Figure 12: Male-female correlation between the relative risks for high-risk census tracts, adjusted for social class.

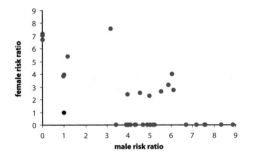

Nodular Sclerosis Hodgkin Lymphoma

ICDO-2 Code Anatomic Site: C 0–80
ICDO-2 Code Histology: 9661, 9663–9667
Age: All
Male Cases: 1473
Female Cases: 1386

Background

Hodgkin lymphoma is comprised of several malignancies having a common characteristic—an identical malignant cell of origin present in small numbers within a tumor mass largely comprised of other cell types. Each of these malignancies shows a characteristic microscopic appearance and occurs in a characteristic pattern in the population. In contrast to other types of Hodgkin lymphoma, nodular sclerosis is more common among young adults, occurs with equal frequency among males and females, and has historically occurred more commonly among persons of higher social class. Being linked to higher social class in childhood, this form of Hodgkin lymphoma has repeatedly been observed to occur more commonly among those with fewer siblings. That observation has led to the belief that the malignancy is caused by one of the many viruses to which we are usually, but not always, exposed in childhood. Polio and hepatitis viruses, among others, more commonly cause symptoms if they are acquired at an older age. In contrast to mixed cell Hodgkin lymphoma, Epstein-Barr Virus (EBV) is not seen within the malignant cells of the nodular sclerosis tumor. However, nodular sclerosis Hodgkin lymphoma has been observed more often following infectious mononucleosis, which is caused when EBV first infects in adolescence or young adulthood. Because early protection against one childhood virus is likely to mean early protection against all childhood viruses, some suspect that a different, as yet undetermined, childhood virus is responsible for this variant of Hodgkin. Heritable factors also play a role in etiology, although probably by means of increasing susceptibility to infection. Hodgkin lymphoma in general has been linked to woodworking occupations, but thisassociation has not been found consistently or linked to one cell type.

Local Pattern

Nodular sclerosis Hodgkin lymphoma is no more common in Los Angeles County than in other parts of the country. It occurs with equal frequency in men and women and is much more common in young adulthood. Incidence is especially high among whites, and is higher among African-Americans than among Latinos. The disease is especially common among those of high social class. This malignancy has been increasing in frequency over time in the county as a whole, although the trend among residents of high-risk census tracts has been more stable. Figure 6 shows a slight nonrandom excess of census tracts with unexpectedly few or unexpectedly many cases. Figure 6 shows a modest nonrandom tendency for census tracts with fewer or more than the expected number of cases. Although there are no contiguous census tracts and only a few census tracts with high risk for both males and females, the census tracts at high risk for this malignancy are generally among those of high social class, and that pattern disappears after adjustment.

Thumbnail Interpretation

The pattern seen in Los Angeles County is consistent with the generally accepted causal hypothesis, namely that the malignancy results when an older child or young adult first contracts a particular infection.

Figure 1: Age-adjusted incidence rate by place.

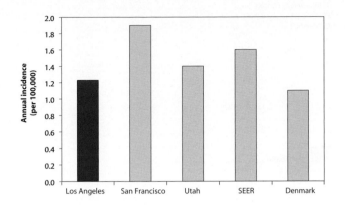

Figure 2: Age-adjusted incidence rate over the period.

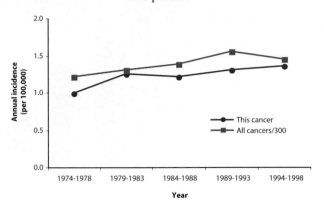

Figure 3: Age-adjusted incidence rate by age and race/ethnicity.

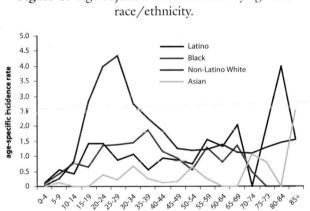

Figure 4: Age-adjusted incidence rate by social class.

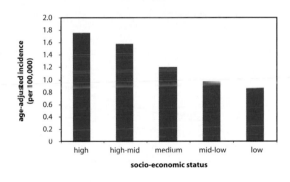

Figure 5: Distribution of the relative risk values for all census tracts.

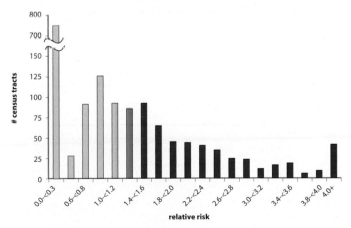

Figure 6: Census tracts by the number of cases per tract.

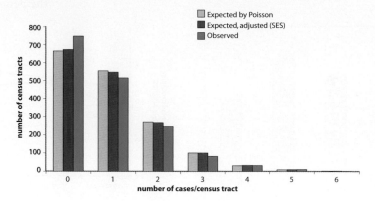

Figure 7a and b: Census tracts at high risk by the number of cases. (a) Unadjusted and (b) adjusted for social class.

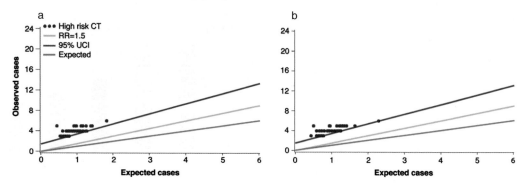

Figure 8: Risk over the period for high-risk census tracts relative to all census tracts.

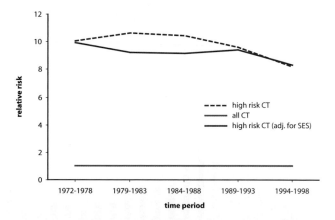

Nodular Sclerosis Hodgkin Lymphoma: Female

Figure 1: Age-adjusted incidence rate by place.

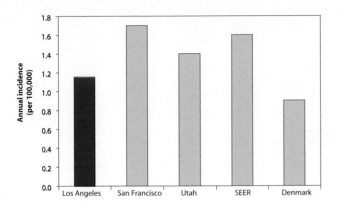

Figure 2: Age-adjusted incidence rate over the period.

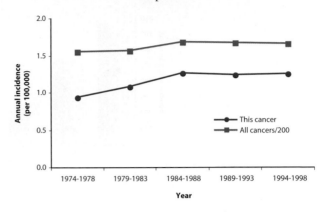

Figure 3: Age-adjusted incidence rate by age and race/ethnicity.

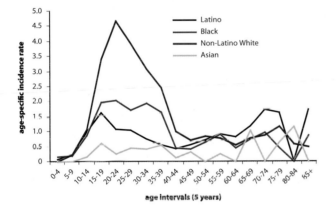

Figure 4: Age-adjusted incidence rate by social class.

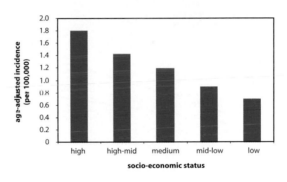

Figure 5: Distribution of the relative risk values for all census tracts.

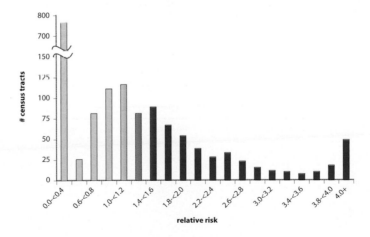

Figure 6: Census tracts by the number of cases per tract.

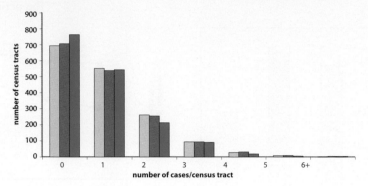

Figure 7a and b: Census tracts at high risk by the number of cases. (a) Unadjusted and (b) adjusted for social class.

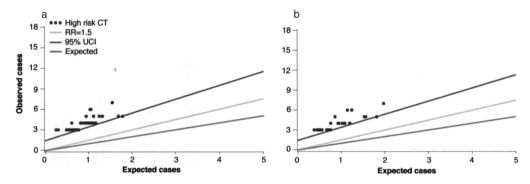

Figure 8: Risk over the period for high-risk census tracts relative to all census tracts.

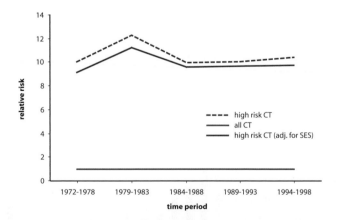

Figure 9: Map of census tracts at high risk.

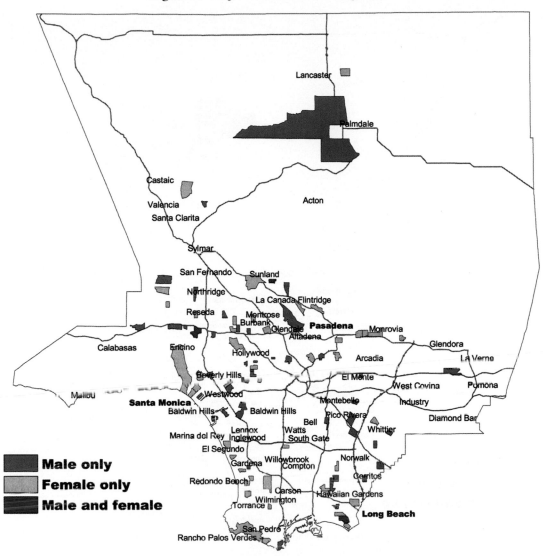

Figure 10: Male-female correlation between the relative risks for high-risk census tracts.

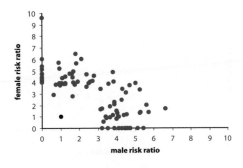

Figure 11: Map of census tracts at high risk, adjusted for social class.

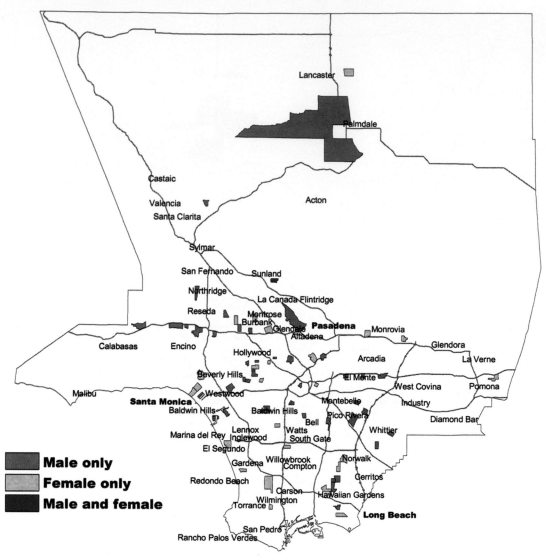

Figure 12: Male-female correlation between the relative risks for high-risk census tracts, adjusted for social class.

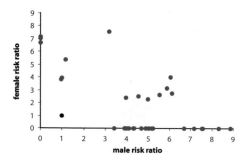

Other Hodgkin Lymphoma

ICDO-2 Code Anatomic Site: C 0–80
ICDO-2 Code Histology: 9650–9651, 9653–9655, 9657–9659, 9662
Age: All
Male Cases: 762
Female Cases: 467

Background

Hodgkin lymphoma is comprised of several malignancies having a common characteristic—an identical malignant cell of origin present in small numbers within a tumor mass largely comprised of other cell types. Each of these malignancies shows a characteristic microscopic appearance and occurs in a characteristic pattern in the population. The less common forms of Hodgkin lymphoma are less well characterized, but some features are also consistent with infection by a childhood virus. Hodgkin lymphoma in general has been linked to woodworking occupations, but this association has not been found consistently.

adulthood, occur more frequently at older ages, and do not occur more commonly in any particular racial/ethnic group. Among males in the county as a whole, they have been decreasing in frequency over time, although this tendency is less obvious among the residents of high-risk census tracts. Incidence is not consistently linked to social class. Figure 6 shows a moderate nonrandom excess of census tracts with unexpectedly few or unexpectedly many cases. Census tracts at high risk are almost all based on males and are distributed widely around the county in no obvious pattern. No census tract stands out on the basis of a particularly high number of excess cases.

Local Pattern

The less common Hodgkin lymphomas have occurred with about equal frequency in Los Angeles County and other registry populations and are slightly more common among males. These tumors begin to occur in young

Thumbnail Interpretation

The preponderance among males is consistent with the pattern of other lymphomas. No systematic pattern of geographical occurrence is apparent, and therefore no local source of causation can be proposed.

Figure 1: Age-adjusted incidence rate by place.

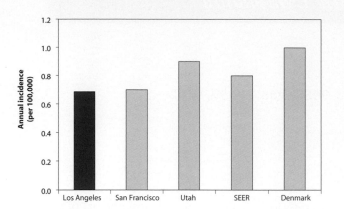

Figure 2: Age-adjusted incidence rate over the period.

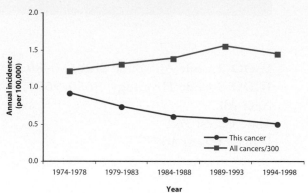

Figure 3: Age-adjusted incidence rate by age and race/ethnicity.

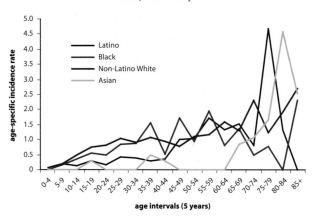

Figure 4: Age-adjusted incidence rate by social class.

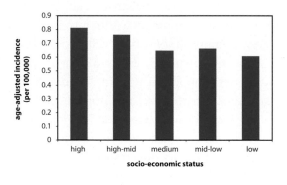

Figure 5: Distribution of the relative risk values for all census tracts.

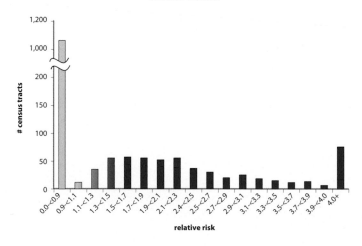

Figure 6: Census tracts by the number of cases per tract.

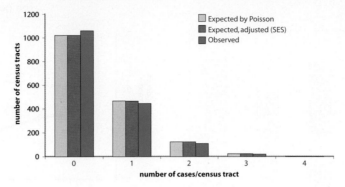

Figure 7a and b: Census tracts at high risk by the number of cases. (a) Unadjusted and (b) adjusted for social class.

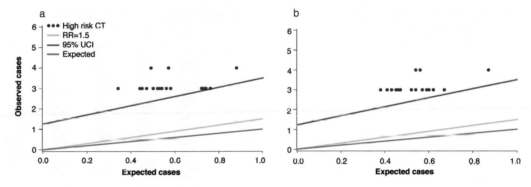

Figure 8: Risk over the period for high-risk census tracts relative to all census tracts.

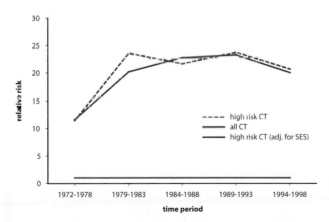

Figure 1: Age-adjusted incidence rate by place.

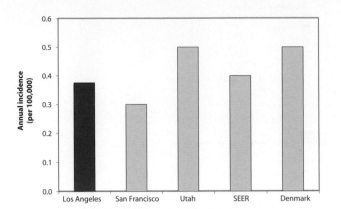

Figure 2: Age-adjusted incidence rate over the period.

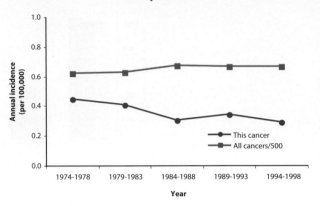

Figure 3: Age-adjusted incidence rate by age and race/ethnicity.

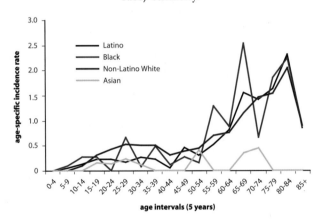

Figure 4: Age-adjusted incidence rate by social class.

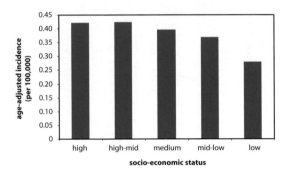

Figure 5: Distribution of the relative risk values for all census tracts.

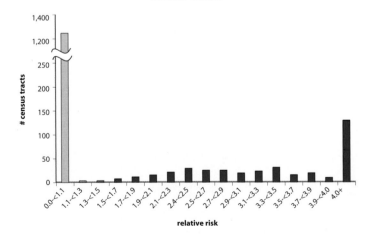

Figure 6: Census tracts by the number of cases per tract.

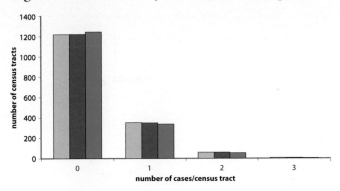

Figure 7a and b: Census tracts at high risk by the number of cases. (a) Unadjusted and (b) adjusted for social class.

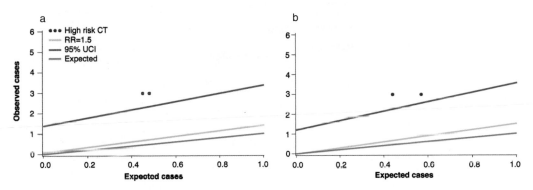

Figure 8: Risk over the period for high risk census tracts relative to all census tracts.

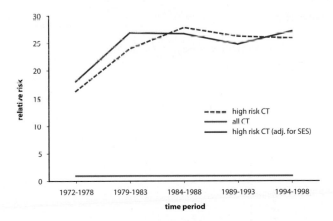

Figure 9: Map of census tracts at high risk.

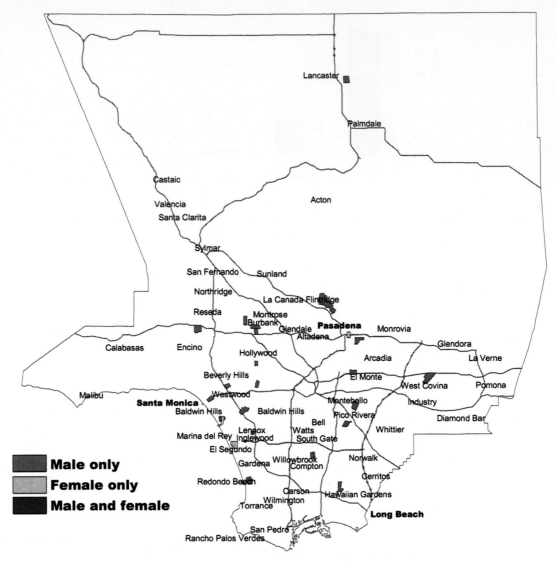

Figure 10: Male-female correlation between the relative risks for high-risk census tracts.

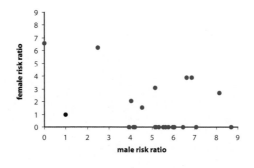

Figure 11: Map of census tracts at high risk, adjusted for social class.

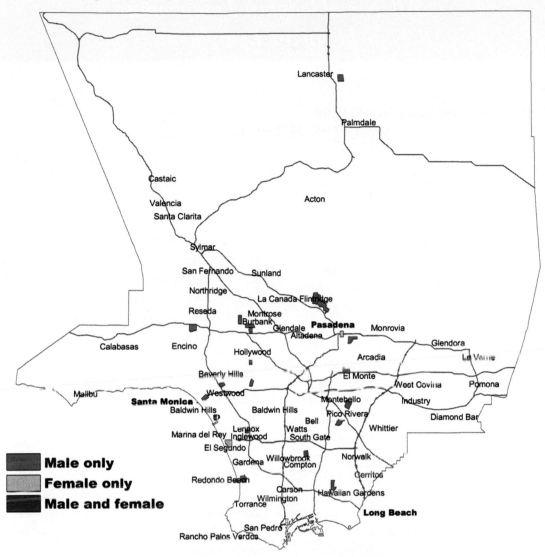

Male only
Female only
Male and female

Figure 12: Male female correlation between the relative risks for high risk census tracts, adjusted for social class.

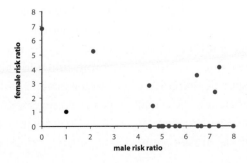

All Non-Hodgkin Lymphomas
(All Non-Hodgkin Lymphoma Types)

ICDO-2 Code Anatomic Site: C 0–80
ICDO-2 Code Histology: 9590–9596, 9670–9723
Age: All
Male Cases: 15001
Female Cases: 12375

Background

This category of non-Hodgkin lymphoma is now known to consist of a large number of conditions with different histological and clinical characteristics and presumably different causation. They vary greatly by grade (degree of functional differentiation), with those of high grade being particularly aggressive. The method of classifying these component conditions has evolved with advancing technology, but is still far from perfect, and few well-crafted studies have distinguished between the subgroups. To combine the best of new knowledge with historical consistency, we have created a classification that combines elements of the older standard Working Formulation with the newer international classification.

As a group, non-Hodgkin lymphomas have been increasing in frequency for the past 50 years. This increase is partly a result of better diagnosis, and more recently, large increases have occurred as a result of AIDS. Other known causes include immunosuppressive drugs given as chemotherapy or prior to transplantation, ionizing radiation, autoimmune disease, and specific hereditary syndromes of immunodeficiency. Certain occupational and personal exposures, such as to black hair dye, benzidine, dioxins, and herbicides have also been reported as causes. In some circumstances infections have been strongly suspected. These include Epstein-Barr Virus (EBV), the bacterium *H. pylori*, human T-cell lymphotropic virus type-1 (HTLV-1), and possibly hepatitis C virus (HCV). Other suspected causes include agricultural pesticides, repeated blood transfusions (possibly a source of unspecified infection), and excessive exposure to solar radiation.

Local Pattern

Non-Hodgkin lymphomas are nearly twice as common among men as among women, and occur with roughly equal frequency in all areas of the country including Los Angeles County. Incidence begins to increase in young adulthood. Whites and Latinos, especially women, are at higher risk than African-Americans and those of Asian origin. There is increased risk among both men and women of higher social class. Incidence among men in the county as a whole has increase over time, both among young and old, although the increase in the former group has been less dramatic in recent years. A lower rate of increase among women in the county as a whole prevailed throughout the entire period. In both men and women, a trend over time was observed to be initially higher, then lower, among the residents of high-risk census tracts. Figure 6 shows a substantial nonrandom excess of tracts with unexpectedly few or unexpectedly many cases. Many census tracts met the high-risk criteria, and in some, the level of risk was very high. Although some of these are scattered around the county, a very large aggregate of high-risk census tracts extends from Westwood across West Hollywood to the Silver Lake district, and this pattern is only slightly modified by adjustment for social class.

Thumbnail Interpretation

Non-Hodgkin's lymphomas are diverse in histological detail, doubtless reflecting causal diversity. Immune deficiency as a result of AIDS has resulted in excess risk to members of the gay community, and that largely explains the geographical and demographic pattern of occurrence in Los Angeles County. The reason why whites of higher social class find themselves at higher risk is unknown, but may at least partly be due to genetic determinants.

All Non-Hodgkin Lymphomas
(All Non-Hodgkin Lymphoma Types): Male

Figure 1: Age-adjusted incidence rate by place.

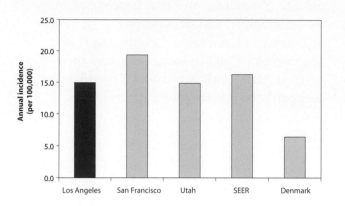

Figure 2: Age-adjusted incidence rate over the period.

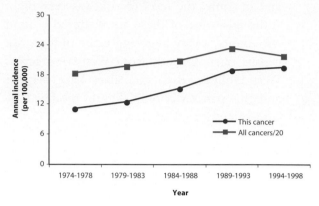

Figure 3: Age-adjusted incidence rate by age and race/ethnicity.

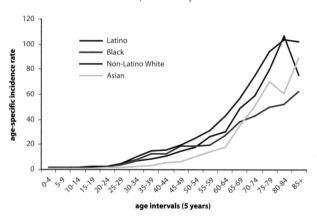

Figure 4: Age-adjusted incidence rate by social class.

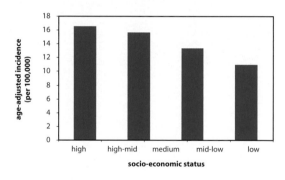

Figure 5: Distribution of the relative risk values for all census tracts.

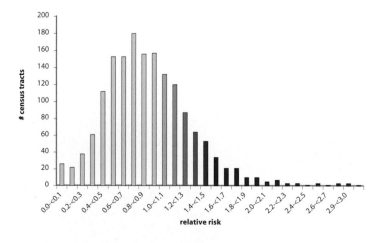

Figure 6: Census tracts by the number of cases per tract.

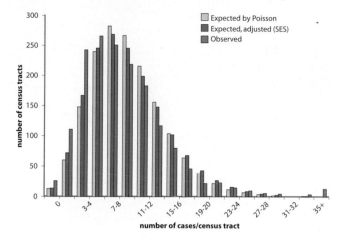

Figure 7a and b: Census tracts at high risk by the number of cases. (a) Unadjusted and (b) adjusted for social class.

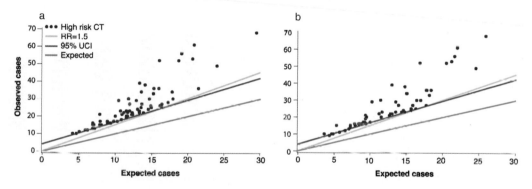

Figure 8: Risk over the period for high-risk census tracts relative to all census tracts.

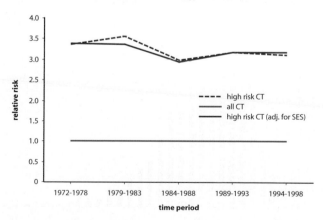

All Non-Hodgkin Lymphomas
(All Non-Hodgkin Lymphoma Types): Female

Figure 1: Age-adjusted incidence rate by place.

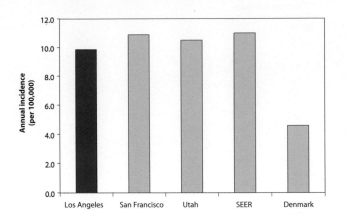

Figure 2: Age-adjusted incidence rate over the period.

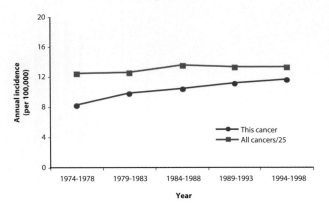

Figure 3: Age-adjusted incidence rate by age and race/ethnicity.

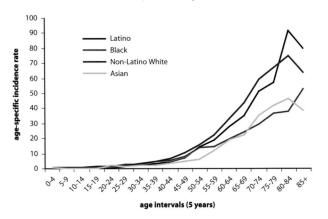

Figure 4: Age-adjusted incidence rate by social class.

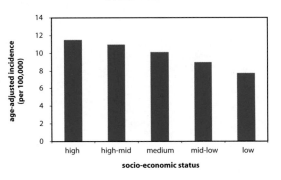

Figure 5: Distribution of the relative risk values for all census tracts.

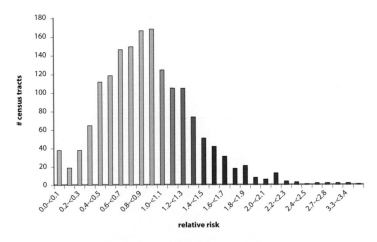

All Non-Hodgkin Lymphomas
(All Non-Hodgkin Lymphoma Types): Female

Figure 6: Census tracts by the number of cases per tract.

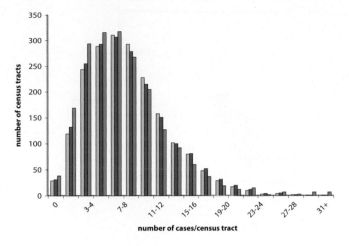

Figure 7a and b: Census tracts at high risk by the number of cases. (a) Unadjusted and (b) adjusted for social class.

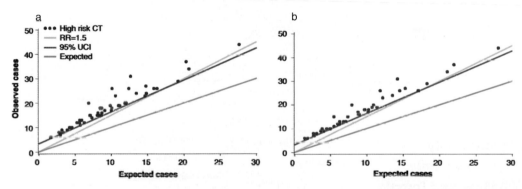

Figure 8: Risk over the period for high-risk census tracts relative to all census tracts.

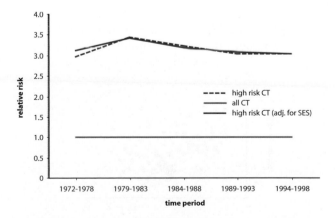

Figure 9: Map of census tracts at high risk.

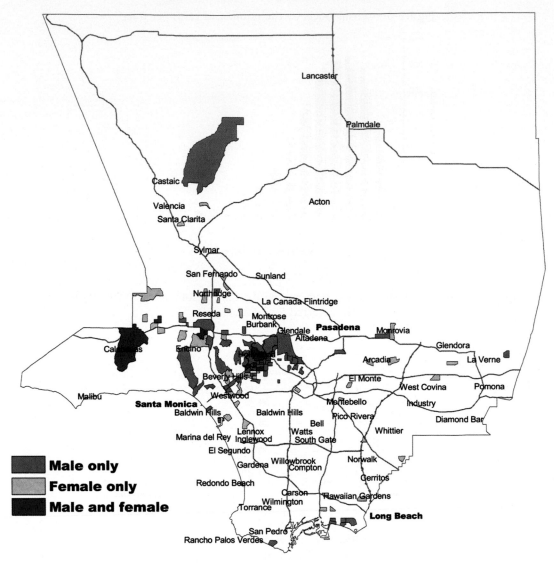

Figure 10: Male-female correlation between the relative risks for high-risk census tracts.

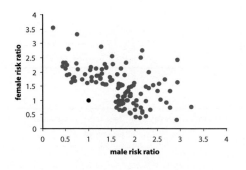

Figure 11: Map of census tracts at high risk, adjusted for social class.

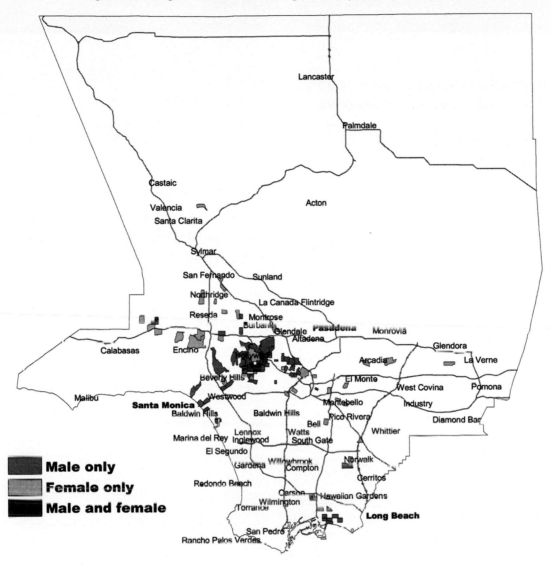

Figure 12: Male-female correlation between the relative risks for high-risk census tracts, adjusted for social class.

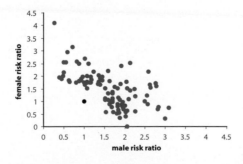

Follicular Non-Hodgkin Lymphoma

ICDO-2 Code Anatomic Site: C 0–80
ICDO-2 Code Histology: 9690–9698
Age: All
Male Cases: 2354
Female Cases: 2643

Background

See the previous background discussion above for all non-Hodgkin lymphomas. These relatively common low-grade non-Hodgkin lymphomas are of unknown causation, but in contrast to most other forms, they tend to occur in seemingly healthy persons. They have been described to occur in excess among farmers (with the specific causal exposures unknown), among smokers, among chronic users of dark hair dye, and among persons with certain rare genetic syndromes.

Local Pattern

Follicular non-Hodgkin lymphomas occur with about equal frequency in men and women, and at about the same rate in Los Angeles County as in other populations including San Francisco. Incidence begins to increase in young adulthood. Whites are at highest risk, followed by Latinos, and the disease occurs preferentially among those of higher social class. Incidence in the county as a whole has been stable in men but has decreased among women; it has been stable among the residents of high-risk census tracts. Figure 6 shows only a slight nonrandom excess of census tracts with unexpectedly few or unexpectedly many cases. The census tracts at high-risk are scattered around the higher social class regions of the county, and that tendency is reduced somewhat after adjustment for social class. No census tract stands out on the basis of a particularly high number of excess cases.

Thumbnail Interpretation

The reason for the higher incidence among whites of higher social class is unknown. No systematic pattern of geographical occurrence is apparent, and therefore no local source of causation can be identified.

Figure 1: Age-adjusted incidence rate by place.

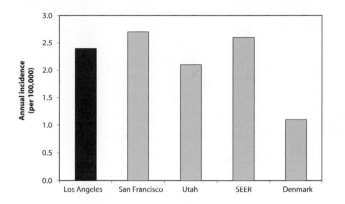

Figure 2: Age-adjusted incidence rate over the period.

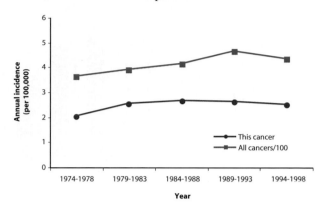

Figure 3: Age-adjusted incidence rate by age and race/ethnicity.

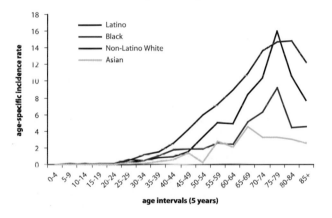

Figure 4: Age-adjusted incidence rate by social class.

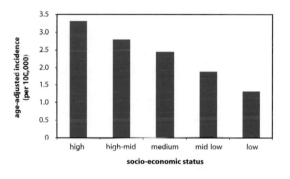

Figure 5: Distribution of the relative risk values for all census tracts.

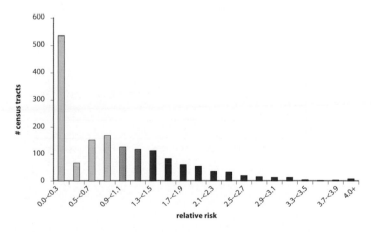

Figure 6: Census tracts by the number of cases per tract.

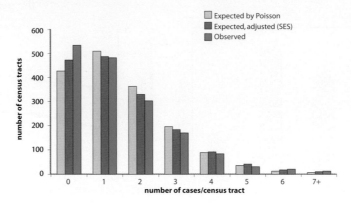

Figure 7a and b: Census tracts at high risk by the number of cases. (a) Unadjusted and (b) adjusted for social class.

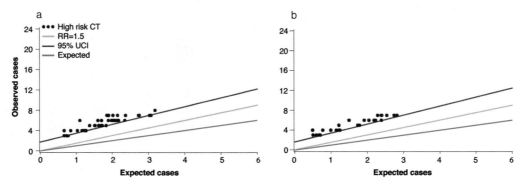

Figure 8: Risk over the period for high-risk census tracts relative to all census tracts.

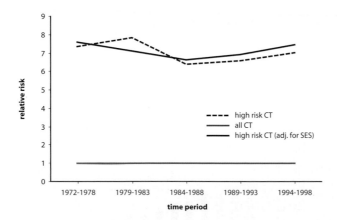

Figure 1: Age-adjusted incidence rate by place.

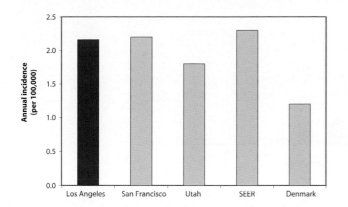

Figure 2: Age-adjusted incidence rate over the period.

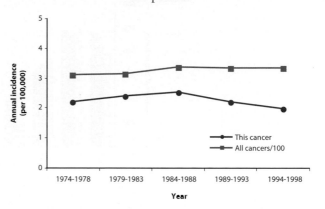

Figure 3: Age-adjusted incidence rate by age and race/ethnicity.

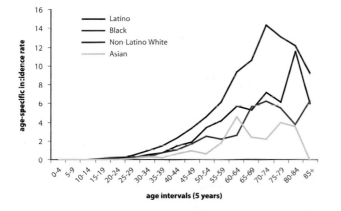

Figure 4: Age-adjusted incidence rate by social class.

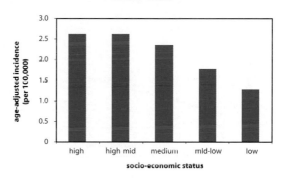

Figure 5: Distribution of the relative risk values for all census tracts.

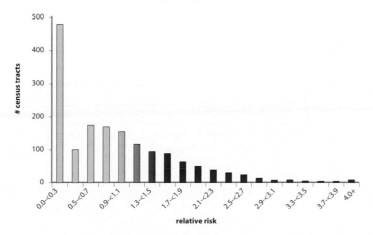

Figure 6: Census tracts by the number of cases per tract.

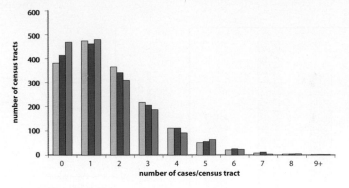

Figure 7a and b: Census tracts at high risk by the number of cases. (a) Unadjusted and (b) adjusted for social class.

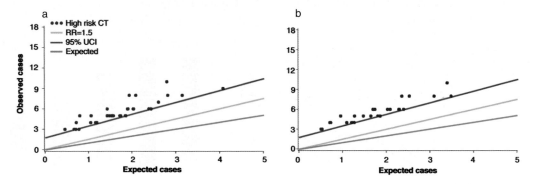

Figure 8: Risk over the period for high-risk census tracts relative to all census tracts.

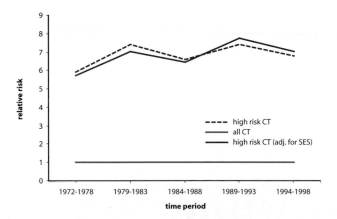

Figure 9: Map of census tracts at high risk.

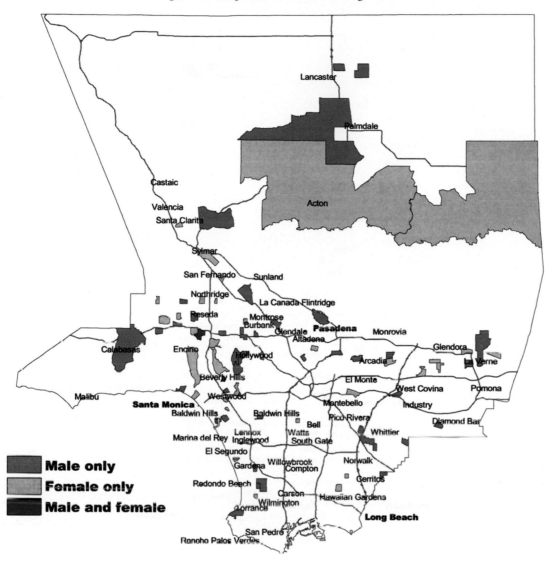

Figure 10: Male-female correlation between the relative risks for high-risk census tracts.

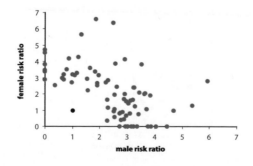

Figure 11: Map of census tracts at high risk, adjusted for social class.

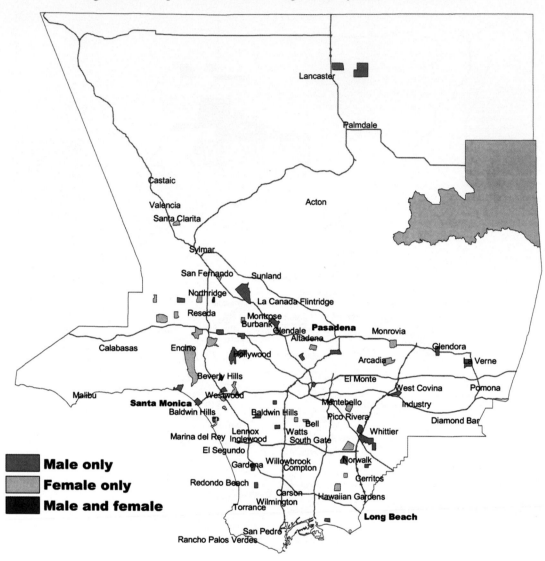

Figure 12: Male-female correlation between the relative risks for high-risk census tracts, adjusted for social class.

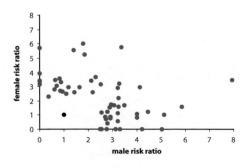

Diffuse Large B-Cell Non-Hodgkin Lymphoma

ICDO-2 Code Anatomic Site: C 0–80
ICDO-2 Code Histology: 9680–9683, 9688
Age: All
Male Cases: 3378
Female Cases: 3008

Background

See the previous background discussion for all non-Hodgkin lymphomas. This is the most common form of non-Hodgkin lymphoma, and the malignancies are usually of intermediate grade. The causes are unknown. Few studies have distinguished between this subgroup and those other lymphomas consisting of smaller B cells. This form of non-Hodgkin lymphoma has been linked to AIDS, to immunosuppressive drugs prior to transplantation, and to the exposures characteristic of farmers.

Americans, who are at somewhat lower risk. These malignancies are more common among those of higher social class and are increasing in frequency over time and among residents of the county as a whole as well as among residents of high-risk census tracts. Figure 6 shows a moderate nonrandom excess of census tracts with unexpectedly few or unexpectedly many cases. High-risk census tracts are scattered around the county, but a large aggregate of census tracts is seen in West Hollywood and Beverly Hills. The latter confluence is unchanged after adjustment for social class.

Local Pattern

Diffuse large B-cell non-Hodgkin lymphoma is slightly more common among men than women, and is less common in Los Angeles County than in San Francisco. By age, incidence gradually increases throughout adulthood. The racial/ethnicity groups are all at roughly equal risk, except African-

Thumbnail Interpretation

The higher risk in San Francisco and the higher risk among men suggest that these lymphomas commonly occur because of the immunosuppression produced by AIDS. The presence of high-risk census tracts in and around West Hollywood is consistent with that suspicion.

Figure 1: Age-adjusted incidence rate by place.

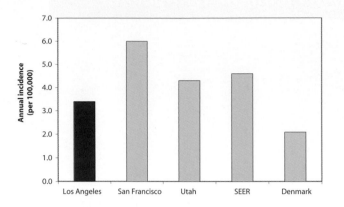

Figure 2: Age-adjusted incidence rate over the period.

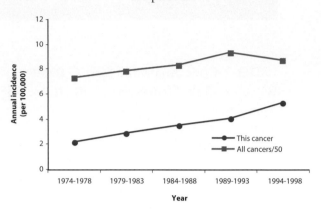

Figure 3: Age-adjusted incidence rate by age and race/ethnicity.

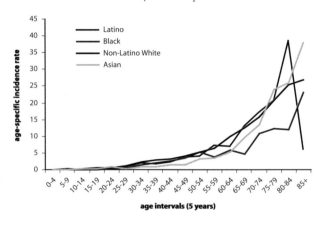

Figure 4: Age-adjusted incidence rate by social class.

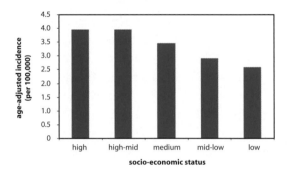

Figure 5: Distribution of the relative risk values for all census tracts.

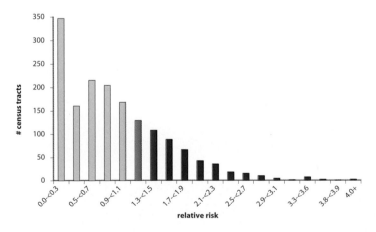

Figure 6: Census tracts by the number of cases per tract.

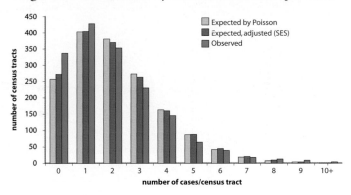

Figure 7a and b: Census tracts at high risk by the number of cases. (a) Unadjusted and (b) adjusted for social class.

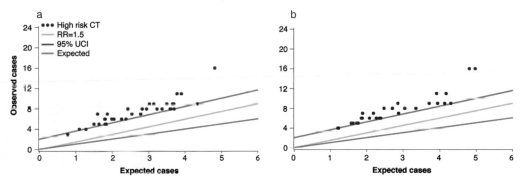

Figure 8: Risk over the period for high-risk census tracts relative to all census tracts.

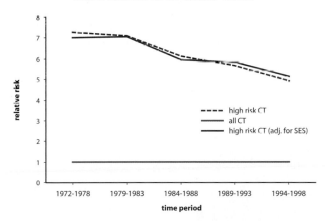

Figure 1: Age-adjusted incidence rate by place.

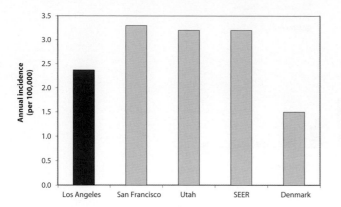

Figure 2: Age-adjusted incidence rate over the period.

Figure 3: Age-adjusted incidence rate by age and race/ethnicity.

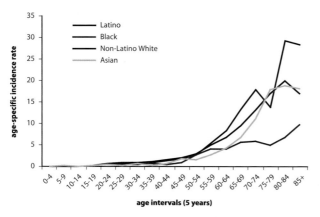

Figure 4: Age-adjusted incidence rate by social class.

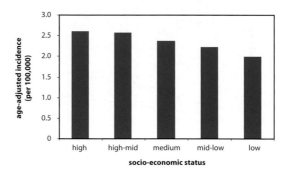

Figure 5: Distribution of the relative risk values for all census tracts.

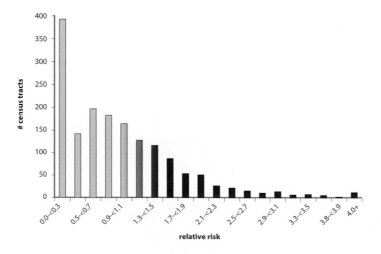

Figure 6: Census tracts by the number of cases per tract.

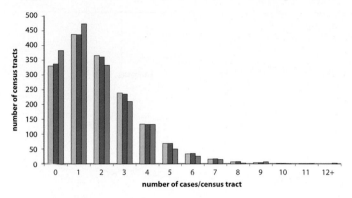

Figure 7a and b: Census tracts at high risk by the number of cases. (a) Unadjusted and (b) adjusted for social class.

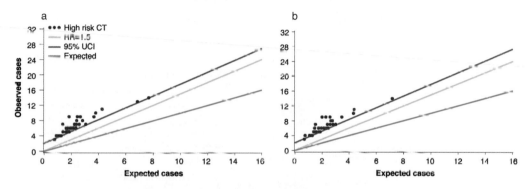

Figure 8: Risk over the period for high-risk census tracts relative to all census tracts.

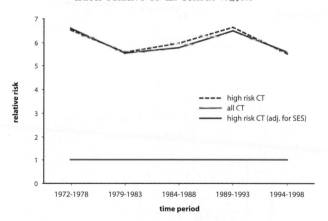

Figure 9: Map of census tracts at high risk.

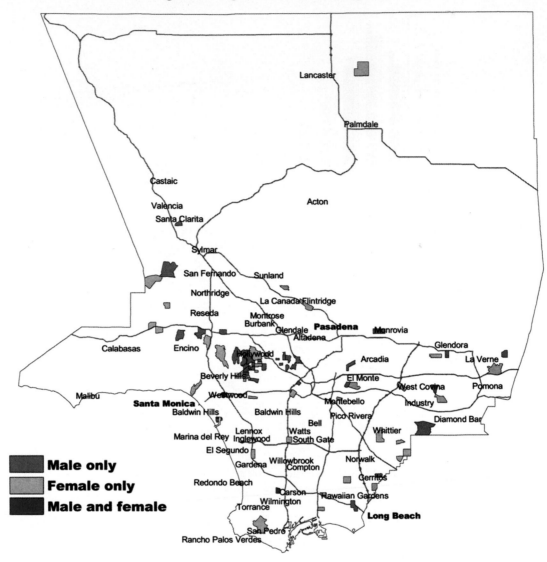

Figure 10: Male-female correlation between the relative risks for high-risk census tracts.

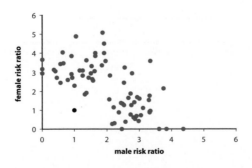

Figure 11: Map of census tracts at high risk, adjusted for social class.

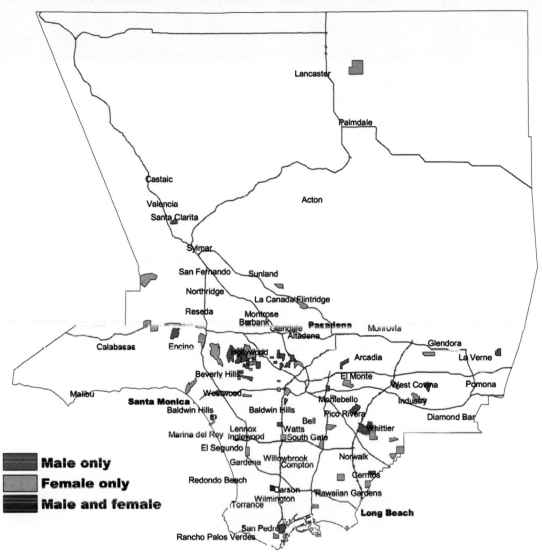

Figure 12: Male-female correlation between the relative risks for high-risk census tracts, adjusted for social class.

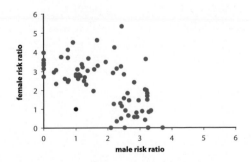

Diffuse Small Lymphocytic or Plasmacytic B-Cell Non-Hodgkin Lymphoma

ICDO-2 Code Anatomic Site: C 0–80
ICDO-2 Code Histology: 9670–9672
Age: All
Male Cases: 2357
Female Cases: 2018

Background

See the previous background discussion for all non-Hodgkin lymphomas. Few studies have distinguished between this category, which itself contains multiple subgroups, and the lymphomas composed of larger B cells. Inherited chromosomal abnormalities are sometimes associated with this low-grade form of lymphoma. It also has been reported to occur more often in agricultural settings. Nonetheless, the causes are unknown.

Local Pattern

This variety of non-Hodgkin lymphoma occurs somewhat more frequently among men than women, and occurs with roughly equal frequency among residents of Los Angeles County and other regions of North America and Europe. Whites are at slightly higher risk in comparison with the members of other racial/ethnicity groups. These malignancies are more common among those of higher social class. They have decreased in frequency over time, both in the county as a whole and among the residents of high-risk census tracts. Figure 6 shows only a slight nonrandom excess of census tracts with unexpectedly few or unexpectedly many cases. Before and after adjustment for social class, high-risk census tracts are scattered around the county in the high social class regions, but without contiguity. No census tract stands out on the basis of a particularly high number of excess cases.

Thumbnail Interpretation

The reason for the higher incidence among whites of higher social class and the decrease over time are unknown. No systematic pattern of geographical occurrence is apparent, and therefore speculation about local sources of causation is precluded.

Figure 1: Age-adjusted incidence rate by place.

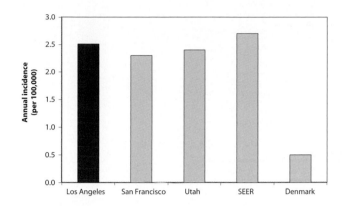

Figure 2: Age-adjusted incidence rate over the period.

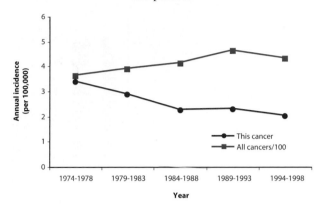

Figure 3: Age-adjusted incidence rate by age and race/ethnicity.

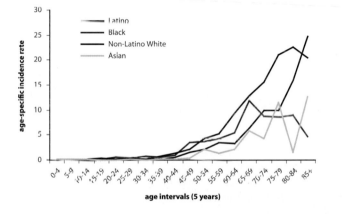

Figure 4: Age-adjusted incidence rate by social class.

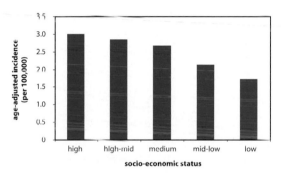

Figure 5: Distribution of the relative risk values for all census tracts.

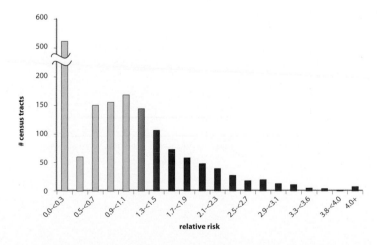

Figure 6: Census tracts by the number of cases per tract.

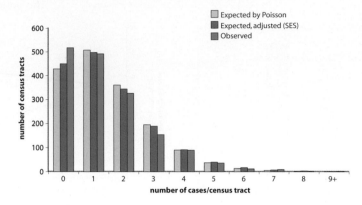

Figure 7a and b: Census tracts at high risk by the number of cases. (a) Unadjusted and (b) adjusted for social class.

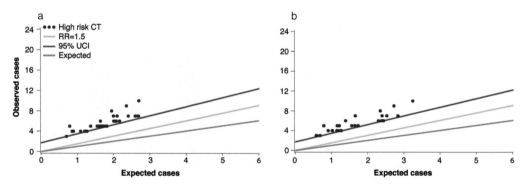

Figure 8: Risk over the period for high-risk census tracts relative to all census tracts.

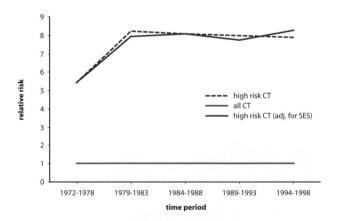

Diffuse Small Lymphocytic or Plasmacytic B-Cell Non-Hodgkin Lymphoma: Female

Figure 1: Age-adjusted incidence rate by place.

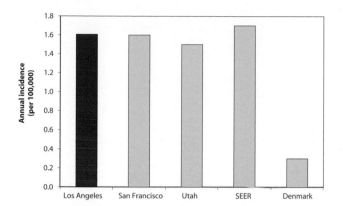

Figure 2: Age-adjusted incidence rate over the period.

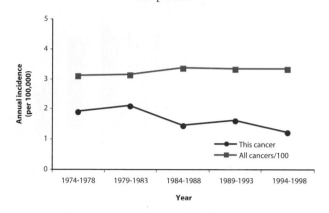

Figure 3: Age-adjusted incidence rate by age and race/ethnicity.

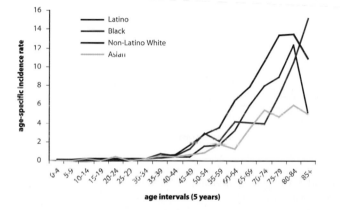

Figure 4: Age-adjusted incidence rate by social class.

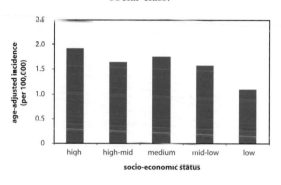

Figure 5: Distribution of the relative risk values for all census tracts.

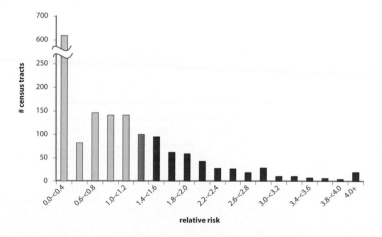

Figure 6: Census tracts by the number of cases per tract.

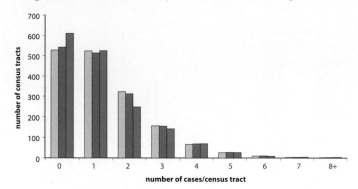

Figure 7a and b: Census tracts at high risk by the number of cases. (a) Unadjusted and (b) adjusted for social class.

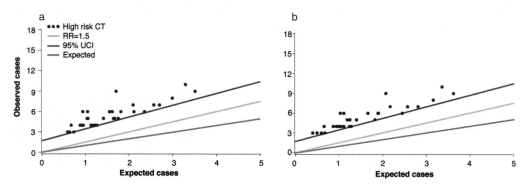

Figure 8: Risk over the period for high-risk census tracts relative to all census tracts.

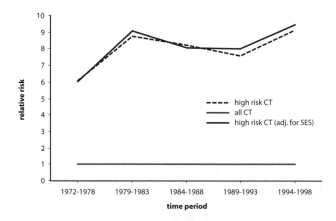

Figure 9: Map of census tracts at high risk.

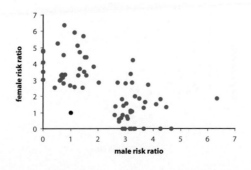

Figure 10: Male-female correlation between the relative risks for high-risk census tracts.

Diffuse Small Lymphocytic or Plasmacytic B-Cell Non-Hodgkin Lymphoma

Figure 11: Map of census tracts at high risk, adjusted for social class.

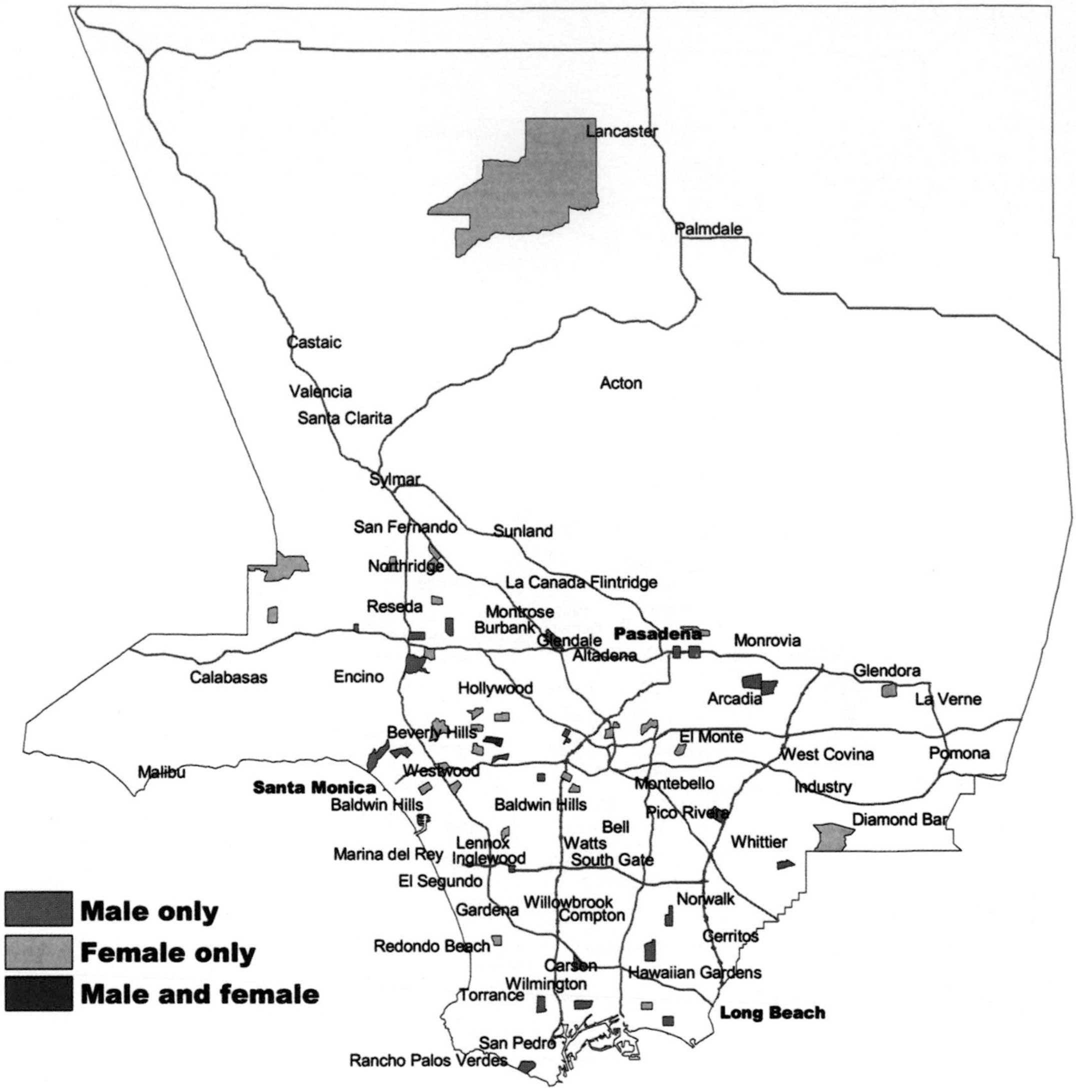

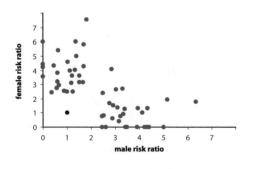

Figure 12: Male-female correlation between the relative risks for high-risk census tracts, adjusted for social class.

Diffuse Mixed B-Cell Non-Hodgkin Lymphoma

ICDO-2 Code Anatomic Site: C 0–80
ICDO-2 Code Histology: 9673–9677
Age: All
Male Cases: 974
Female Cases: 927

Background

See the previous background discussion for all non-Hodgkin lymphomas. These non-Hodgkin lymphomas are composed of both small and large B lymphocytes, but the category also includes certain newly described entities such as mantle-cell lymphoma. This class of intermediate-grade lymphomas of relatively uniform histology has never been studied as a separate entity.

Local Pattern

This class of non-Hodgkin lymphoma occurs slightly more often in men, and is seen with equal frequency in Los Angeles County and other regions. Incidence begins to increase in middle age. The malignancy is more common among those of higher social class and is less common among African-Americans and Asian-Americans, especially women. It has occurred with relatively constant frequency, both in the county as a whole and among the residents of high-risk census tracts. Figure 6 shows only a slight nonrandom excess of census tracts with unexpectedly few or unexpectedly many cases. High-risk census tract are widely scattered, but there does exist a complex of adjacent census tracts at high risk extending from Westwood to West Hollywood. Census tracts at high risk of males tend to appear to the northwest of those at high risk to women. This pattern is only slightly modified by adjustment for social class.

Thumbnail Interpretation

The pattern of this group of lymphomas is not similar to that of diffuse large B-cell or high-grade non-Hodgkin lymphomas, because neither the predominance among males, the higher risk in San Francisco, nor the trend over time are as extreme. Moreover, while there is some geographical localization, the pattern of high-risk census tracts is not the pattern related to AIDS, and that is especially true of census tracts at high risk of female cases. However, the cases occurring in high-risk census tracts appear demographically similar to cases of follicular lymphoma, which has no geographical concentration. Although some of the tumors in these census tracts may have occurred in persons with AIDS, there is a preponderance of female cases in some of them, and other unknown causes appear to be responsible.

Figure 1: Age-adjusted incidence rate by place.

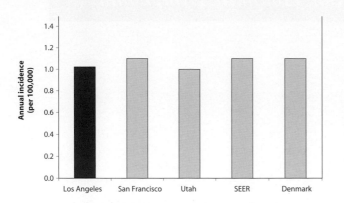

Figure 2: Age-adjusted incidence rate over the period.

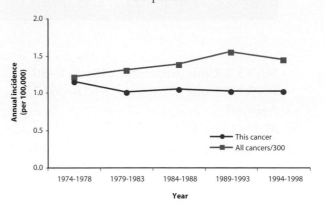

Figure 3: Age-adjusted incidence rate by age and race/ethnicity.

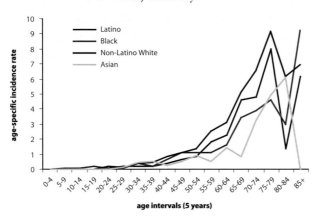

Figure 4: Age-adjusted incidence rate by social class.

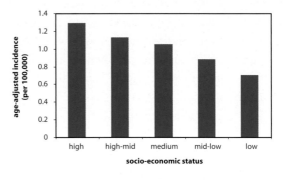

Figure 5: Distribution of the relative risk values for all census tracts.

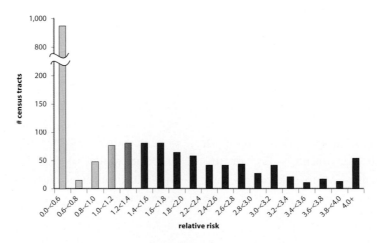

Figure 6: Census tracts by the number of cases per tract.

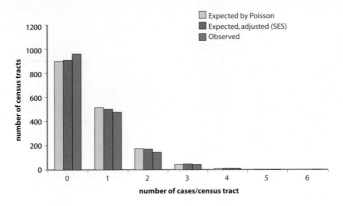

Figure 7a and b: Census tracts at high risk by the number of cases. (a) Unadjusted and (b) adjusted for social class.

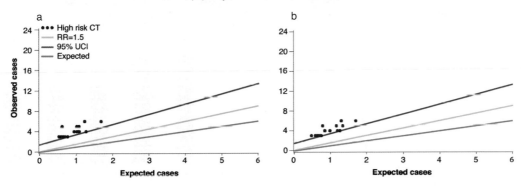

Figure 8: Risk over the period for high-risk census tracts relative to all census tracts.

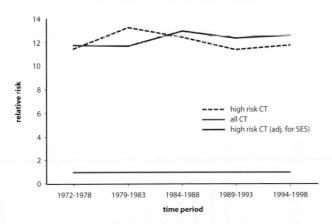

Figure 1: Age-adjusted incidence rate by place.

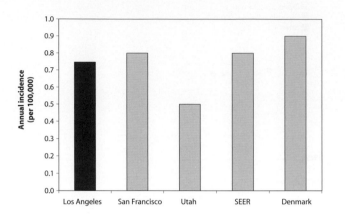

Figure 2: Age-adjusted incidence rate over the period.

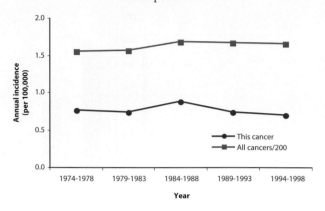

Figure 3: Age-adjusted incidence rate by age and race/ethnicity.

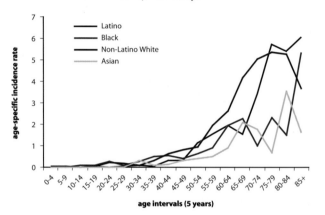

Figure 4: Age-adjusted incidence rate by social class.

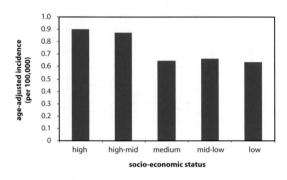

Figure 5: Distribution of the relative risk values for all census tracts.

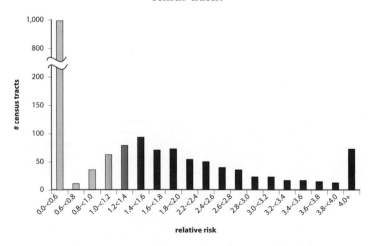

Figure 6: Census tracts by the number of cases per tract.

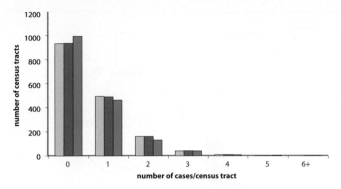

Figure 7a and b: Census tracts at high risk by the number of cases. (a) Unadjusted and (b) adjusted for social class.

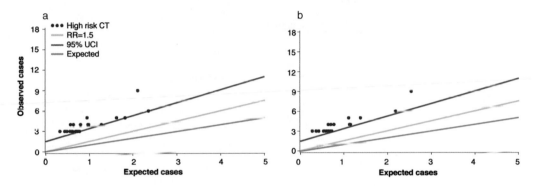

Figure 8: Risk over the period for high-risk census tracts relative to all census tracts.

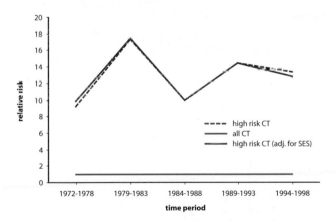

Figure 9: Map of census tracts at high risk.

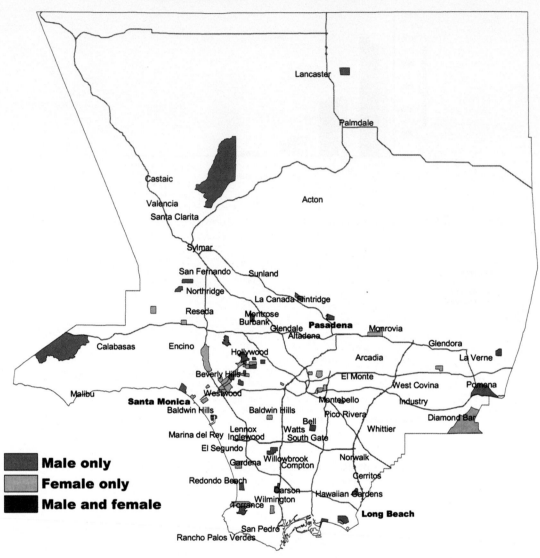

Figure 10: Male-female correlation between the relative risks for high-risk census tracts.

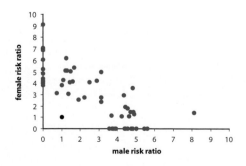

Figure 11: Map of census tracts at high risk, adjusted for social class.

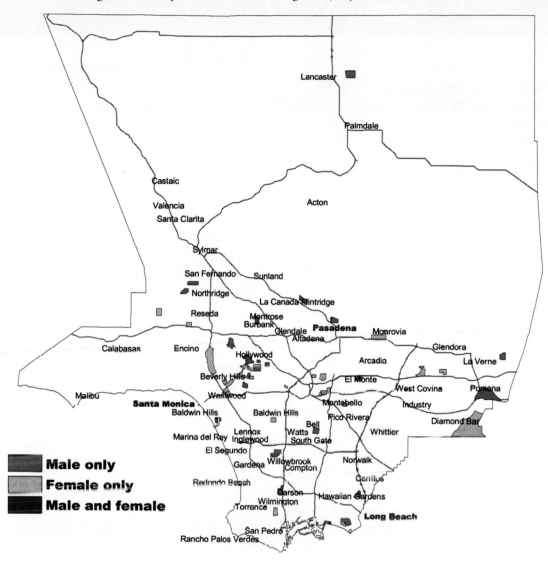

Male only
Female only
Male and female

Figure 12: Male-female correlation between the relative risks for high-risk census tracts, adjusted for social class.

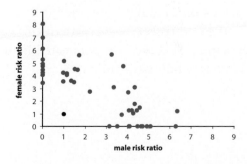

High-Grade Non-Hodgkin Lymphoma

ICDO-2 Code Anatomic Site: C 0–80
ICDO-2 Code Histology: 9684–9685, 9687
Age: All
Male Cases: 2054
Female Cases: 920

Background

See the previous background discussion for all non-Hodgkin lymphomas. These lymphomas commonly appear in persons with immunodeficiency, such as those with AIDS or immunosuppression for organ transplantation. They are associated with and perhaps caused by specific infections, most notably the ubiquitous Epstein-Barr Virus (EBV). One of these tumors, called African Burkitt's lymphoma, is clearly caused by EBV infection among persons who have had repeated malaria infections and may have deficient immune protection on that basis. High-grade non-Hodgkin lymphomas may also occur among persons with rare genetic syndromes, and some studies have suggested a link with cigarette smoking.

Local Pattern

High-grade lymphomas occur twice as frequently among men as among women, and occur more often in Los Angeles County and especially in San Francisco than in other regions of the country. They appear as commonly in all races and all social classes, but risk is higher among young adult males and among the elderly. These malignancies have increased in frequency in the county as a whole with time, although that increase has recently reversed, both in the county as a whole and among the residents of census tracts at high risk. Figure 6 shows a moderate nonrandom excess of census tracts with unexpectedly few or unexpectedly many cases. Residents in some census tracts have experienced very high levels of risk. Whether or not adjusted for social class, a large cluster of contiguous high-risk census tracts, mostly reflecting risk to young men, is centered on West Hollywood.

Thumbnail Interpretation

The increased frequency of this class of non-Hodgkin lymphomas among men, the high incidence in San Francisco, the increased frequency among young men and the elderly, and the presence of high-risk census tracts in and around West Hollywood all indicate that the pattern of occurrence of this malignancy is dominated by immunodeficiency as a result of AIDS.

Figure 1: Age-adjusted incidence rate by place.

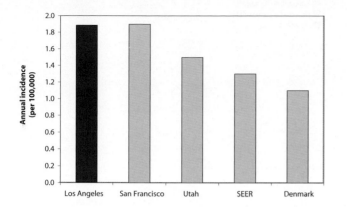

Figure 2: Age-adjusted incidence rate over the period.

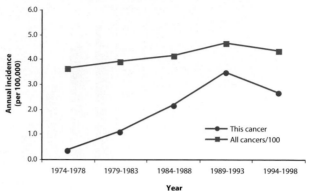

Figure 3: Age-adjusted incidence rate by age and race/ethnicity.

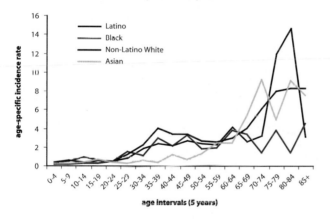

Figure 4: Age-adjusted incidence rate by social class.

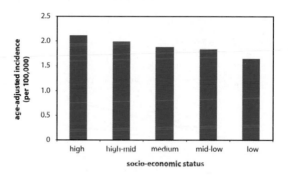

Figure 5: Distribution of the relative risk values for all census tracts.

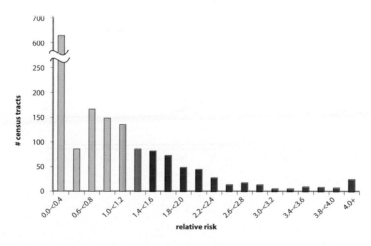

Figure 6: Census tracts by the number of cases per tract.

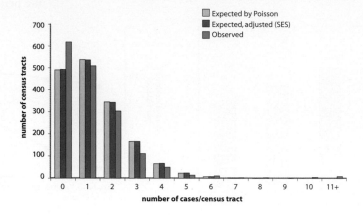

Figure 7a and b: Census tracts at high risk by the number of cases. (a) Unadjusted and (b) adjusted for social class.

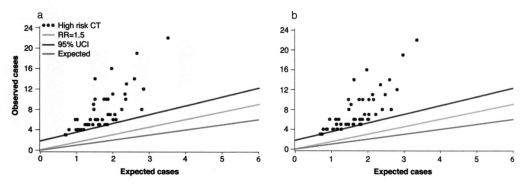

Figure 8: Risk over the period for high-risk census tracts relative to all census tracts.

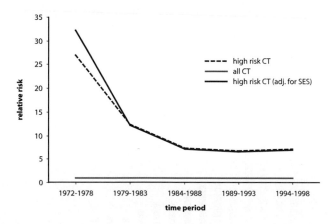

Figure 1: Age-adjusted incidence rate by place.

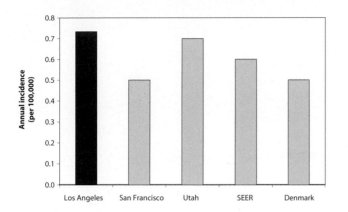

Figure 2: Age-adjusted incidence rate over the period.

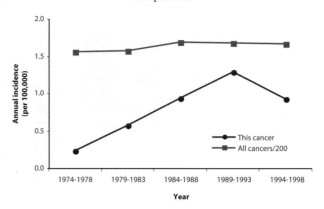

Figure 3: Age-adjusted incidence rate by age and race/ethnicity.

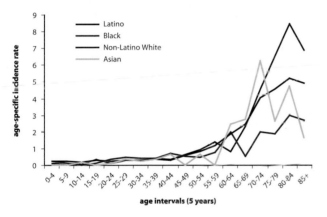

Figure 4: Age-adjusted incidence rate by social class.

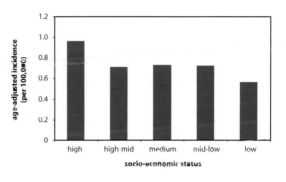

Figure 5: Distribution of the relative risk values for all census tracts.

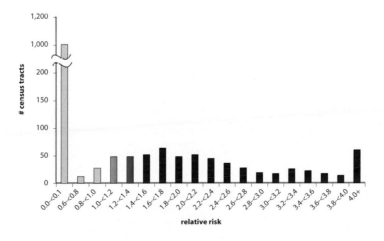

Figure 6: Census tracts by the number of cases per tract.

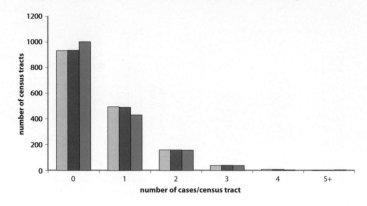

Figure 7a and b: Census tracts at high risk by the number of cases. (a) Unadjusted and (b) adjusted for social class.

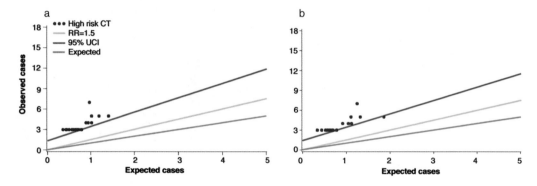

Figure 8: Risk over the period for high-risk census tracts relative to all census tracts.

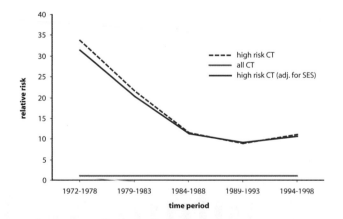

Figure 9: Map of census tracts at high risk.

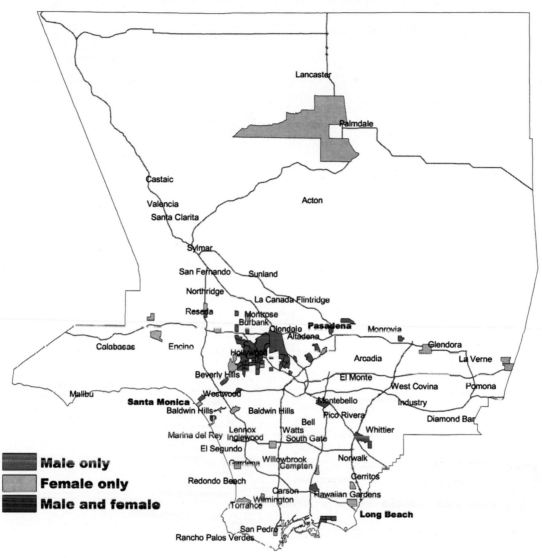

Figure 10: Male-female correlation between the relative risks for high-risk census tracts.

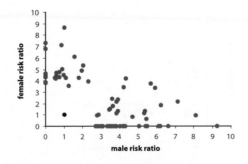

Figure 11: Map of census tracts at high risk, adjusted for social class.

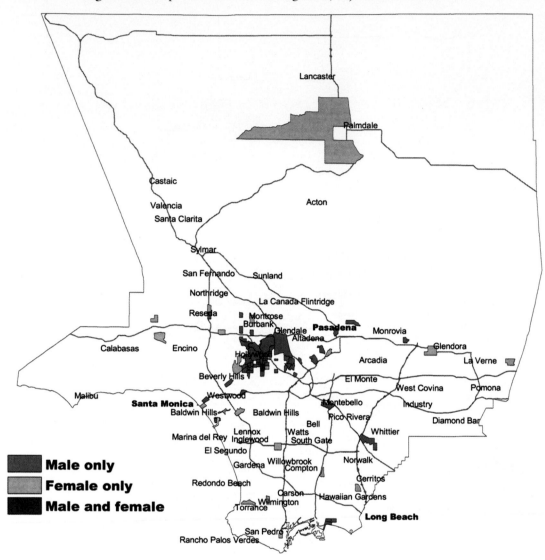

Figure 12: Male-female correlation between the relative risks for high-risk census tracts, adjusted for social class.

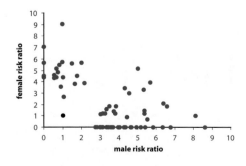

T-Cell Non-Hodgkin Lymphoma

ICDO-2 Code Anatomic Site: C 0–80
ICDO-2 Code Histology: 9700–9709, 9712–9714, 9717
Age: All
Male Cases: 604
Female Cases: 427

Background

See the previous background discussion for all non-Hodgkin lymphomas. These lymphomas occur among persons with immunodeficiency, and are especially common in persons with heritable chronic autoimmune conditions, such as celiac disease of the bowel. They also occur in patients whose immune systems have been disarmed prior to organ transplantation. They have been thought to result from certain infectious agents, especially HTLV-1, and possibly Epstein-Barr Virus (EBV) and they occur commonly in persons with HIV/AIDS.

Local Pattern

T-cell non-Hodgkin lymphoma is about twice as common among men as among women. It occurs more commonly in San Francisco than in Los Angeles County, and begins to increase in frequency in middle age. Among young adults, African-Americans are more often affected than are the members of other racial/ethnic groups. Incidence is relatively independent of social class. Incidence has been relatively constant over time among the residents of high-risk census tracts, but has been increasing markedly in the county as a whole. Figure 6 shows only a slight nonrandom excess of census tracts with unexpectedly few or unexpectedly many cases. High-risk census tracts are scattered over the county, with no apparent determinant other than chance. No census tract stands out on the basis of a particularly high number of excess cases.

Thumbnail Interpretation

The reasons for the higher incidence among men and the increase over time are unknown. There is no geographical pattern of the type that would be expected on the basis of a higher risk among those with AIDS. No systematic pattern of geographical occurrence is apparent, and therefore no local source of causation can be speculated upon.

Figure 1: Age-adjusted incidence rate by place.

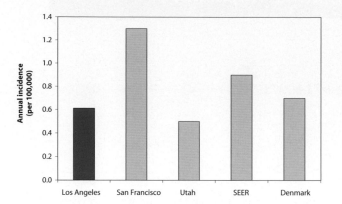

Figure 2: Age-adjusted incidence rate over the period.

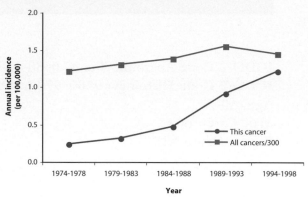

Figure 3: Age-adjusted incidence rate by age and race/ethnicity.

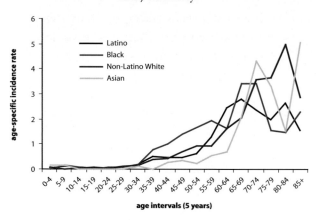

Figure 4: Age-adjusted incidence rate by social class.

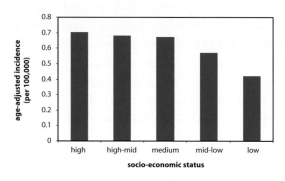

Figure 5: Distribution of the relative risk values for all census tracts.

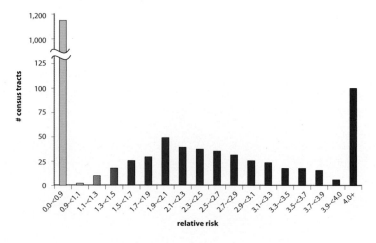

Figure 6: Census tracts by the number of cases per tract.

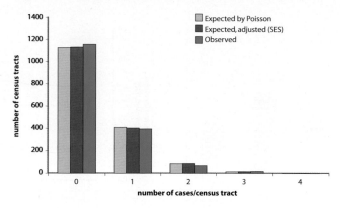

Figure 7a and b: Census tracts at high risk by the number of cases. (a) Unadjusted and (b) adjusted for social class.

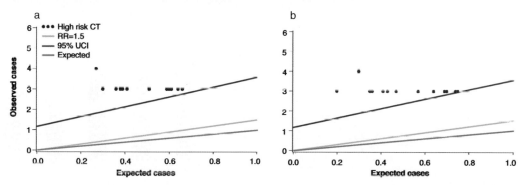

Figure 8: Risk over the period for high-risk census tracts relative to all census tracts.

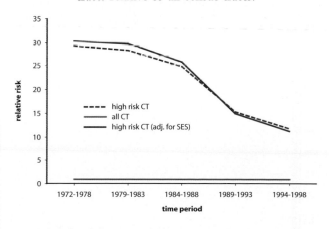

T-Cell Non-Hodgkin Lymphoma: Female

Figure 1: Age-adjusted incidence rate by place.

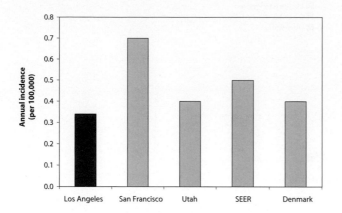

Figure 2: Age-adjusted incidence rate over the period.

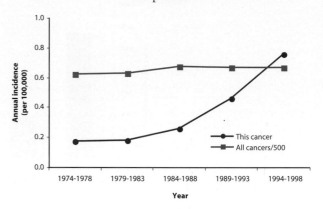

Figure 3: Age-adjusted incidence rate by age and race/ethnicity.

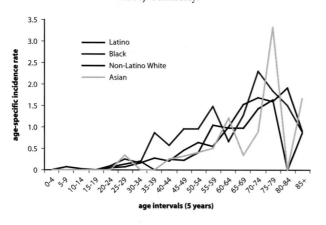

Figure 4: Age-adjusted incidence rate by social class.

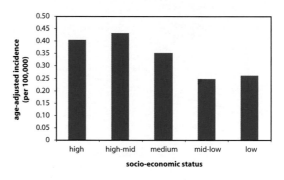

Figure 5: Distribution of the relative risk values for all census tracts.

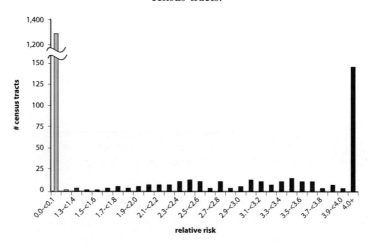

Figure 6: Census tracts by the number of cases per tract.

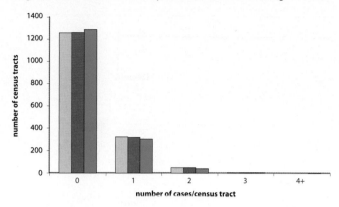

Figure 7a and b: Census tracts at high risk by the number of cases. (a) Unadjusted and (b) adjusted for social class.

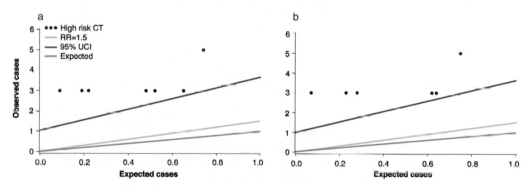

Figure 8: Risk over the period for high-risk census tracts relative to all census tracts.

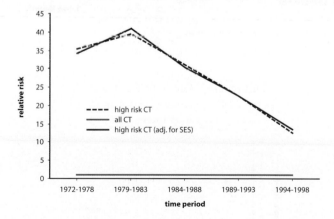

Figure 9: Map of census tracts at high risk.

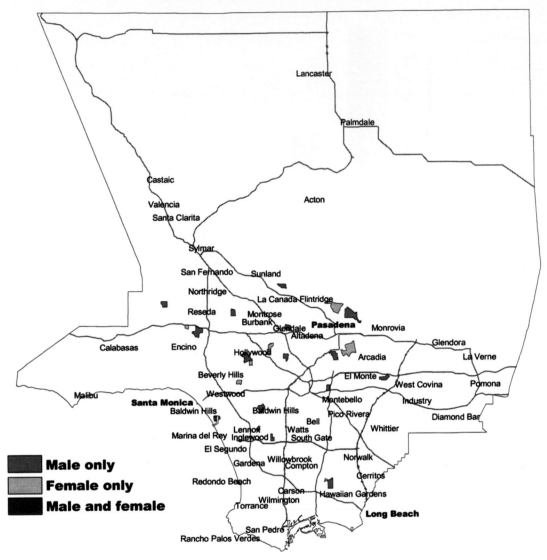

Figure 10: Male-female correlation between the relative risks for high-risk census tracts.

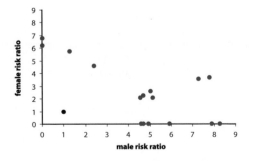

Figure 11: Map of census tracts at high risk, adjusted for social class.

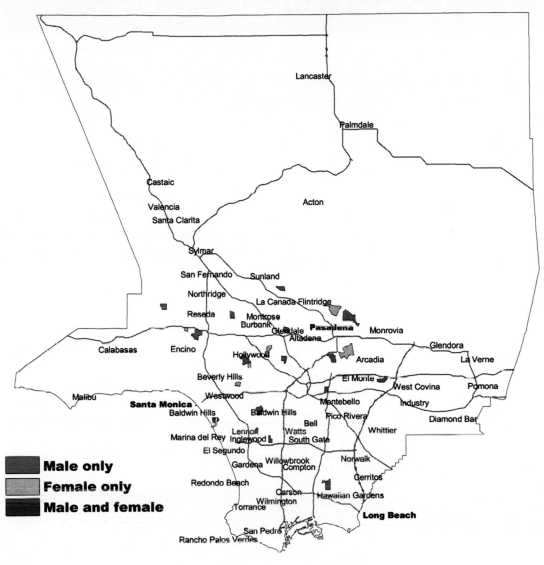

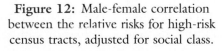

Figure 12: Male-female correlation between the relative risks for high-risk census tracts, adjusted for social class.

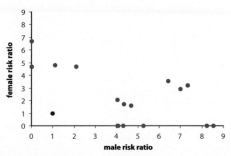

Maltoma Non-Hodgkin Lymphoma

ICDO-2 Code Anatomic Site: C 0–80
ICDO-2 Code Histology: 9710–9711, 9715–9716
Age: All
Male Cases: 253
Female Cases: 302

Background

See the previous background discussion above for all non-Hodgkin lymphomas. Maltomas (lymphomas occurring in mucosa-associated lymphoid tissue, MALT) are low-grade B-cell lymphomas, usually of the gastrointestinal tract. They are presumed to result from infectious agents. The category has only been recognized in the past 15 years or so. Among the infectious agents that have been implicated at different anatomic sites are the bacteria *Helicobacter pylori* and *Campylobacter jejuni*, the spirochete *Borrelia burgdorfia*, and possibly the hepatitis C virus.

Local Pattern

Incidence of this lymphoma is roughly similar among men and women, and does not vary markedly between Los Angeles County and other regions of the country. The rate begins to rise in young adulthood, and Asian-Americans and older Latinos are at higher risk. Risk does not vary by social class. As expected from the recent establishment of the classification, the occurrence of lymphomas coded to maltoma increased greatly in the last decade or so. Figure 6 shows no non-random excess of census tracts with unexpectedly few or unexpectedly many cases. Only a few geographically disparate census tracts fulfil the high-risk criteria, and none do so for both men and women. No census tract stands out on the basis of a particularly high number of excess cases.

Thumbnail Interpretation

The reason for the racial/ethnic disparity is unknown, although *H. pylori* is more common in developing societies. No systematic pattern of geographical occurrence is yet apparent, and therefore no local source of causation can be proposed.

Figure 1: Age-adjusted incidence rate by place.

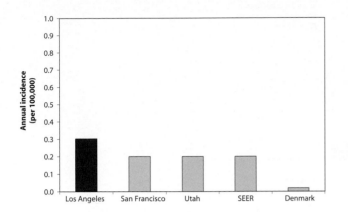

Figure 2: Age-adjusted incidence rate over the period.

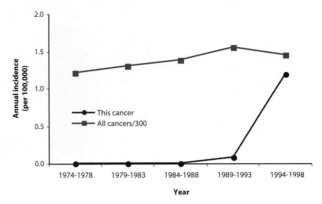

Figure 3: Age-adjusted incidence rate by age and race/ethnicity.

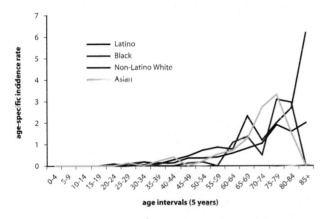

Figure 4: Age-adjusted incidence rate by social class.

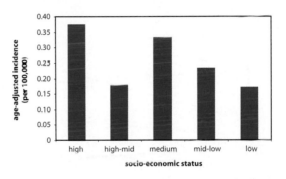

Figure 5: Distribution of the relative risk values for all census tracts.

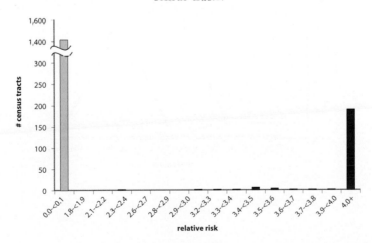

Figure 6: Census tracts by the number of cases per tract.

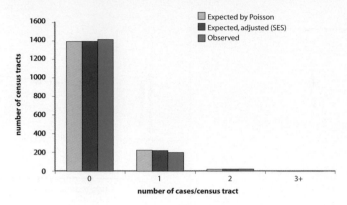

Figure 7a and b: Census tracts at high risk by the number of cases. (a) Unadjusted and (b) adjusted for social class.

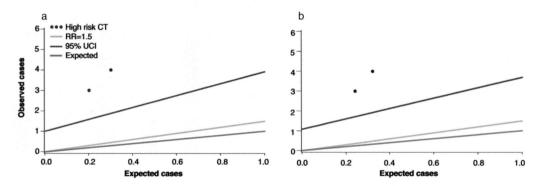

Figure 8: Risk over the period for high-risk census tracts relative to all census tracts.

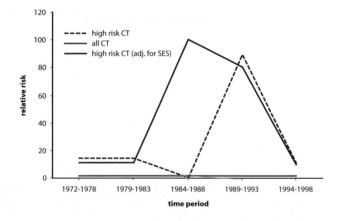

Maltoma Non-Hodgkin Lymphoma: Female

Figure 1: Age-adjusted incidence rate by place.

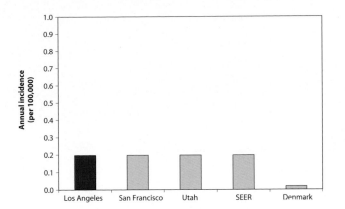

Figure 2: Age-adjusted incidence rate over the period.

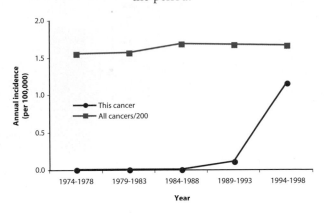

Figure 3: Age-adjusted incidence rate by age and race/ethnicity.

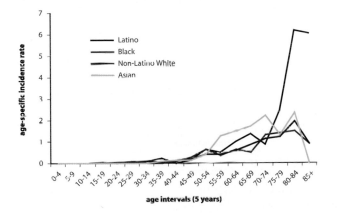

Figure 4: Age-adjusted incidence rate by social class.

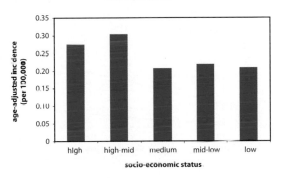

Figure 5: Distribution of the relative risk values for all census tracts.

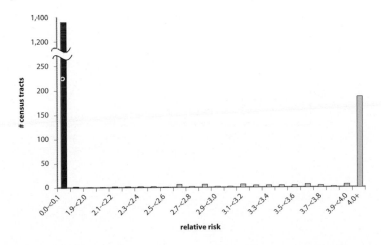

Figure 6: Census tracts by the number of cases per tract.

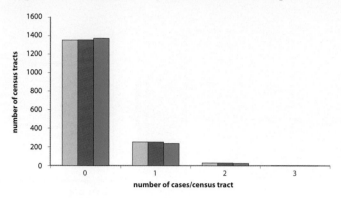

Figure 7a and b: Census tracts at high risk by the number of cases. (a) Unadjusted and (b) adjusted for social class.

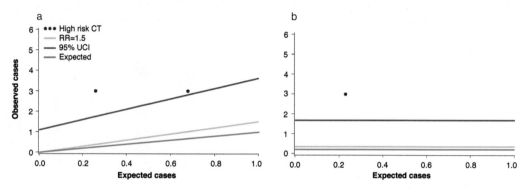

Figure 8: Risk over the period for high-risk census tracts relative to all census tracts.

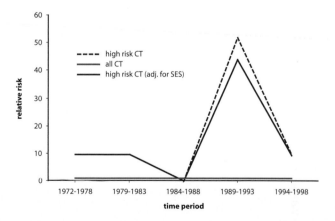

Figure 9: Map of census tracts at high risk.

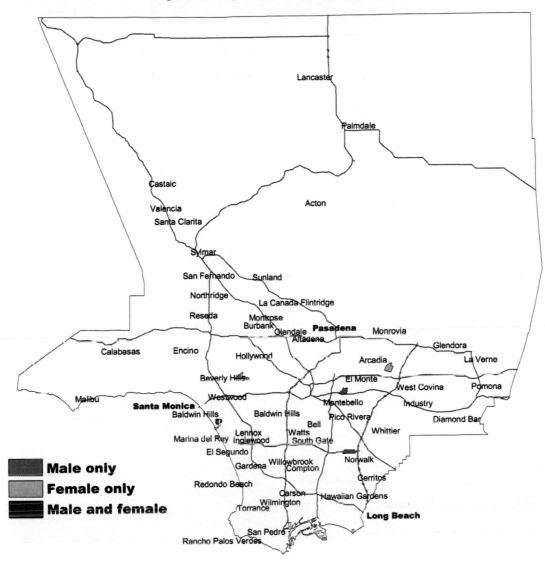

Male only
Female only
Male and female

Figure 10: Male-female correlation between the relative risks for high-risk census tracts.

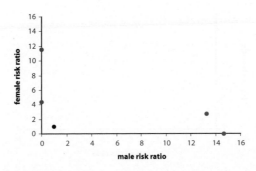

Figure 11: Map of census tracts at high risk, adjusted for social class.

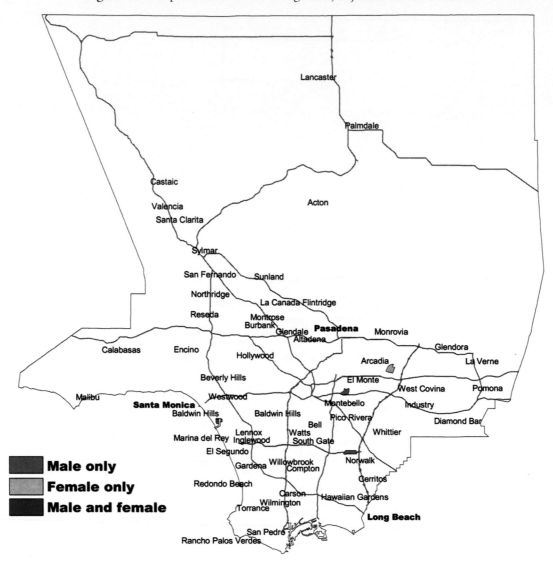

Figure 12: Male-female correlation between the relative risks for high-risk census tracts, adjusted for social class.

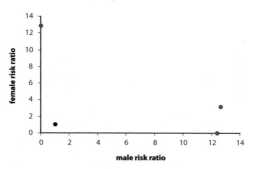

Other Non-Hodgkin Lymphoma

ICDO-2 Code Anatomic Site: C 0–80
ICDO-2 Code Histology: 9590–9595, 9677, 9686, 9720–9723
Age: All
Male Cases: 3027
Female Cases: 2130

Background

See the previous background discussion for all non-Hodgkin lymphomas. This group of non-Hodgkin lymphomas includes miscellaneous rare subtypes as well as many individual cases that have been nonspecifically coded, some of which are surely identical to the non-Hodgkin lymphomas placed in other categories.

Local Pattern

As a group, these non-Hodgkin lymphomas are almost twice as common in males, and occur more commonly in San Francisco than in Los Angeles County or other areas of the country. All race/ethnicity groups and social classes are at roughly equal risk. Incidence begins to increase in young adulthood. Over time, incidence has increased, although the trend subsequently reversed, both in the county as a whole and among the residents of high-risk census tracts. Figure 6 shows a slight nonrandom excess of census tracts with unexpectedly few or unexpectedly many cases. Census tracts at high risk are widely distributed, and clusters of contiguous census tracts appear in West Hollywood and the Silver Lake district.

Thumbnail Interpretation

Even though the pattern of occurrence is not identical to that of diffuse large B-cell non-Hodgkin lymphomas or high-grade non-Hodgkin lymphomas, the higher risk in San Francisco, the higher risk among men, and the presence of high-risk census tracts in and around West Hollywood all suggest that this class of lymphomas includes a substantial number that have occurred preferentially among persons with AIDS.

Figure 1: Age-adjusted incidence rate by place.

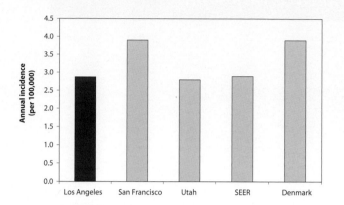

Figure 2: Age-adjusted incidence rate over the period.

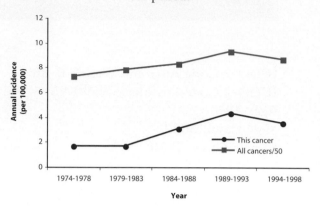

Figure 3: Age-adjusted incidence rate by age and race/ethnicity.

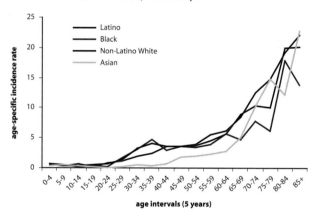

Figure 4: Age-adjusted incidence rate by social class.

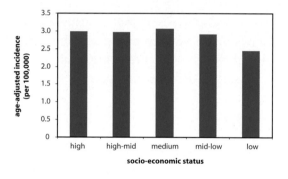

Figure 5: Distribution of the relative risk values for all census tracts.

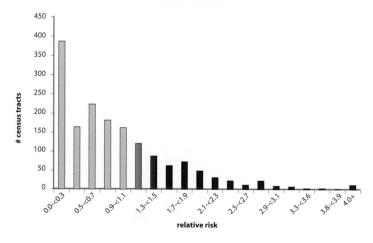

Figure 6: Census tracts by the number of cases per tract.

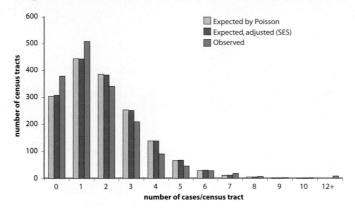

Figure 7a and b: Census tracts at high risk by the number of cases. (a) Unadjusted and (b) adjusted for social class.

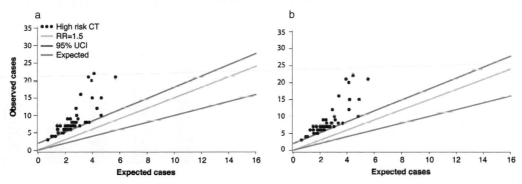

Figure 8: Risk over the period for high-risk census tracts relative to all census tracts.

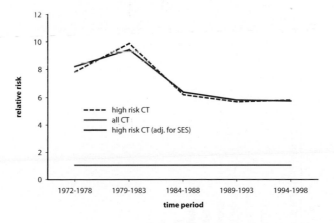

Figure 1: Age-adjusted incidence rate by place.

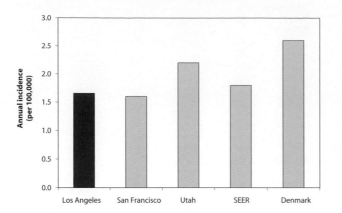

Figure 2: Age-adjusted incidence rate over the period.

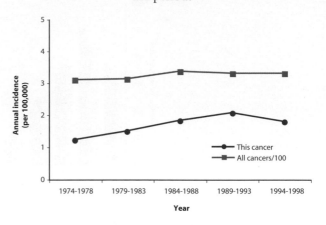

Figure 3: Age-adjusted incidence rate by age and race/ethnicity.

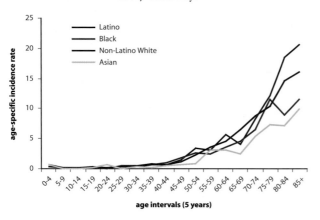

Figure 4: Age-adjusted incidence rate by social class.

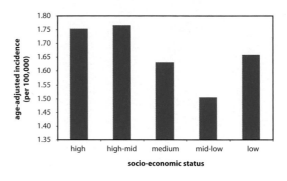

Figure 5: Distribution of the relative risk values for all census tracts.

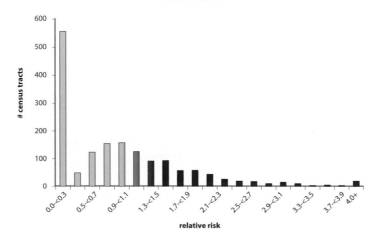

Figure 6: Census tracts by the number of cases per tract.

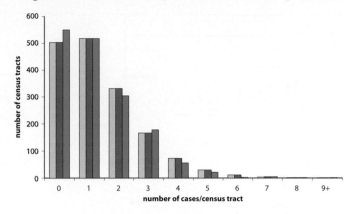

Figure 7a and b: Census tracts at high risk by the number of cases. (a) Unadjusted and (b) adjusted for social class.

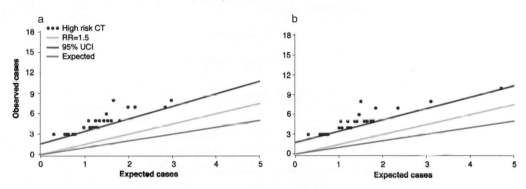

Figure 8: Risk over the period for high-risk census tracts relative to all census tracts.

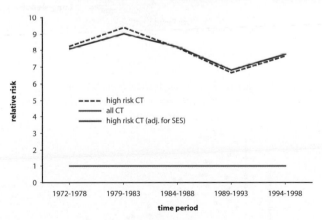

Figure 9: Map of census tracts at high risk.

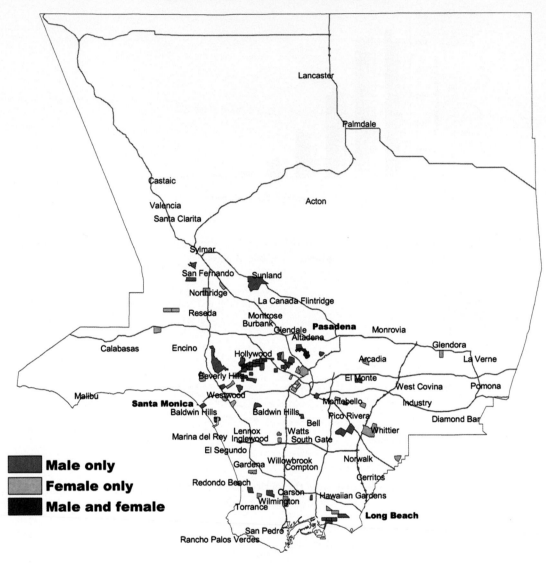

Male only
Female only
Male and female

Figure 10: Male-female correlation between the relative risks for high-risk census tracts.

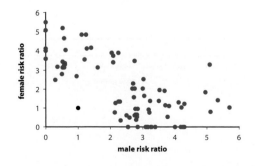

Figure 11: Map of census tracts at high risk, adjusted for social class.

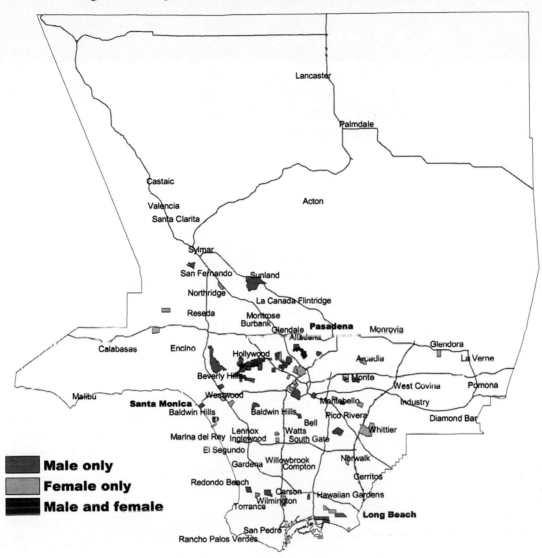

Figure 12: Male-female correlation between the relative risks for high-risk census tracts, adjusted for social class.

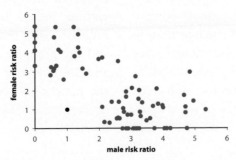

Central Nervous System Non-Hodgkin Lymphomas (All Non-Hodgkin Lymphoma Types)

ICDO-2 Code Anatomic Site: C 70–72
ICDO-2 Code Histology: 9590–9596, 9670–9723
Age: All
Male Cases: 739
Female Cases: 220

Background

See the previous background discussion above for all non-Hodgkin lymphomas. Non-Hodgkin lymphomas of the brain and spinal cord are generally high-grade tumors, often deriving from cells of the immune system normally found in the central nervous system. A large proportion of them are known to occur in persons with AIDS or in the elderly. Like other AIDS-related lymphomas, they have recently become less common among patients who have received the more effective modern antiviral therapy. Early in the period, many central nervous system lymphomas were never biopsied and are therefore of unknown histology.

Local Pattern

These lymphomas occur three to four times as often among men as among women, and more frequently in San Francisco than in Los Angeles County or elsewhere. They increase in frequency in young adulthood, at which age they are most common, and occur in persons of all races and levels of social class. Incidence in the county as a whole had increased in recent decades, but that increase has now flattened out, and this is particularly evident among residents of high-risk census tracts. Figure 6 shows a moderate nonrandom excess of census tracts with unexpectedly few or unexpectedly many cases. A small number of census tracts are seen to be at very high risk. No census tracts meet the high-risk criteria for women. High-risk census tracts are concentrated in contiguous census tracts in West Hollywood, Silver Lake, and Long Beach. The geographical pattern is unchanged after adjustment for social class.

Thumbnail Interpretation

A substantial proportion of the lymphomas occurring in the central nervous system are of the high-grade immature cell variety. Thus the background information, the increased frequency of these lymphomas among men, the high incidence in San Francisco, the frequency among young men, and the presence of high-risk census tracts in and around West Hollywood all suggest that the pattern of occurrence of this malignancy is dominated by susceptibility engendered by AIDS.

Central Nervous System Non-Hodgkin Lymphomas (All Non-Hodgkin Lymphoma Types): Male

Figure 1: Age-adjusted incidence rate by place.

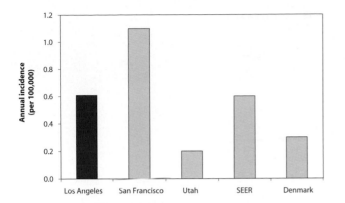

Figure 2: Age-adjusted incidence rate over the period.

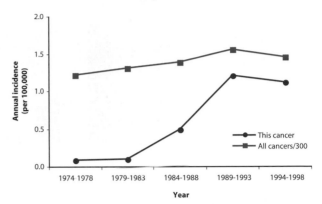

Figure 3: Age-adjusted incidence rate by age and race/ethnicity.

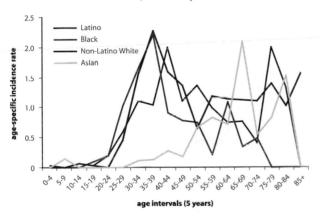

Figure 4: Age-adjusted incidence rate by social class.

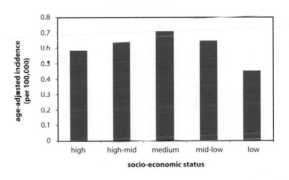

Figure 5: Distribution of the relative risk values for all census tracts.

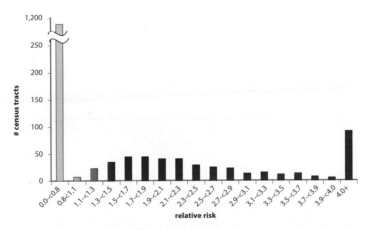

Figure 6: Census tracts by the number of cases per tract.

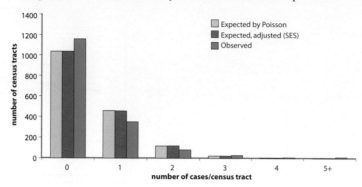

Figure 7a and b: Census tracts at high risk by the number of cases. (a) Unadjusted and (b) adjusted for social class.

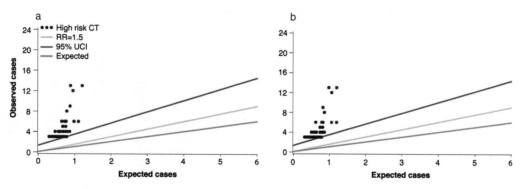

Figure 8: Risk over the period for high-risk census tracts relative to all census tracts.

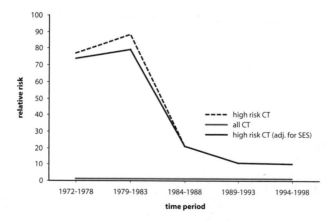

Central Nervous System Non-Hodgkin Lymphomas
(All Non-Hodgkin Lymphoma Types): Female

Figure 1: Age-adjusted incidence rate by place.

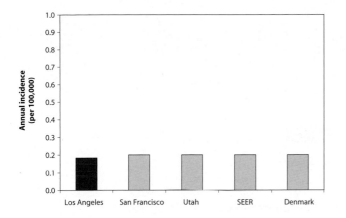

Figure 2: Age-adjusted incidence rate over the period.

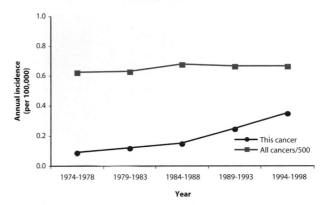

Figure 3: Age-adjusted incidence rate by age and race/ethnicity.

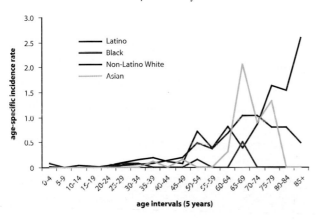

Figure 4: Age-adjusted incidence rate by social class.

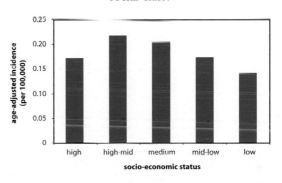

Figure 5: Distribution of the relative risk values for all census tracts.

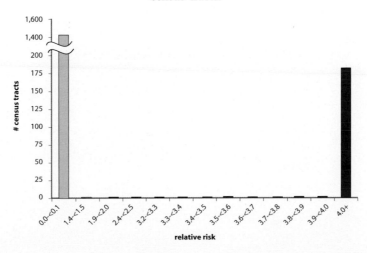

Figure 6: Census tracts by the number of cases per tract.

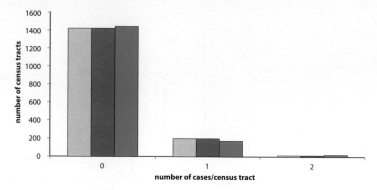

Figure 7a and b: Census tracts at high risk by the number of cases. (a) Unadjusted and (b) adjusted for social class.

There were no high risk census tracts.

Figure 8: Risk over the period for high-risk census tracts relative to all census tracts.

There were no high risk census tracts.

Central Nervous System Non-Hodgkin Lymphomas
(All Non-Hodgkin Lymphoma Types)

Figure 9: Map of census tracts at high risk.

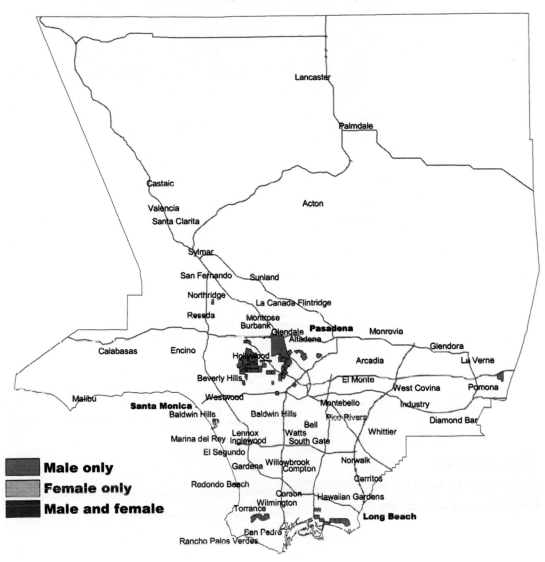

Figure 10: Male-female correlation between the relative risks for high-risk census tracts.

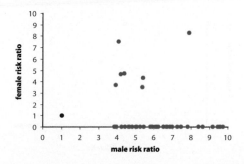

Central Nervous System Non-Hodgkin Lymphomas
(All Non-Hodgkin Lymphoma Types)

Figure 11: Map of census tracts at high risk, adjusted for social class.

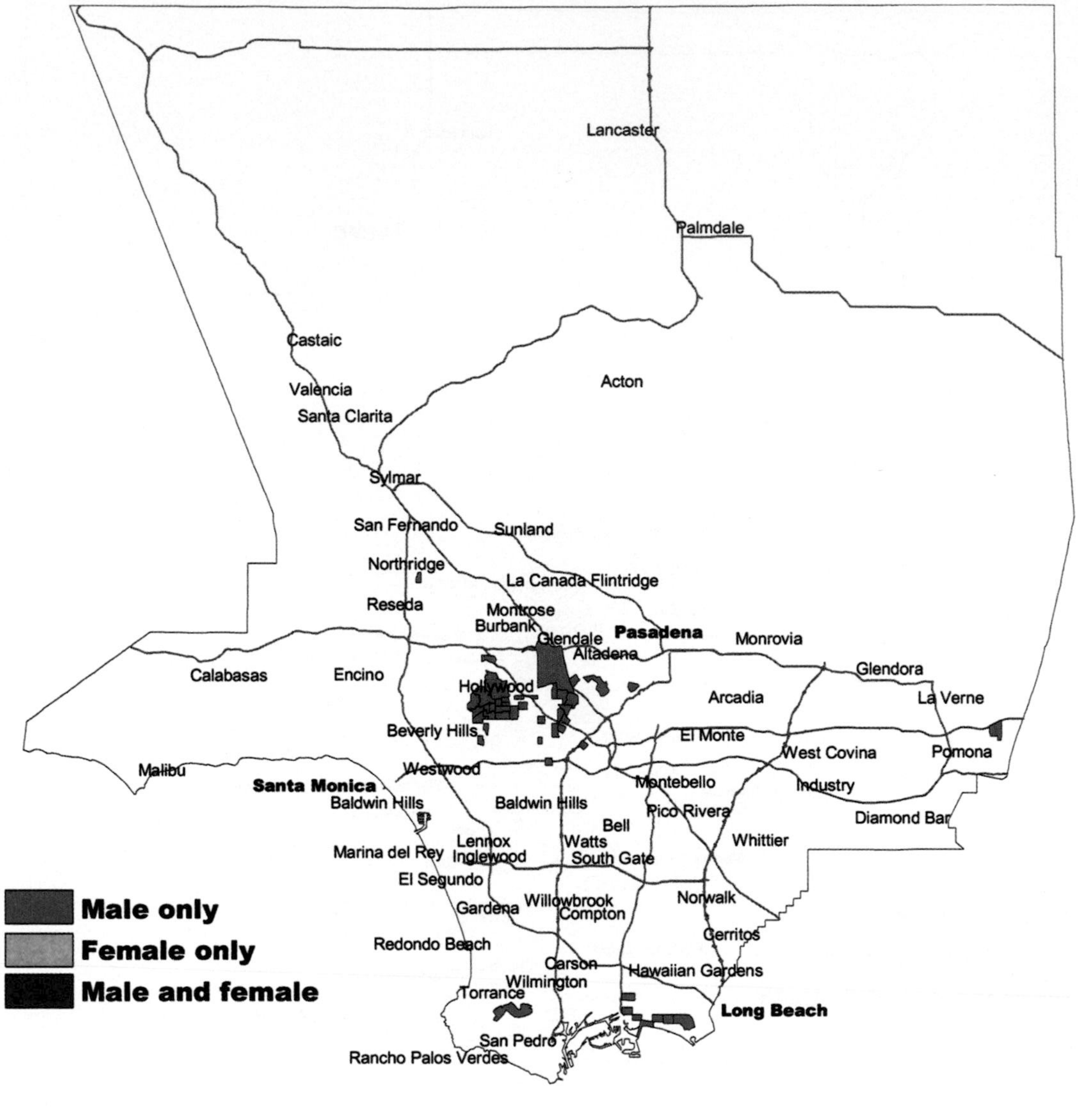

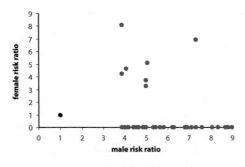

Figure 12: Male-female correlation between the relative risks for high-risk census tracts, adjusted for social class.

Stomach Non-Hodgkin Lymphomas (All Non-Hodgkin Lymphoma Types)

ICDO-2 Code Anatomic Site: C 16
ICDO-2 Code Histology: 9590–9596, 9670–9723
Age: All
Male Cases: 858
Female Cases: 762

Background

See the previous background discussion for all non-Hodgkin lymphomas. Most non-Hodgkin lymphomas of the stomach now are either diffuse large B-cell lymphomas or are placed into the maltoma group. The bacterium *Helicobacter pylori* is thought to be the predominant cause, although other predisposing factors are likely to be important. High-grade lymphomas of the stomach also occur, and their causes are presumed similar to lymphomas of the same histological types occurring elsewhere.

Local Pattern

Stomach lymphomas are more common in men than in women, and occur in all races, with roughly equal frequency in Los Angeles County and other areas. There is a tendency for higher risk in persons of lower social class.

These tumors increased in frequency in the county as a whole in the last several decades, especially among men, although incidence among residents of high-risk census tracts was more constant. Figure 6 shows a slight non-random excess of census tracts with unexpectedly few or unexpectedly many cases. With and without adjustment for census tract, the census tracks at high-risk are scattered over the county in no obvious pattern and few contiguous pairs. No census tract stands out on the basis of a particularly high number of excess cases.

Thumbnail Interpretation

The reasons for the increase over time and the higher incidence among men and those of lower social class are unknown. No systematic pattern of geographical occurrence is apparent, and therefore no local source of causation can be proposed.

Stomach Non-Hodgkin Lymphomas
(All Non-Hodgkin Lymphoma Types): Male

Figure 1: Age-adjusted incidence rate by place.

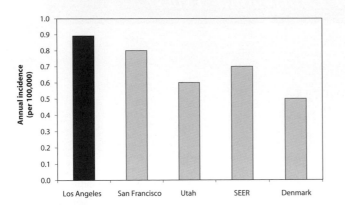

Figure 2: Age-adjusted incidence rate over the period.

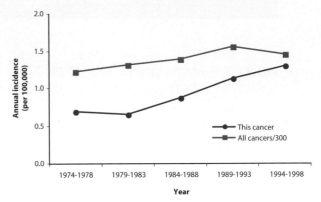

Figure 3: Age-adjusted incidence rate by age and race/ethnicity.

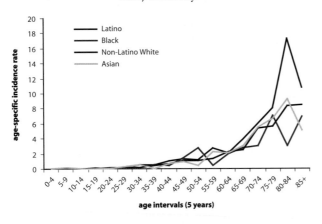

Figure 4: Age-adjusted incidence rate by social class.

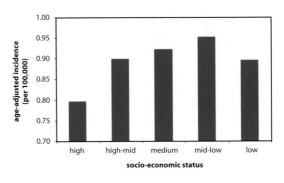

Figure 5: Distribution of the relative risk values for all census tracts.

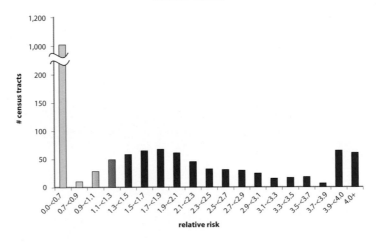

Stomach Non-Hodgkin Lymphomas
(All Non-Hodgkin Lymphoma Types): Male

Figure 6: Census tracts by the number of cases per tract.

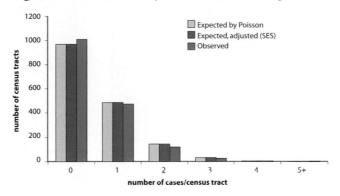

Figure 7a and b: Census tracts at high risk by the number of cases. (a) Unadjusted and (b) adjusted for social class.

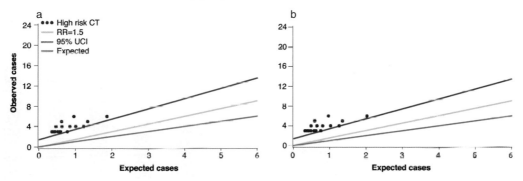

Figure 8: Risk over the period for high-risk census tracts relative to all census tracts.

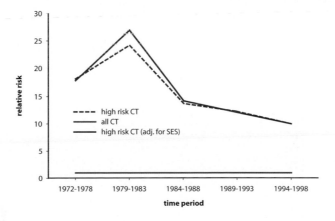

Stomach Non-Hodgkin Lymphomas
(All Non-Hodgkin Lymphoma Types): Female

Figure 1: Age-adjusted incidence rate by place.

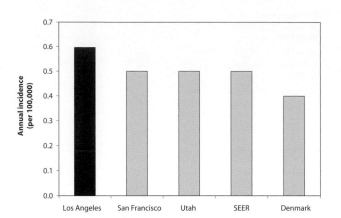

Figure 2: Age-adjusted incidence rate over the period.

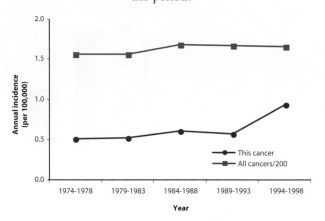

Figure 3: Age-adjusted incidence rate by age and race/ethnicity.

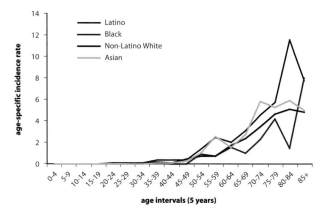

Figure 4: Age-adjusted incidence rate by social class.

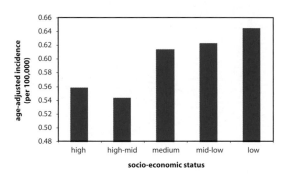

Figure 5: Distribution of the relative risk values for all census tracts.

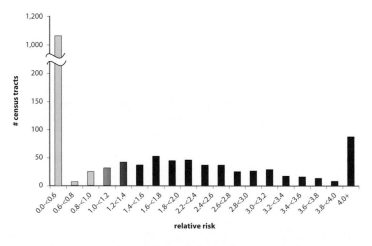

Figure 6: Census tracts by the number of cases per tract.

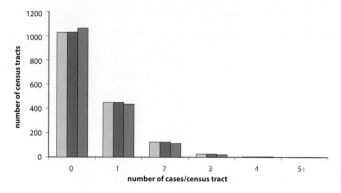

Figure 7a and b: Census tracts at high risk by the number of cases. (a) Unadjusted and (b) adjusted for social class.

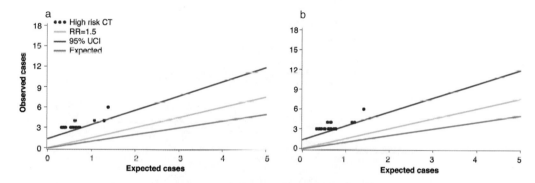

Figure 8: Risk over the period for high-risk census tracts relative to all census tracts.

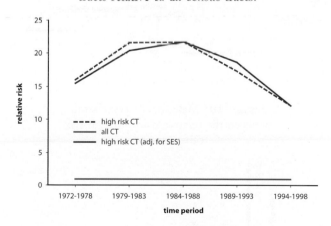

Figure 9: Map of census tracts at high risk.

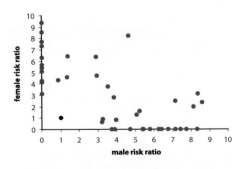

Male only
Female only
Male and female

Figure 10: Male-female correlation
between the relative risks for high-risk
census tracts.

Stomach Non-Hodgkin Lymphomas
(All Non-Hodgkin Lymphoma Types)

Figure 11: Map of census tracts at high risk, adjusted for social class.

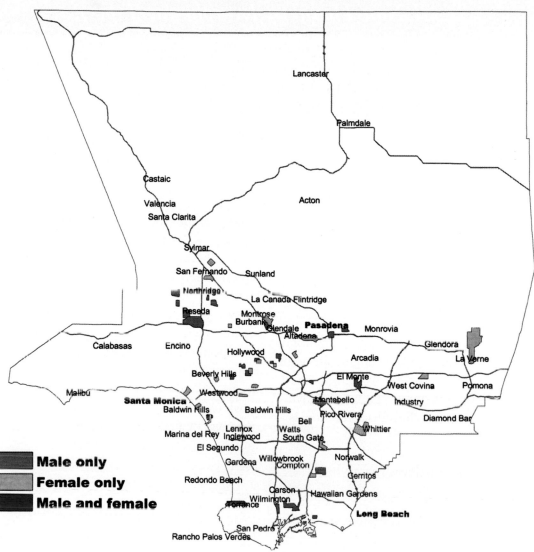

Figure 12: Male-female correlation between the relative risks for high-risk census tracts, adjusted for social class.

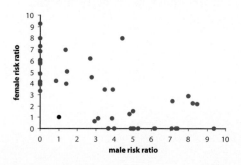

Other Gastrointestinal Non-Hodgkin Lymphomas (All Non-Hodgkin Lymphoma Types)

ICDO-2 Code Anatomic Site: C 15, 17–20
ICDO-2 Code Histology: 9590–9596, 9670–9723
Age: All
Male Cases: 614
Female Cases: 345

Background

See the previous background discussion for all non-Hodgkin lymphomas. This category was separated from non-Hodgkin lymphomas of the stomach because most of these are histologically different. Some lymphomas of the bowel have occurred more commonly in persons of Middle Eastern origin, although the reasons why are obscure. As a group, these tumors include maltomas, high-grade lymphomas, and large B-cell diffuse lymphomas.

Local Pattern

Lymphomas of the lower gastrointestinal tract are about twice as common in men as women and occur as commonly elsewhere as in Los Angeles County. Incidence in different racial groups is roughly similar, and among men there is some tendency for increased risk among those of higher social class. These malignancies begin to increase in frequency in young adulthood. Their occurrence in the county as a whole has increased over time, especially among men, although the rate among men living in high-risk census tracts appears to have been more constant. Figure 6 shows only a slight nonrandom excess of census tracts with unexpectedly few or unexpectedly many cases. Before and after adjustment for social class, the high-risk census tracts are scattered over the county in no obvious pattern. No census tract stands out on the basis of a particularly high relative risk and number of excess cases.

Thumbnail Interpretation

The reason for the higher incidence over time and among men and those of higher social class is unknown. No systematic pattern of geographical occurrence is apparent, and therefore no local source of causation can be proposed.

Figure 1: Age-adjusted incidence rate by place.

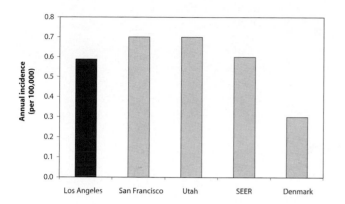

Figure 2: Age-adjusted incidence rate over the period.

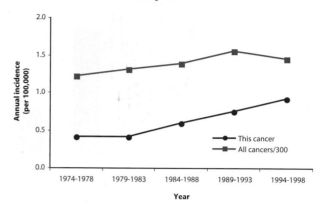

Figure 3: Age-adjusted incidence rate by age and race/ethnicity.

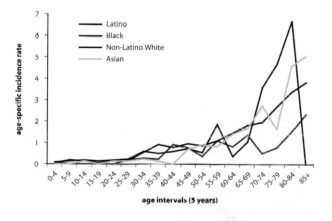

Figure 4: Age-adjusted incidence rate by social class.

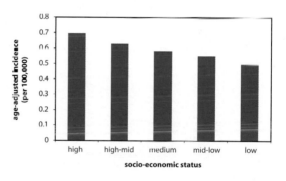

Figure 5: Distribution of the relative risk values for all census tracts.

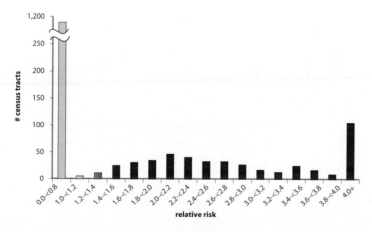

Figure 6: Census tracts by the number of cases per tract.

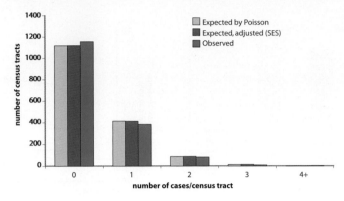

Figure 7a and b: Census tracts at high risk by the number of cases. (a) Unadjusted and (b) adjusted for social class.

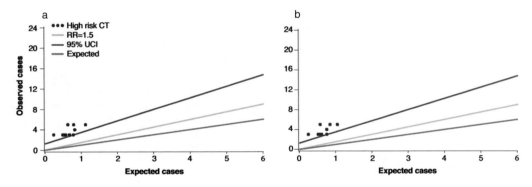

Figure 8: Risk over the period for high-risk census tracts relative to all census tracts.

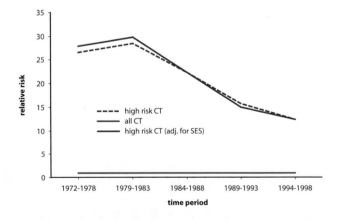

Figure 1: Age-adjusted incidence rate by place.

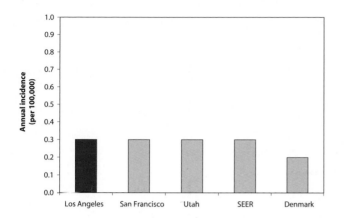

Figure 2: Age-adjusted incidence rate over the period.

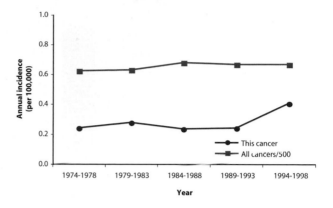

Figure 3: Age-adjusted incidence rate by age and race/ethnicity.

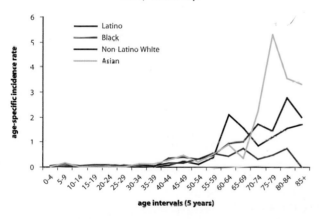

Figure 4: Age-adjusted incidence rate by social class.

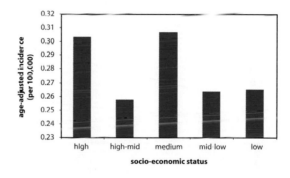

Figure 5: Distribution of the relative risk values for all census tracts.

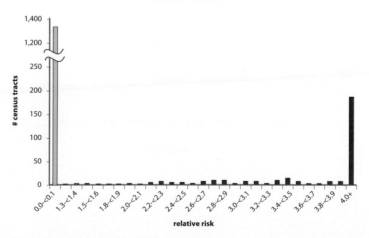

Figure 6: Census tracts by the number of cases per tract.

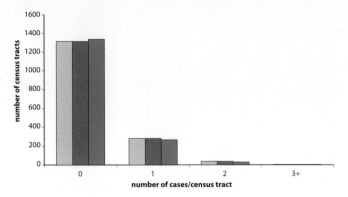

Figure 7a and b: Census tracts at high risk by the number of cases. (a) Unadjusted and (b) adjusted for social class.

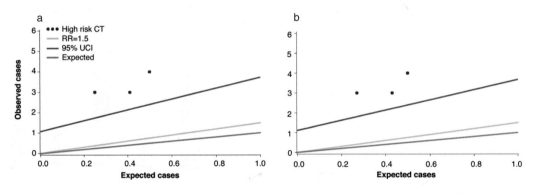

Figure 8: Risk over the period for high-risk census tracts relative to all census tracts.

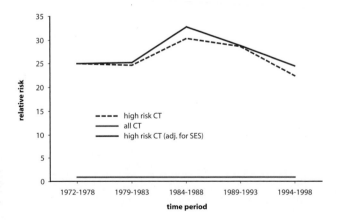

Other Gastrointestinal Non-Hodgkin Lymphomas
(All Non-Hodgkin Lymphoma Types)

Figure 9: Map of census tracts at high risk.

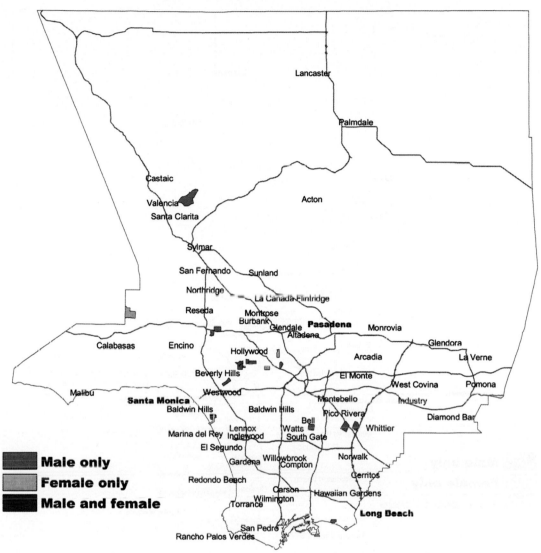

Figure 10: Male-female correlation between the relative risks for high-risk census tracts.

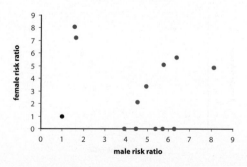

Other Gastrointestinal Non-Hodgkin Lymphomas
(All Non-Hodgkin Lymphoma Types)

Figure 11: Map of census tracts at high risk, adjusted for social class.

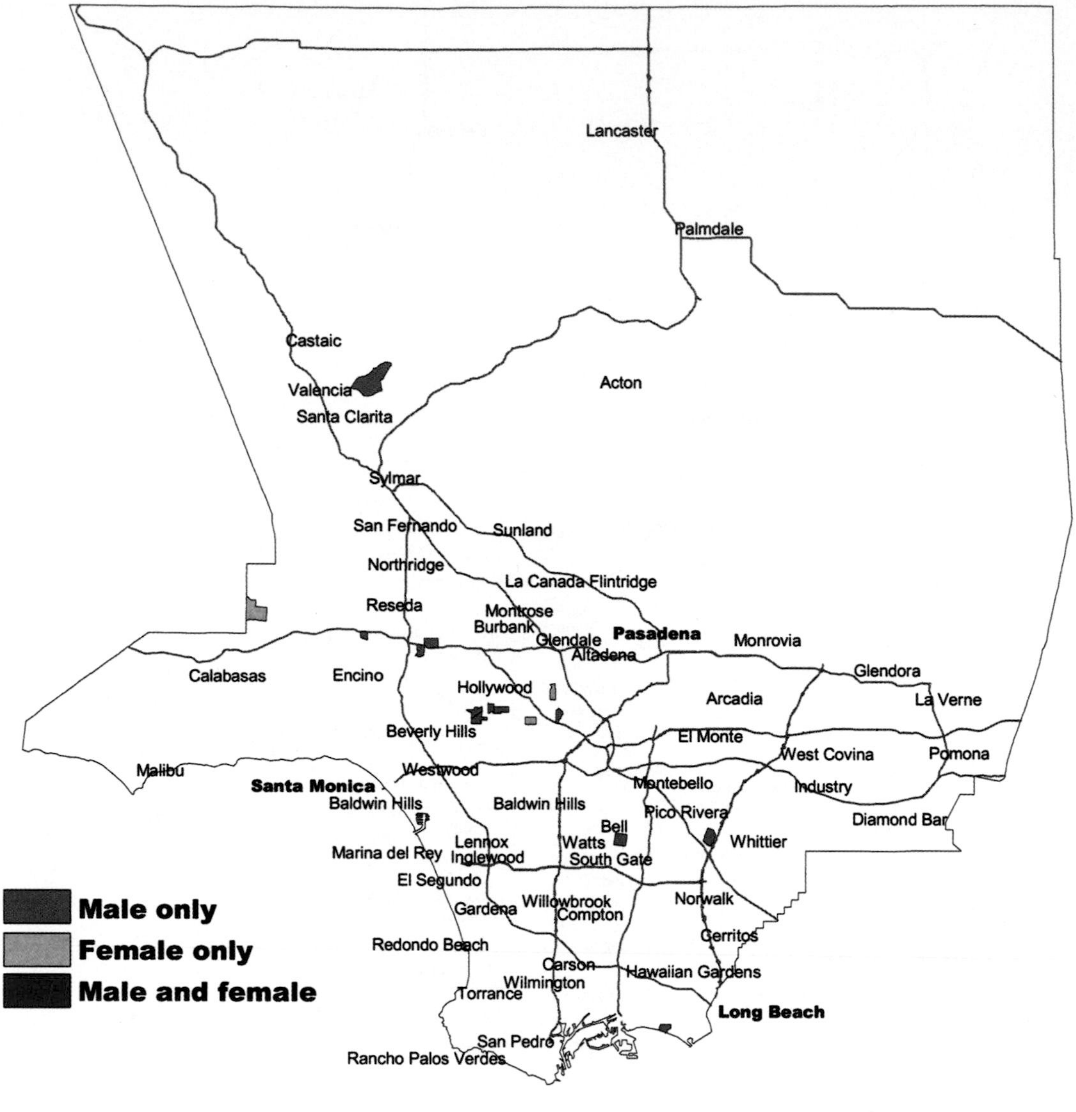

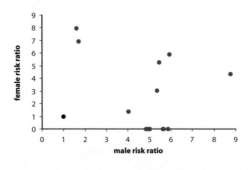

Figure 12: Male-female correlation between the relative risks for high-risk census tracts, adjusted for social class.

Lymph Node/Other Organ Non-Hodgkin Lymphomas (All Non-Hodgkin Lymphoma Types)

ICDO-2 Code Anatomic Site: C 0–14, 21–69, 73–80
ICDO-2 Code Histology: 9590–9596, 9670–9723
Age: All
Male Cases: 12790
Female Cases: 11048

Background

See the previous background discussion for all non-Hodgkin lymphomas. Non-Hodgkin lymphomas generally first appear in a lymph node. This group therefore includes the majority of these malignancies. The causes are as described above for all lymphomas.

Local Pattern

Nodal lymphomas occur about half-again as often in men as women, but with roughly equal frequency in Los Angeles County compared to other regions. Incidence begins to increase in young adulthood. African-Americans and Asian-Americans, especially women, are at lower risk than whites and Latinos, and there is a clear tendency for high risk to be associated with those of higher social class. Until recently, incidence was increasing, both in the county as a whole and among the residents of high-risk census tracts. Figure 6 shows a moderate nonrandom excess of census tracts with unexpectedly few or unexpectedly many cases. High-risk census tracts are scattered over the county, with a large aggregate of contiguous census tracts in a band from Santa Monica through West Hollywood.

Thumbnail Interpretation

This class of lymphomas includes a very large number of those not assigned to any organ system, and on that basis one would expect it to represent an amalgam of tumors of variable histology. The frequency among young men and the presence of high-risk census tracts in and around West Hollywood all suggest that many of these malignancies are determined by susceptibility due to AIDS.

Lymph Node/Other Organ Non-Hodgkin Lymphomas
(All Non-Hodgkin Lymphoma Types): Male

Figure 1: Age-adjusted incidence rate by place.

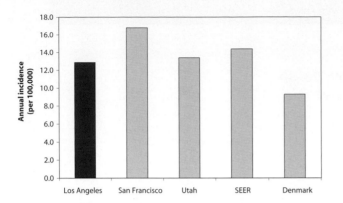

Figure 2: Age-adjusted incidence rate over the period.

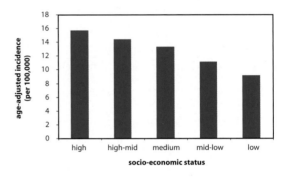

Figure 3: Age-adjusted incidence rate by age and race/ethnicity.

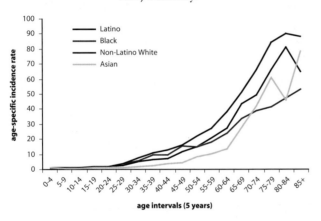

Figure 4: Age-adjusted incidence rate by social class.

Figure 5: Distribution of the relative risk values for all census tracts.

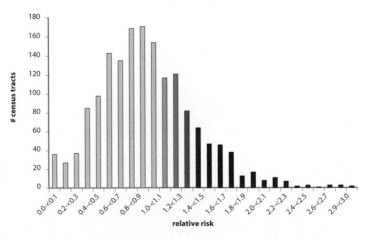

Figure 6: Census tracts by the number of cases per tract.

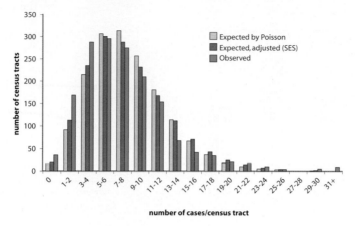

Figure 7a and b: Census tracts at high risk by the number of cases. (a) Unadjusted and (b) adjusted for social class.

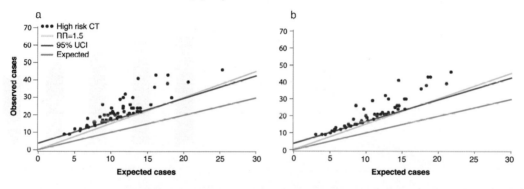

Figure 8: Risk over the period for high-risk census tracts relative to all census tracts.

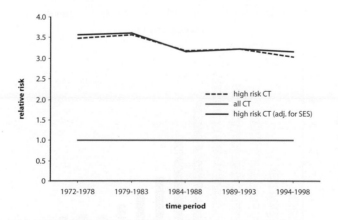

Lymph Node/Other Organ Non-Hodgkin Lymphomas
(All Non-Hodgkin Lymphoma Types): Female

Figure 1: Age-adjusted incidence rate by place.

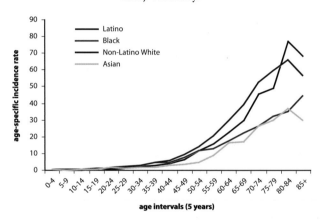

Figure 2: Age-adjusted incidence rate over the period.

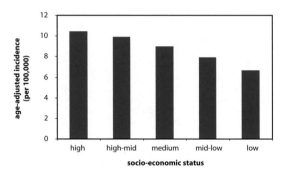

Figure 3: Age-adjusted incidence rate by age and race/ethnicity.

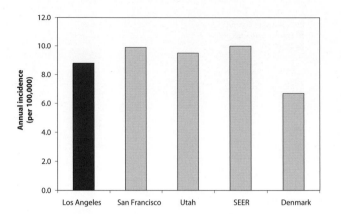

Figure 4: Age-adjusted incidence rate by social class.

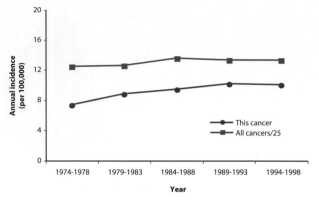

Figure 5: Distribution of the relative risk values for all census tracts.

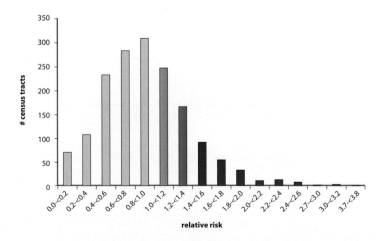

Figure 6: Census tracts by the number of cases per tract.

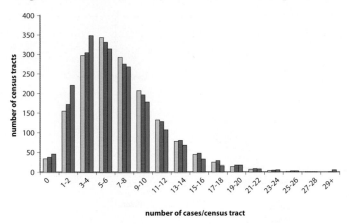

Figure 7a and b: Census tracts at high risk by the number of cases. (a) Unadjusted and (b) adjusted for social class.

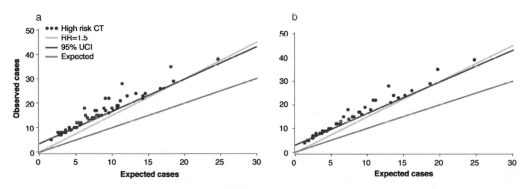

Figure 8: Risk over the period for high-risk census tracts relative to all census tracts.

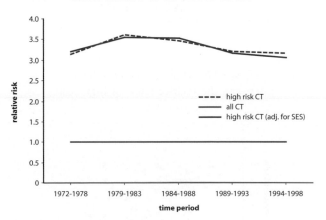

Lymph Node/Other Organ Non-Hodgkin Lymphomas
(All Non-Hodgkin Lymphoma Types)

Figure 9: Map of census tracts at high risk.

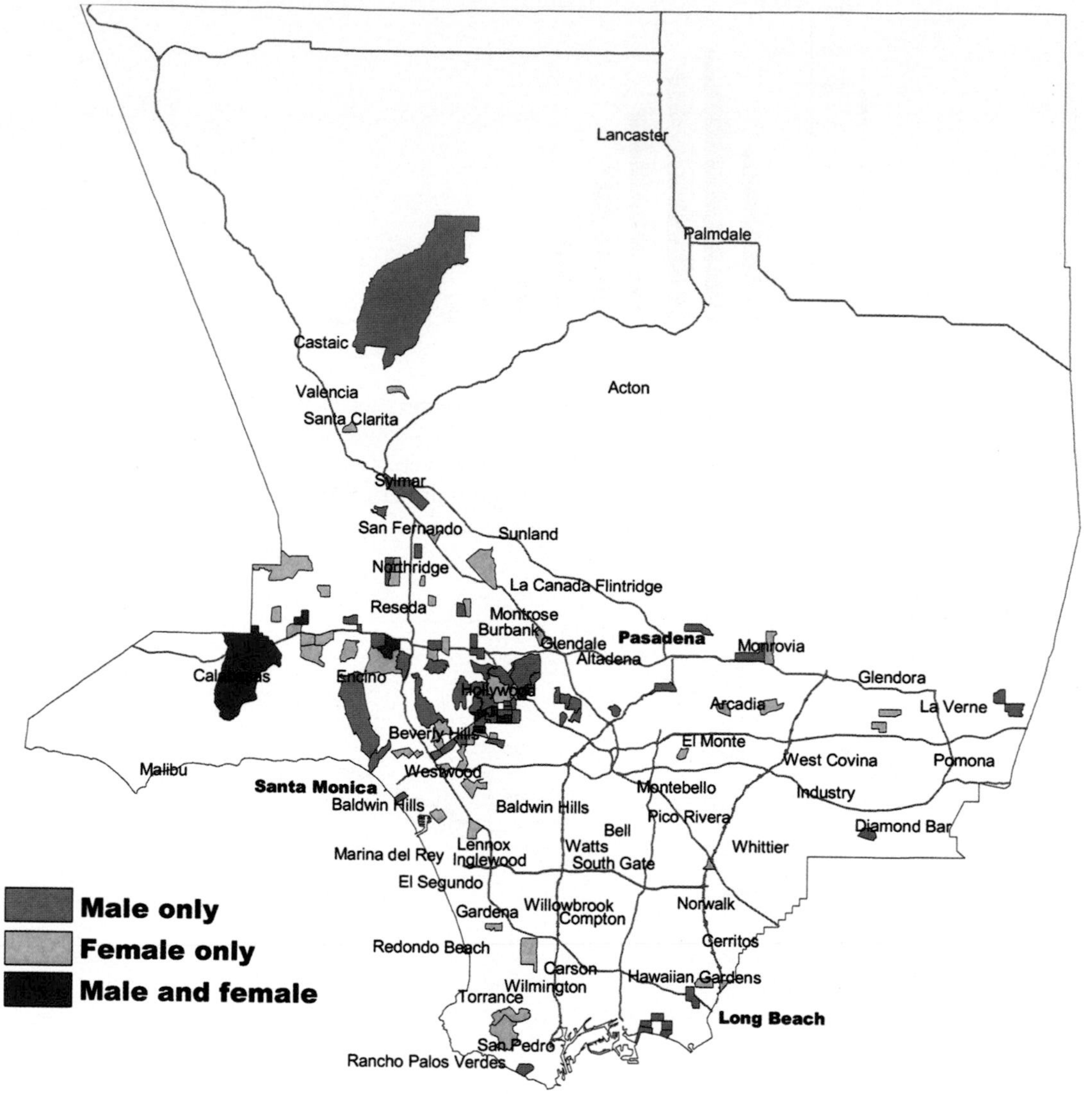

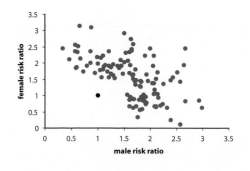

Figure 10: Male-female correlation between the relative risks for high-risk census tracts.

Figure 11: Map of census tracts at high risk, adjusted for social class.

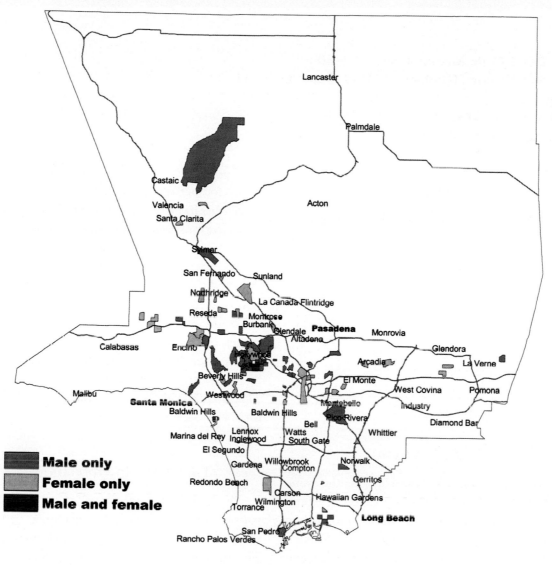

Male only
Female only
Male and female

Figure 12: Male-female correlation between the relative risks for high-risk census tracts, adjusted for social class.

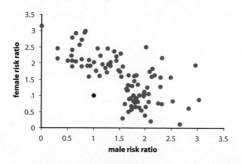

Multiple Myeloma

ICDO-2 Code Anatomic Site: C 0–80
ICDO-2 Code Histology: 9731–9732, 9760–9764
Age: All
Male Cases: 4737
Female Cases: 4305

Background

Despite many suspicions and many studies, no causes of multiple myeloma have been clearly established. Among those that have been suggested, with substantial inconsistency between studies, are chronic antigenic stimulation, ionizing radiation, pesticide and/or other agricultural exposures, and, among women, repeated use of black hair dyes. Genetic determinants are suggested by the large ethnic differences in risk.

Local Pattern

Myeloma occurs with roughly the same frequency in Los Angeles as it does elsewhere, and is somewhat more common among men than among women. The malignancy is much more common among African-Americans and much less common among Asian-Americans than among whites or Latinos. Rates have gradually increased over the period in the county as a whole. Among the residents of high-risk census tracts, however, they temporarily increased at a more rapid rate, then became stable. Although there is a general ten-dency for persons (especially men) of higher social class to be affected, the rate among those of the lowest social class is highest of all. Figure 6 shows a moderate nonrandom excess of census tracts with unexpectedly few or un-expectedly many cases. Although high-risk census tracts are distributed widely throughout the county, there is a notable concentration of census tracts, some of which are contiguous, in the heavily African-American regions of South Central Los Angeles and Baldwin Hills. This is true both before and after adjustment for social class.

Thumbnail Interpretation

Both the demographic and the geographic pattern of occurrence of myeloma reflect the high incidence among African-Americans. The geographic pattern is notable because it includes a number of census tracts in Baldwin Hills, a community that is predominately African-American, but is not of lower social class. Although infections and other reasons for this high incidence have been postulated, a more likely explanation is that of genetic susceptibility.

Figure 1: Age-adjusted incidence rate by place.

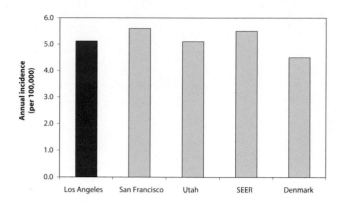

Figure 2: Age-adjusted incidence rate over the period.

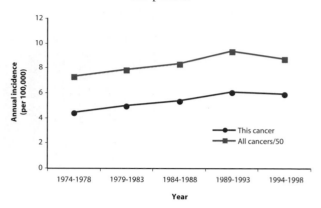

Figure 3: Age-adjusted incidence rate by age and race/ethnicity.

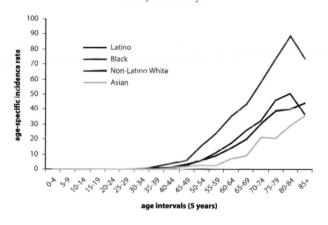

Figure 4: Age-adjusted incidence rate by social class.

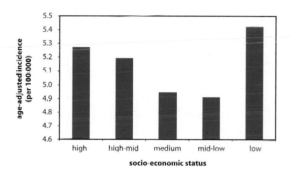

Figure 5: Distribution of the relative risk values for all census tracts.

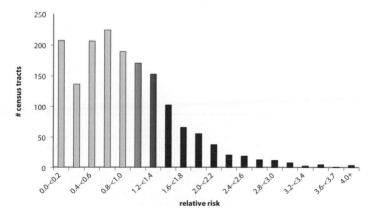

Figure 6: Census tracts by the number of cases per tract.

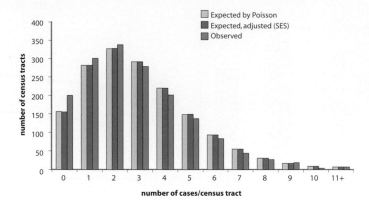

Figure 7a and b: Census tracts at high risk by the number of cases. (a) Unadjusted and (b) adjusted for social class.

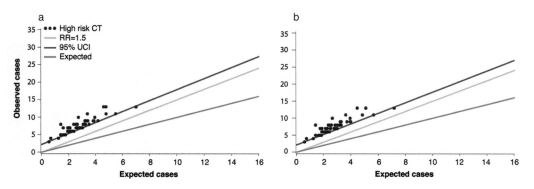

Figure 8: Risk over the period for high-risk census tracts relative to all census tracts.

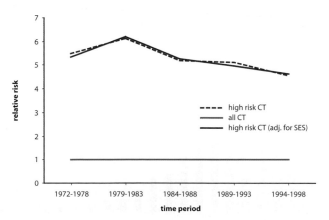

Figure 1: Age-adjusted incidence rate by place.

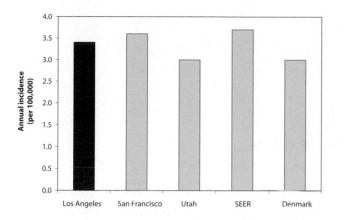

Figure 2: Age-adjusted incidence rate over the period.

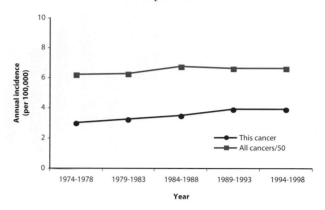

Figure 3: Age-adjusted incidence rate by age and race/ethnicity.

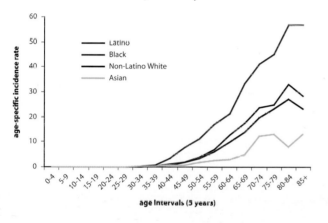

Figure 4: Age-adjusted incidence rate by social class.

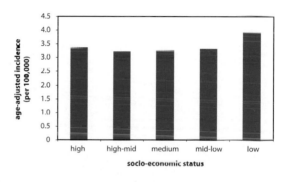

Figure 5: Distribution of the relative risk values for all census tracts.

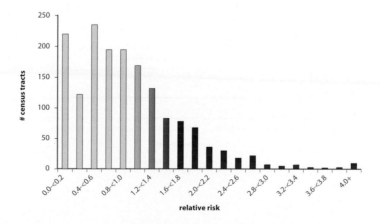

Figure 6: Census tracts by the number of cases per tract.

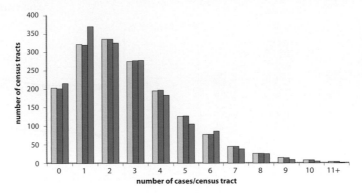

Figure 7a and b: Census tracts at high risk by the number of cases. (a) Unadjusted and (b) adjusted for social class.

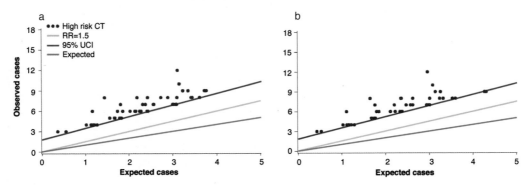

Figure 8: Risk over the period for high-risk census tracts relative to all census tracts.

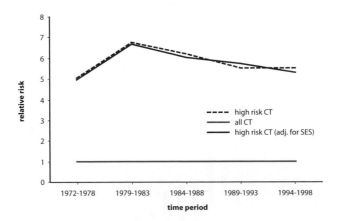

Figure 9: Map of census tracts at high risk.

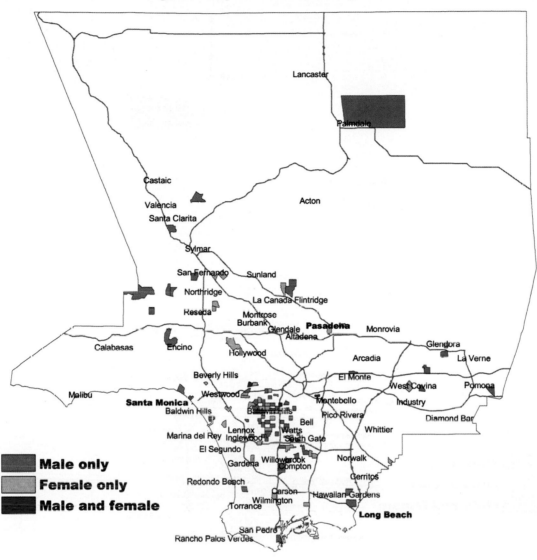

Male only
Female only
Male and female

Figure 10: Male-female correlation between the relative risks for high-risk census tracts.

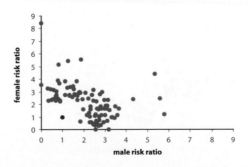

Figure 11: Map of census tracts at high risk, adjusted for social class.

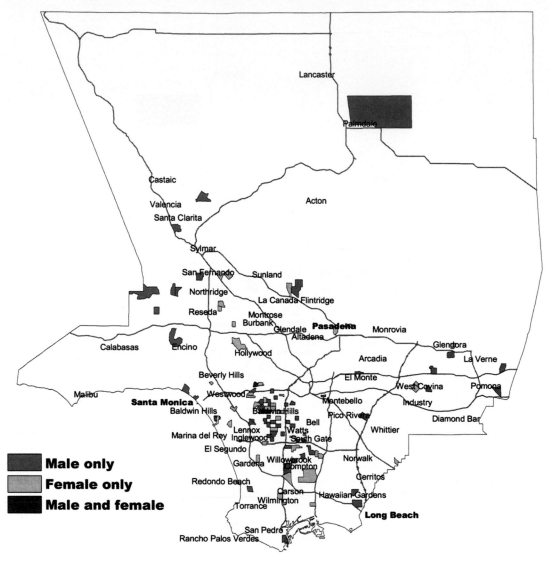

Male only
Female only
Male and female

Figure 12: Male-female correlation
between the relative risks for high-risk
census tracts, adjusted for social class.

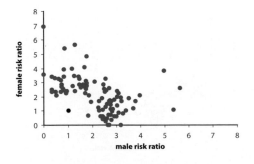

Acute Lymphoblastic Leukemia

ICDO-2 Code Anatomic Site: C 0–80
ICDO-2 Code Histology: 9821–9822, 9824–9826, 9828
Age: All
Male Cases: 1971
Female Cases: 1494

Background

Acute lymphoblastic (lymphocytic, lymphoid) leukemia (ALL) is the most common malignancy between birth and 4 years of age, and for that reason is believed to be caused by exposure to the mother during gestation. The specific causes are not known. Partly on the basis of analogous diseases in animals, the disease has long been thought due to a virus or viruses. An unknown virus is still a leading hypothesis, although the basis is not any evidence of case-to-case transmission, but the observation that incidence of disease increases among the residents of isolated populations after an influx of outsiders likely to carry viruses not already present in the community. Ionizing radiation, as produced by the atomic bomb, has been shown capable of causing cases of this disease as well as other forms of leukemia. ALL occurs more commonly in persons who have certain rare hereditary or congenital conditions. Along with other forms of childhood leukemia, ALL occurs more commonly in persons with Down's syndrome, caused by a chromosome abnormality. The suggestion that it is more common among the children of lower birth weight and those of older mothers also suggests a congenital (not a hereditary) influence. The identical twin of a case is at higher risk of also becoming affected, but this is not strong evidence for a common hereditary cause. Such cases probably occur not because of a hereditary determinant but because leukemic cells may pass from one twin to the other during the mother's pregnancy.

Unspecified pesticides and electromagnetic field exposures have also been suspected of causing ALL, but the evidence is inconsistent. There are also inconsistent reports that various occupational exposures to either the mother or the father might be responsible. ALL among adults has also been observed in association with occupational exposures (rubber, petroleum, agriculture), but these findings are inconsistent.

Local Pattern

ALL occurs at roughly the same frequency in Los Angeles County as in other populations across America and Europe. Incidence decreases from a higher level among toddlers to lower levels among middle-aged adults. Incidence among males is higher than among females. Latinos generally are at highest risk, followed by whites and African-Americans. Incidence does not vary with social class. The number of census tracts meeting the high-risk criteria is not altered by adjustment for social class. Incidence has gradually increased over time in the county as a whole, and to a lesser degree among the residents of high-risk census tracts. Figure 6 shows only a slight nonrandom excess of census tracts with unexpectedly few or unexpectedly many cases. The high-risk census tracts are scattered around the county with no apparent pattern and no apparent explanation for the several small sets of two contiguous census tracts. These are scattered over the county as are the few census tracts in which both males and females are seen to be at high-risk. No census tract stands out on the basis of a particularly high relative risk and number of excess cases.

Thumbnail Interpretation

The pattern of ALL in Los Angeles County provides no evidence of specific causation. No systematic pattern of geographical occurrence is apparent, and therefore no local source of causation can be proposed.

Acute Lymphoblastic Leukemia: Male

Figure 1: Age-adjusted incidence rate by place.

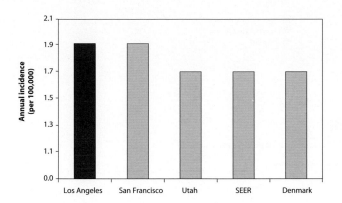

Figure 2: Age-adjusted incidence rate over the period.

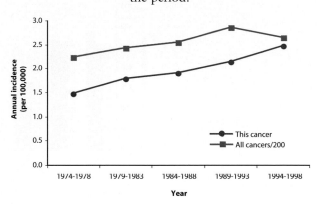

Figure 3: Age-adjusted incidence rate by age and race/ethnicity.

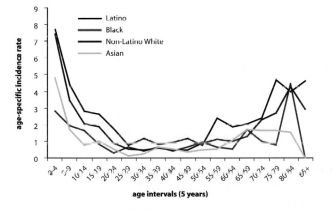

Figure 4: Age-adjusted incidence rate by social class.

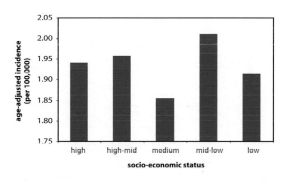

Figure 5: Distribution of the relative risk values for all census tracts.

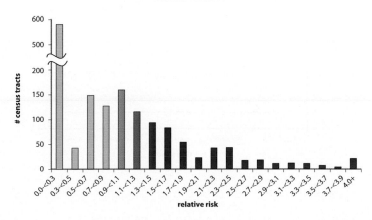

Figure 6: Census tracts by the number of cases per tract.

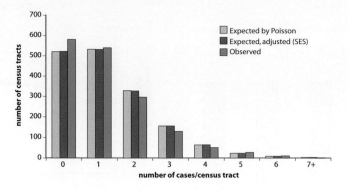

Figure 7a and b: Census tracts at high risk by the number of cases. (a) Unadjusted and (b) adjusted for social class.

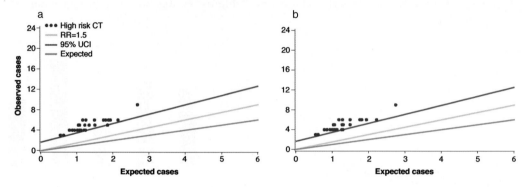

Figure 8: Risk over the period for high-risk census tracts relative to all census tracts.

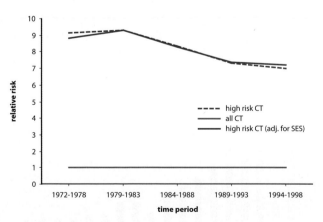

Figure 1: Age-adjusted incidence rate by place.

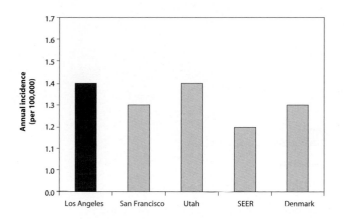

Figure 2: Age-adjusted incidence rate over the period.

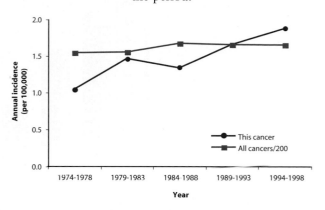

Figure 3: Age-adjusted incidence rate by age and race/ethnicity.

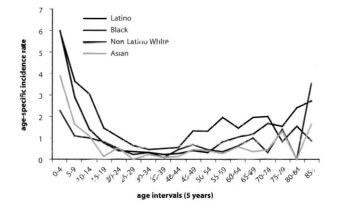

Figure 4: Age-adjusted incidence rate by social class.

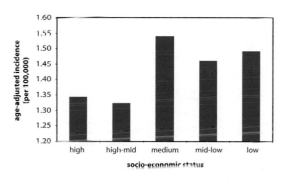

Figure 5: Distribution of the relative risk values for all census tracts.

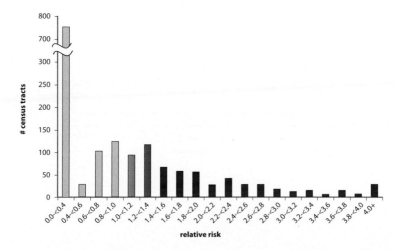

Figure 6: Census tracts by the number of cases per tract.

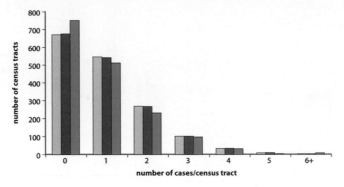

Figure 7a and b: Census tracts at high risk by the number of cases. (a) Unadjusted and (b) adjusted for social class.

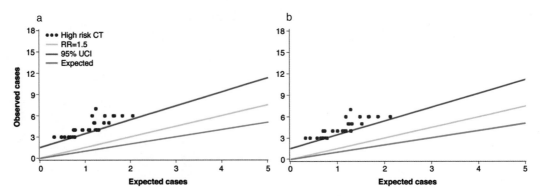

Figure 8: Risk over the period for high-risk census tracts relative to all census tracts.

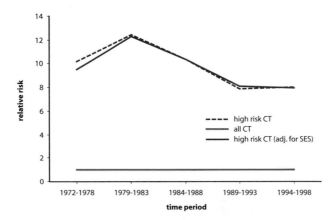

Figure 9: Map of census tracts at high risk.

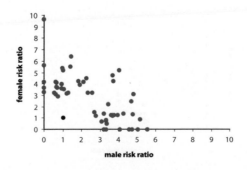

Figure 10: Male-female correlation between the relative risks for high-risk census tracts.

Figure 11: Map of census tracts at high risk, adjusted for social class.

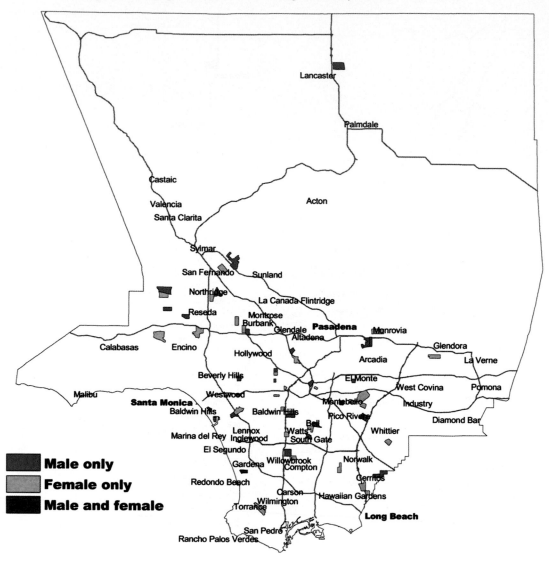

Figure 12: Male-female correlation between the relative risks for high-risk census tracts, adjusted for social class.

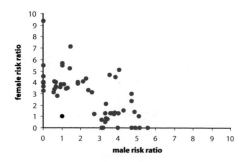

Figure 6: Census tracts by the number of cases per tract.

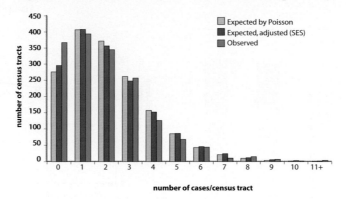

Figure 7a and b: Census tracts at high risk by the number of cases. (a) Unadjusted and (b) adjusted for social class.

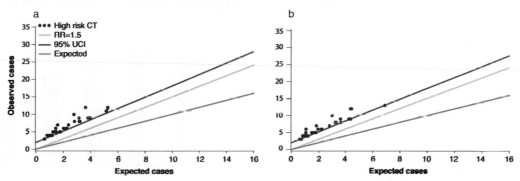

Figure 8: Risk over the period for high-risk census tracts relative to all census tracts.

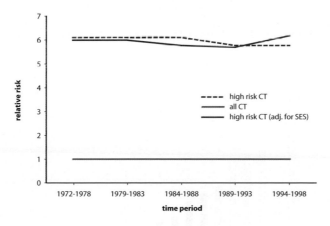

Chronic Lymphocytic Leukemia: Female

Figure 1: Age-adjusted incidence rate by place.

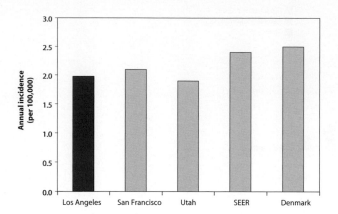

Figure 2: Age-adjusted incidence rate over the period.

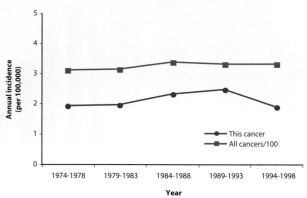

Figure 3: Age-adjusted incidence rate by age and race/ethnicity.

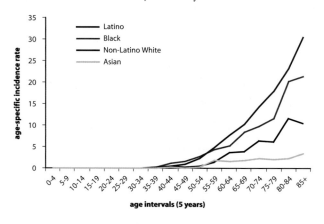

Figure 4: Age-adjusted incidence rate by social class.

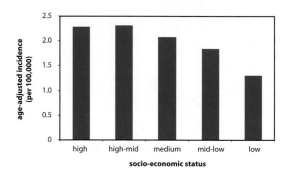

Figure 5: Distribution of the relative risk values for all census tracts.

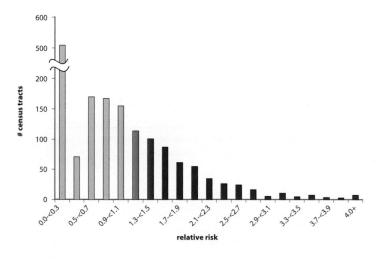

Figure 6: Census tracts by the number of cases per tract.

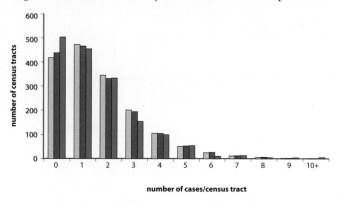

Figure 7a and b: Census tracts at high risk by the number of cases. (a) Unadjusted and (b) adjusted for social class.

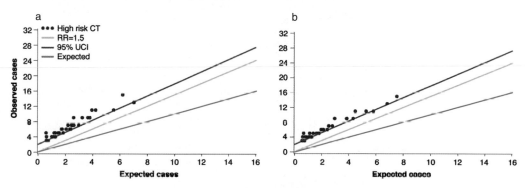

Figure 8: Risk over the period for high-risk census tracts relative to all census tracts.

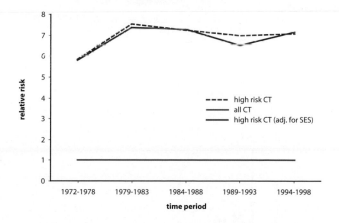

Figure 9: Map of census tracts at high risk.

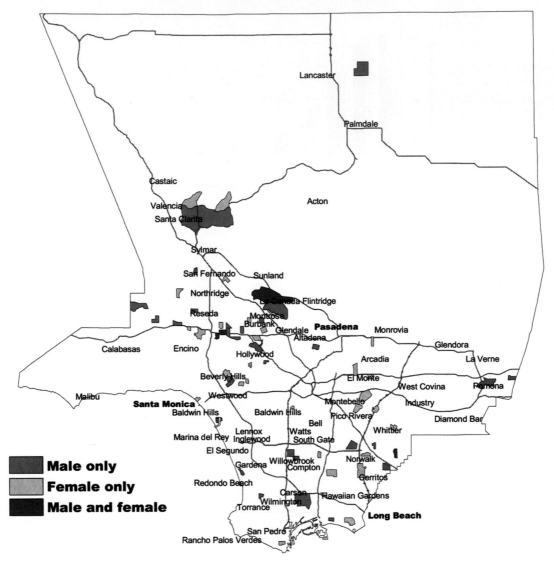

Figure 10: Male-female correlation between the relative risks for high-risk census tracts.

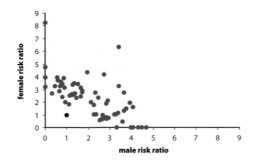

Figure 11: Map of census tracts at high risk, adjusted for social class.

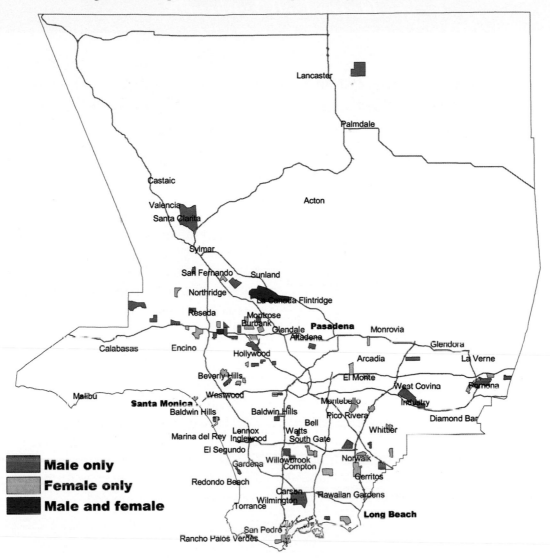

Figure 12: Male-female correlation between the relative risks for high-risk census tracts, adjusted for social class.

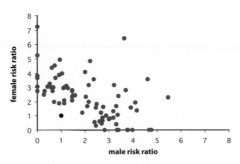

Hairy Cell Leukemia

ICDO-2 Code Anatomic Site: C 0–80
ICDO-2 Code Histology: 9940
Age: All
Male Cases: 336
Female Cases: 88

Background

Some cases of this rare leukemia occur after infection with a specific virus, human T-cell leukemia/lymphoma virus II. Hairy cell leukemia has been reported to be more common among Jewish men. However, the causes of most cases are obscure.

Local Pattern

Hairy cell leukemia occurs with equal frequency in Los Angeles and other areas, and is more common in men than women. It begins to occur more frequently in young adulthood and has occurred at a constant rate over time in both sexes in the county as a whole. It is diagnosed slightly more often among those of upper social class. Figure 6 shows no nonrandom excess of census tracts with unexpectedly few or unexpectedly many cases. The high-risk criteria were fulfilled only by two widely separated census tracts.

Thumbnail Interpretation

The reasons for increased incidence among males and among those of higher social class are unknown. No systematic pattern of occurrence is apparent, and therefore no local source of causation can be proposed.

Hairy Cell Leukemia: Male

Figure 1: Age-adjusted incidence rate by place.

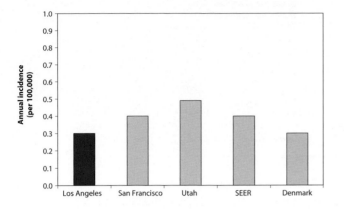

Figure 2: Age-adjusted incidence rate over the period.

There were no high risk census tracts.

Figure 3: Age-adjusted incidence rate by age and race/ethnicity.

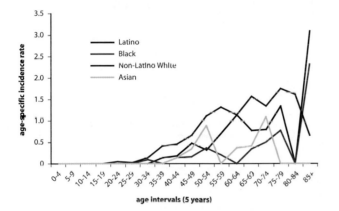

Figure 4: Age-adjusted incidence rate by social class.

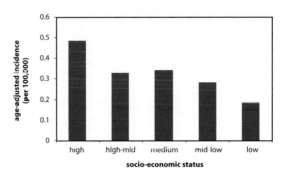

Figure 5: Distribution of the relative risk values for all census tracts.

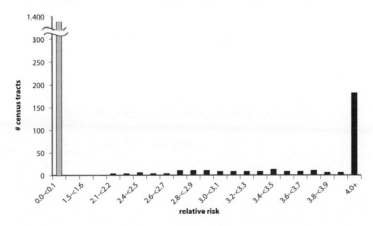

Figure 6: Census tracts by the number of cases per tract.

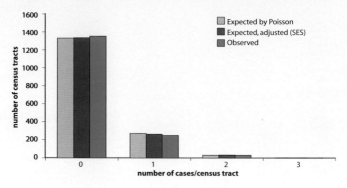

Figure 7a and b: Census tracts at high risk by the number of cases. (a) Unadjusted and (b) adjusted for social class.

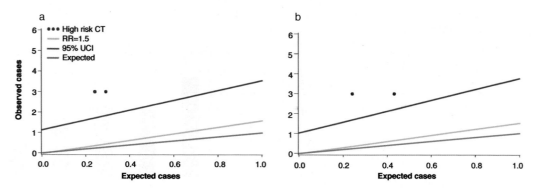

Figure 8: Risk over the period for high-risk census tracts relative to all census tracts.

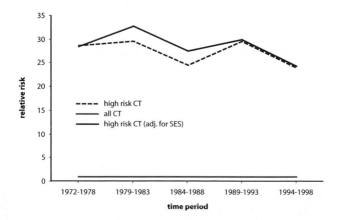

Figure 1: Age-adjusted incidence rate by place.

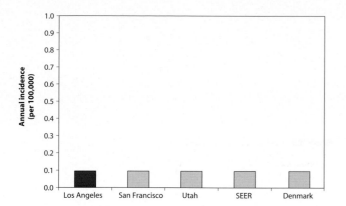

Figure 2: Age-adjusted incidence rate over the period.

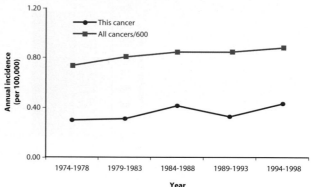

Figure 3: Age-adjusted incidence rate by age and race/ethnicity.

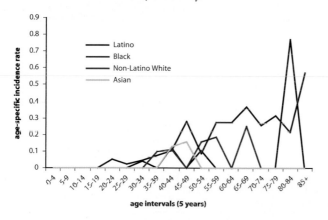

Figure 4: Age-adjusted incidence rate by social class.

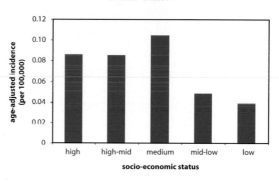

Figure 5: Distribution of the relative risk values for all census tracts.

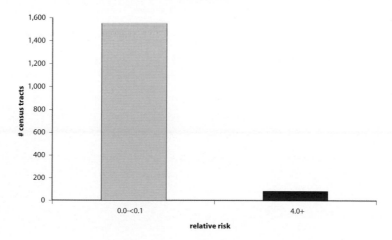

Figure 6: Census tracts by the number of cases per tract.

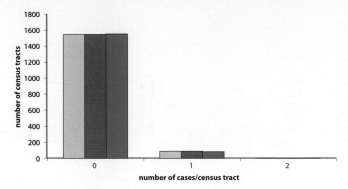

Figure 7a and b: Census tracts at high risk by the number of cases. (a) Unadjusted and (b) adjusted for social class.

There were no high risk census tracts.

Figure 8: Risk over the period for high-risk census tracts relative to all census tracts.

There were no high risk census tracts.

Figure 9: Map of census tracts at high risk.

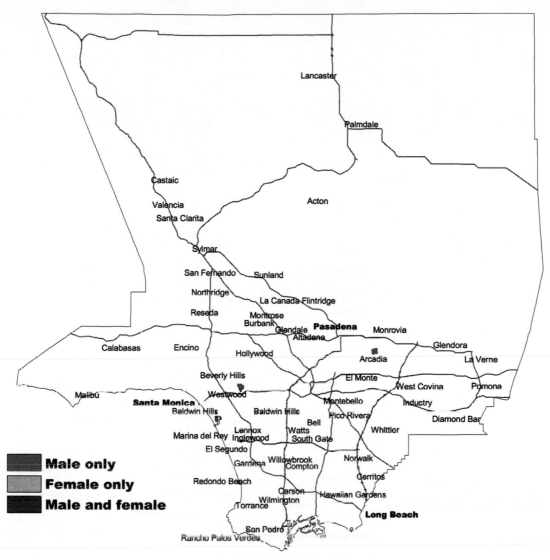

Figure 10: Male-female correlation between the relative risks for high-risk census tracts.

There were no high risk census tracts.

Figure 11: Map of census tracts at high risk, adjusted for social class.

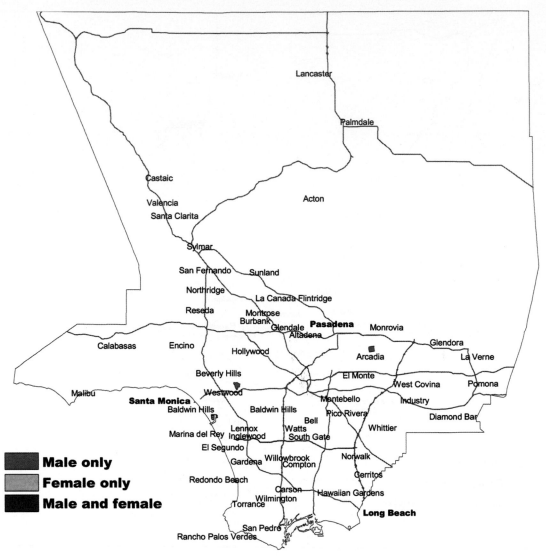

Figure 12: Male-female correlation between the relative risks for high-risk census tracts, adjusted for social class.

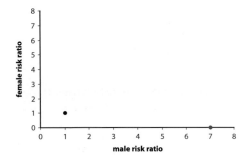

Acute Non-Lymphocytic Leukemia

ICDO-2 Code Anatomic Site: C 0–80
ICDO-2 Code Histology: 9860–9862, 9864–9867, 9870–9874, 9880, 9890–9892, 9894
Age: All
Male Cases: 3751
Female Cases: 3337

Background

Acute non-lymphocytic leukemia (ANLL) is a grouping of several different leukemias. The cells responsible for these malignancies are closely related, and causation of each of the leukemias appears to result from the same set of agents. These leukemias are those most clearly linked to the environment. They are caused by ionizing radiation, either in the form of the atomic bomb or in the form of medical radiotherapy. They can be caused by the powerful chemotherapy that is used, often together with radiation, to treat other forms of cancer. They also have been caused by exposure to benzene, and since benzene is found in cigarette smoke, it is this association that is probably responsible for the fact that this set of leukemias has also been linked to cigarette smoking. ANLL has been suspected, but not proven, to appear more commonly after certain other diverse workplace exposures, such as those of embalmers, foundrymen, underground miners, and hairdressers.

Local Pattern

ANLL is diagnosed less commonly in Utah than in Los Angeles and San Francisco, and is more common among men than among women. It is also more common among older than among young or middle-aged adults, and is more common among whites than among persons of other racial/ethnic groups. ANLL incidence, especially among women, is higher among those of higher social class. Incidence rates have been stable over time, both in the county as a whole and among the residents of high-risk census tracts. Figure 6 shows only a slight nonrandom excess of census tracts with unexpectedly few or unexpectedly many cases. High-risk census tracts are scattered throughout the county, and few are contiguous to one another. No census tract stands out on the basis of a particularly high number of excess cases.

Thumbnail Interpretation

The reason for the higher frequency among men of higher social class is unknown. Benzene is one of the compounds that can be measured at very low dose with modern technology, and it can be identified almost anywhere in the county. No locality is subject to toxic levels from pollution even as high as the level emanating from a lit cigarette. No systematic pattern of geographical occurrence is apparent, and no local source of causation can be proposed.

Figure 1: Age-adjusted incidence rate by place.

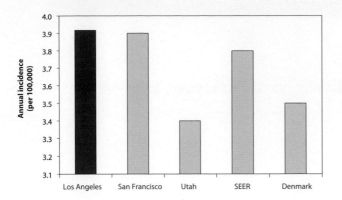

Figure 2: Age-adjusted incidence rate over the period.

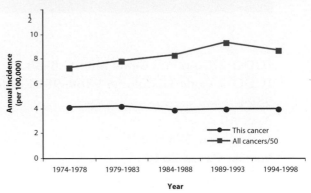

Figure 3: Age-adjusted incidence rate by age and race/ethnicity.

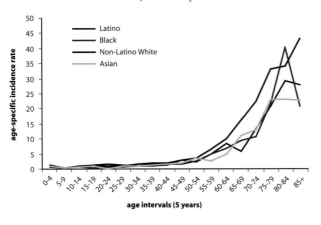

Figure 4: Age-adjusted incidence rate by social class.

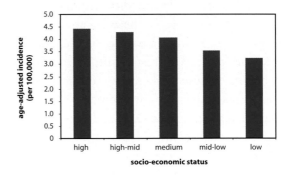

Figure 5: Distribution of the relative risk values for all census tracts.

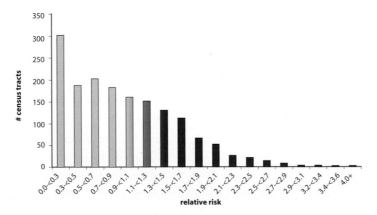

Figure 6: Census tracts by the number of cases per tract.

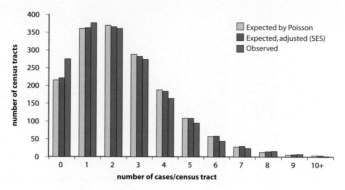

Figure 7a and b: Census tracts at high risk by the number of cases. (a) Unadjusted and (b) adjusted for social class.

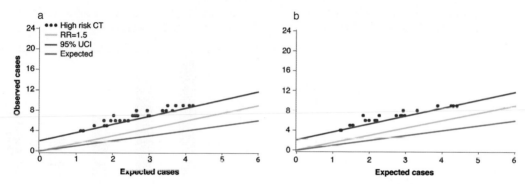

Figure 8: Risk over the period for high-risk census tracts relative to all census tracts.

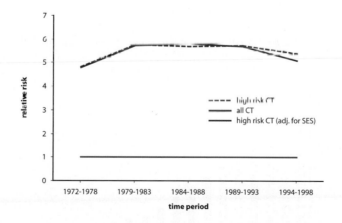

Acute Non-Lymphocytic Leukemia: Female

Figure 1: Age-adjusted incidence rate by place.

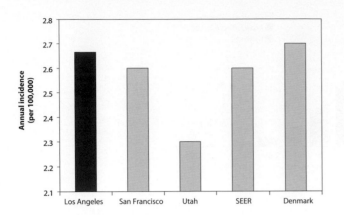

Figure 2: Age-adjusted incidence rate over the period.

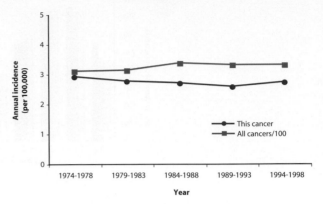

Figure 3: Age-adjusted incidence rate by age and race/ethnicity.

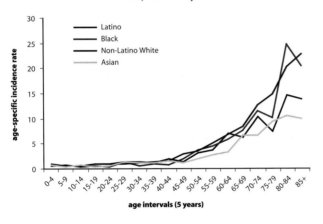

Figure 4: Age-adjusted incidence rate by social class.

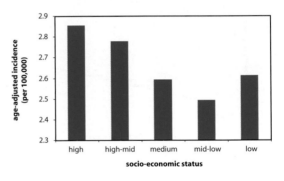

Figure 5: Distribution of the relative risk values for all census tracts.

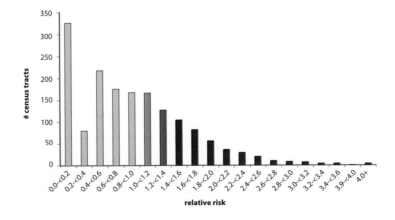

Figure 6: Census tracts by the number of cases per tract.

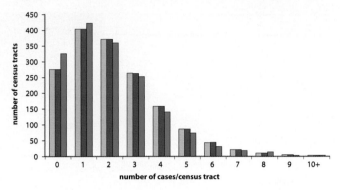

Figure 7a and b: Census tracts at high risk by the number of cases. (a) Unadjusted and (b) adjusted for social class.

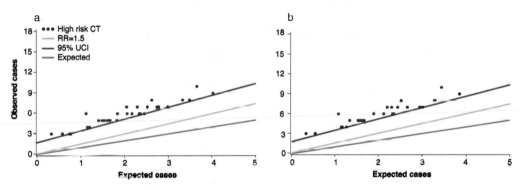

Figure 8: Risk over the period for high-risk census tracts relative to all census tracts.

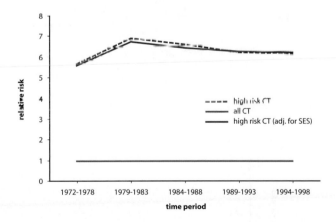

Figure 9: Map of census tracts at high risk.

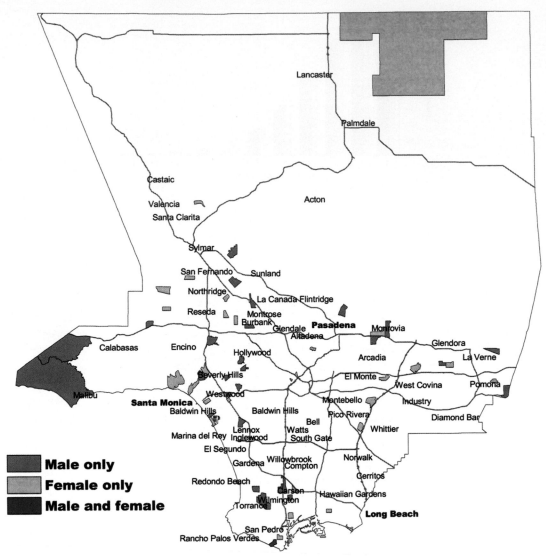

Figure 10: Male-female correlation between the relative risks for high-risk census tracts.

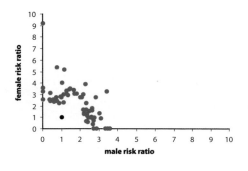

Figure 11: Map of census tracts at high risk, adjusted for social class.

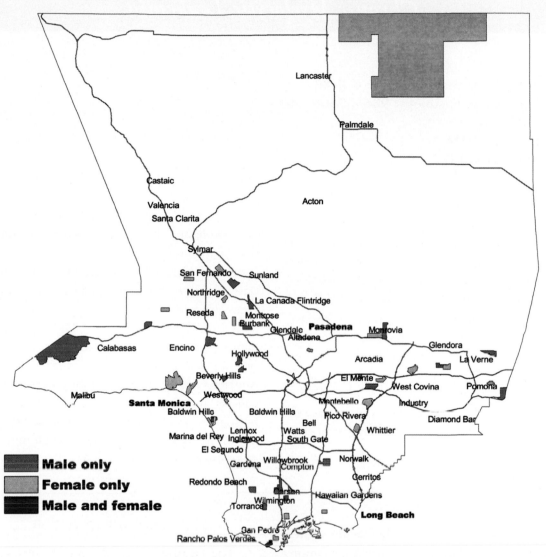

Figure 12: Male-female correlation between the relative risks for high-risk census tracts, adjusted for social class.

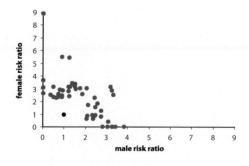

Chronic Myelocytic Leukemia

ICDO-2 Code Anatomic Site: C 0–80
ICDO-2 Code Histology: 9863, 9868, 9893
Age: All
Male Cases: 1924
Female Cases: 1472

Background

Chronic myelocytic (myeloid) leukemia (CML) is marked by the unusual appearance of the affected cells (containing the characteristic "Philadelphia" chromosome). Although many have reported this condition in connection with certain occupational and other environmental causes, individual findings are very inconsistent. Like ANLL, ionizing radiation and treatment with highly toxic drugs used to treat cancer and other chronic diseases are the environmental exposures that have been commonly related to this disease.

Local Pattern

Incidence of CML is no higher among residents of Los Angeles County than among those of other regions of the country. Incidence begins rising in young adulthood, and is generally somewhat higher among men than among women. Incidence levels have been very consistent over time, both in the county as a whole and among the residents of high-risk census tracts. Incidence does not vary according to racial/ethnicity group or social class. Figure 6 shows only a slight nonrandom excess of census tracts with unexpectedly few or unexpectedly many cases. Only a modest number of census tracts meet the high-risk criteria, and they are scattered all over Los Angeles County with few contiguous combinations and few demonstrating high-risk for both males and females. No census tract stands out on the basis of a particularly high number of excess cases.

Thumbnail Interpretation

The reason for the higher incidence among men is unknown. No systematic pattern of geographical occurrence is apparent, and therefore no local source of causation can be proposed.

Figure 1: Age-adjusted incidence rate by place.

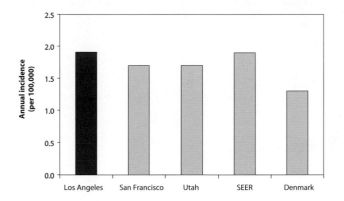

Figure 2: Age-adjusted incidence rate over the period.

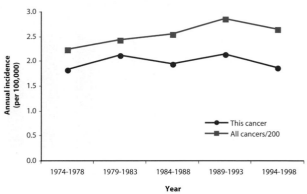

Figure 3: Age-adjusted incidence rate by age and race/ethnicity.

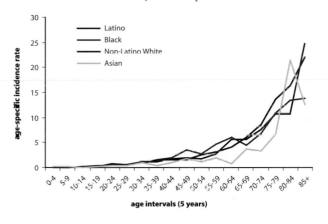

Figure 4: Age-adjusted incidence rate by social class.

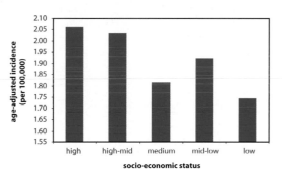

Figure 5: Distribution of the relative risk values for all census tracts.

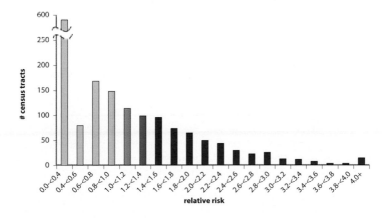

Figure 6: Census tracts by the number of cases per tract.

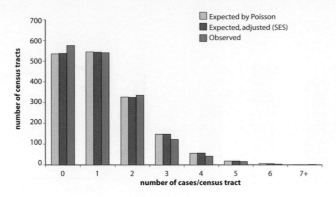

Figure 7a and b: Census tracts at high risk by the number of cases. (a) Unadjusted and (b) adjusted for social class.

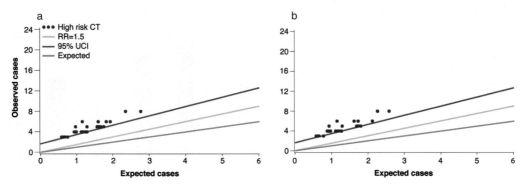

Figure 8: Risk over the period for high-risk census tracts relative to all census tracts.

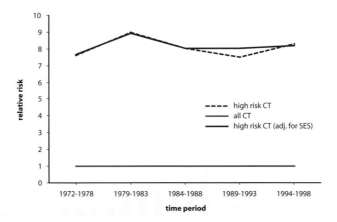

Figure 1: Age-adjusted incidence rate by place.

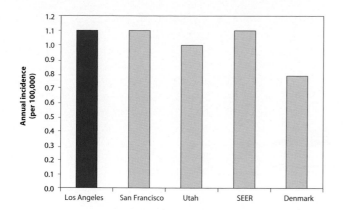

Figure 2: Age-adjusted incidence rate over the period.

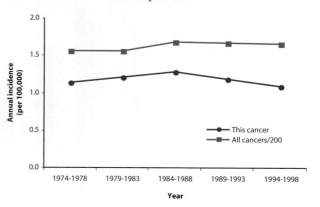

Figure 3: Age-adjusted incidence rate by age and race/ethnicity.

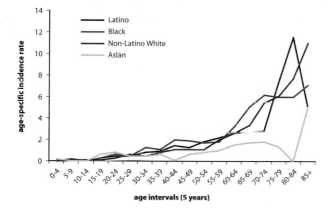

Figure 4: Age-adjusted incidence rate by social class.

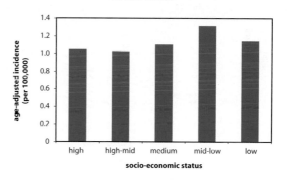

Figure 5: Distribution of the relative risk values for all census tracts.

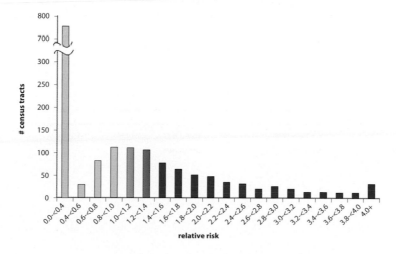

Figure 6: Census tracts by the number of cases per tract.

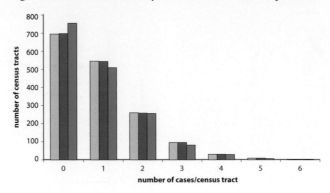

Figure 7a and b: Census tracts at high risk by the number of cases. (a) Unadjusted and (b) adjusted for social class.

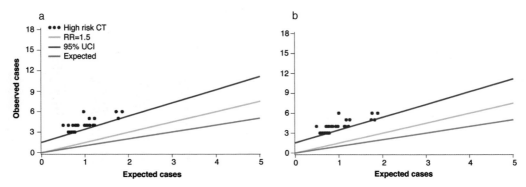

Figure 8: Risk over the period for high-risk census tracts relative to all census tracts.

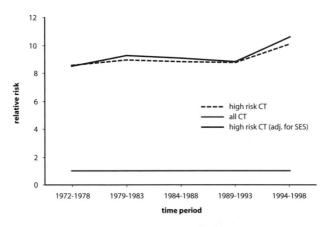

Figure 9: Map of census tracts at high risk.

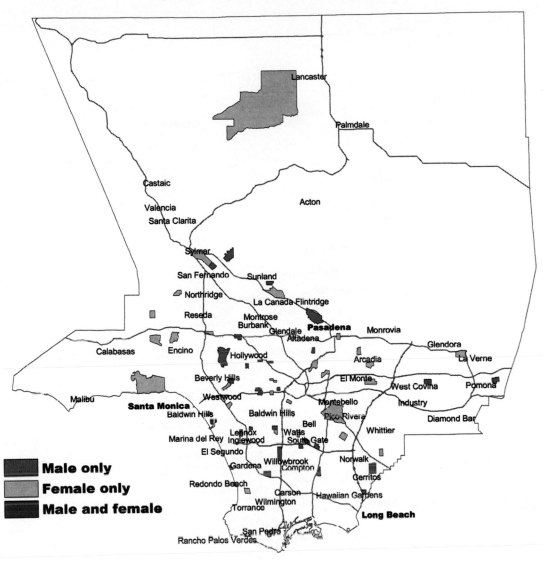

Figure 10: Male-female correlation between the relative risks for high-risk census tracts.

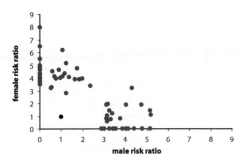

Figure 11: Map of census tracts at high risk, adjusted for social class.

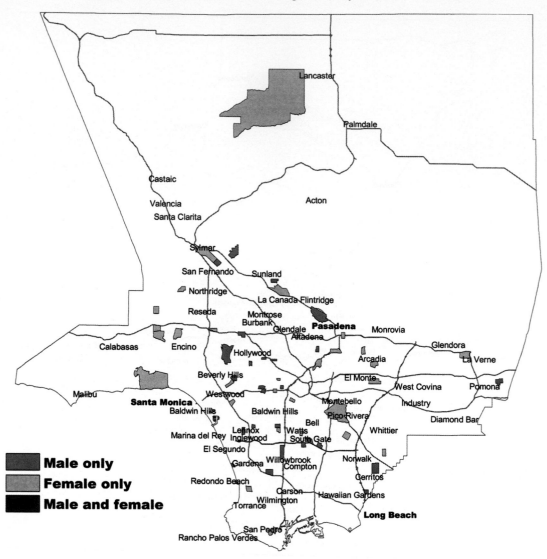

Figure 12: Male-female correlation between the relative risks for high-risk census tracts, adjusted for social class.

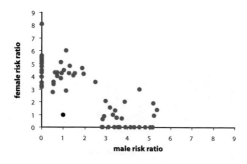

Other Leukemias

ICDO-2 Code Anatomic Site: C 0–80
ICDO-2 Code Histology: 9800–9804, 9827, 9830, 9840–9842, 9850, 9900, 9910, 9930–9932
Age: All
Male Cases: 863
Female Cases: 757

Background

Other leukemias are diverse in cellular origin, in clinical features, and presumably in etiology. This group includes a hodge-podge of many incompletely described cases and some specific entities, such as the adult T-cell leukemias caused by the human T-cell leukemia/lymphoma virus I.

Local Pattern

This miscellaneous group of leukemias is twice as common among men as among women, and as a group, these malignancies appear to be less common among residents of Los Angeles County than among populations in other parts of the country. The malignancies occur with equal frequency among the members of all racial/ethnicity groups and all social classes. Cases in the county as a whole have occurred with roughly constant frequency over time, although residents of high-risk census tracts seem to have experienced higher incidence in the 1970s. Figure 6 shows only a slight nonrandom excess of census tracts with unexpectedly few or unexpectedly many cases. Before and after adjustment for social class, the high-risk census tracts are scattered throughout the county with only a few contiguous pairs and none with high risk for both sexes. No census tract stands out on the basis of a particularly high number of excess cases.

Thumbnail Interpretation

The reason for the higher incidence among men is unknown, although it is characteristic of most leukemias and lymphomas. No systematic pattern of geographical occurrence is apparent, and therefore no local source of causation can be proposed.

Figure 1: Age-adjusted incidence rate by place.

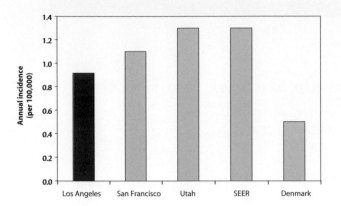

Figure 2: Age-adjusted incidence rate over the period.

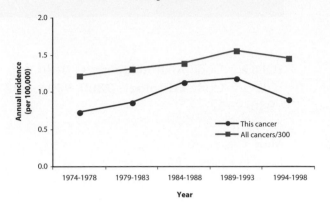

Figure 3: Age-adjusted incidence rate by age and race/ethnicity.

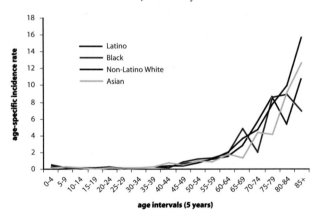

Figure 4: Age-adjusted incidence rate by social class.

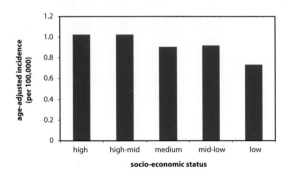

Figure 5: Distribution of the relative risk values for all census tracts.

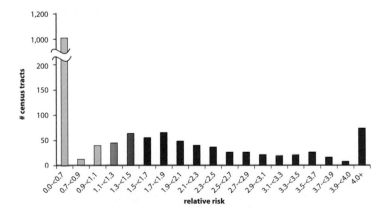

Figure 6: Census tracts by the number of cases per tract.

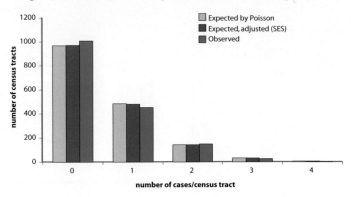

Figure 7a and b: Census tracts at high risk by the number of cases. (a) Unadjusted and (b) adjusted for social class.

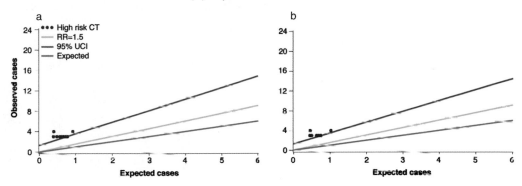

Figure 8: Risk over the period for high-risk census tracts relative to all census tracts.

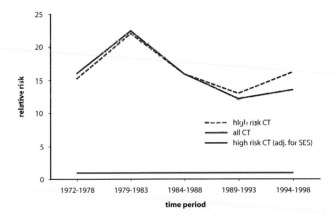

Figure 1: Age-adjusted incidence rate by place.

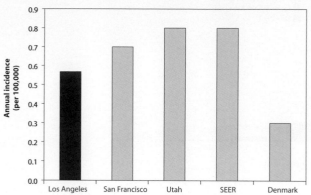

Figure 2: Age-adjusted incidence rate over the period.

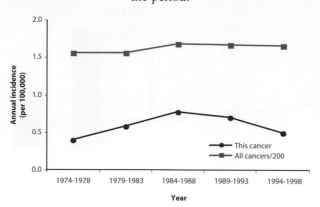

Figure 3: Age-adjusted incidence rate by age and race/ethnicity.

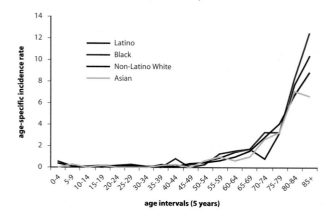

Figure 4: Age-adjusted incidence rate by social class.

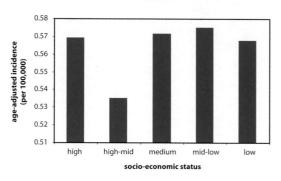

Figure 5: Distribution of the relative risk values for all census tracts.

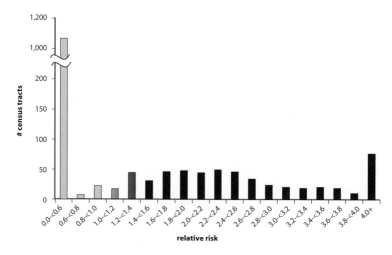

Figure 6: Census tracts by the number of cases per tract.

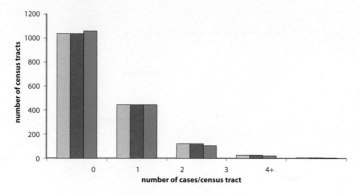

Figure 7a and b: Census tracts at high risk by the number of cases. (a) Unadjusted and (b) adjusted for social class.

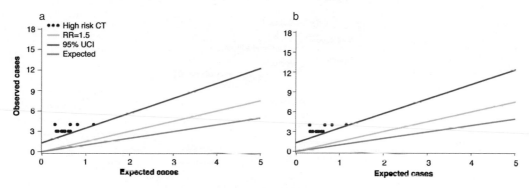

Figure 8: Risk over the period for high-risk census tracts relative to all census tracts.

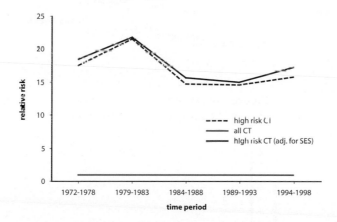

Figure 9: Map of census tracts at high risk.

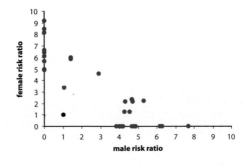

Male only
Female only
Male and female

Figure 10: Male-female correlation between the relative risks for high-risk census tracts.

Figure 11: Map of census tracts at high risk, adjusted for social class.

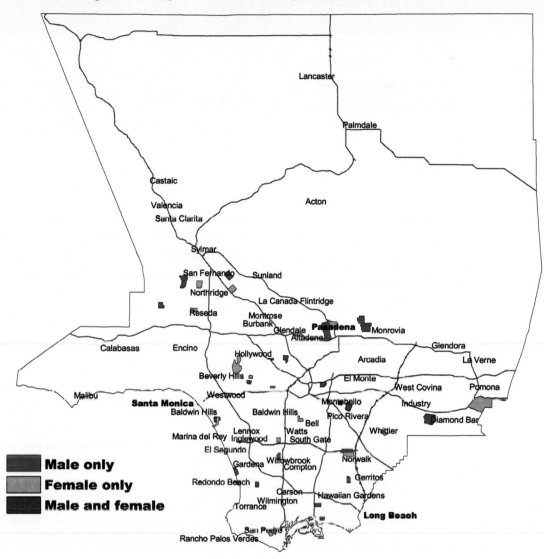

Figure 12: Male-female correlation between the relative risks for high-risk census tracts, adjusted for social class.

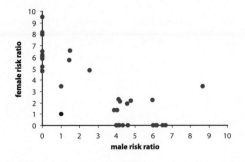

All Malignancies of Infants/Toddlers

ICDO-2 Code Anatomic Site: C 0–80
ICDO-2 Code Histology: All
Age: 0–4
Male Cases: 1968
Female Cases: 1564

Background

Cancers of children under age 5 are diverse in histology, in pattern of occurrence, and probably in etiology, although they tend to be regarded as a group by the public. Although these malignancies are uncommon, their importance transcends their numbers because of their tragic impact. The most common malignancies among pre-school children in the United States are the childhood leukemias, especially acute lymphoblastic leukemia (although lymphomas predominate in some other countries, such as those in sub-Saharan Africa). The second most important group are brain malignancies, followed by neuroblastomas, sarcomas, and rarer neoplasms such as Wilms tumors.

Local Pattern

Malignancies of infants and toddlers occur slightly more often in boys than girls, but are no more common in Los Angeles than in other parts of the country and the developed world. All racial/ethnic groups are at risk, but white and Latino children are affected slightly more often. Incidence has been rather stable over the period in the county as a whole, although it seems to have decreased slightly over the period among the residents of high-risk census tracts. There is little relation to social class. Figure 6 shows only a slight nonrandom excess of census tracts with unexpectedly few or unexpectedly many cases. The census tracts at high risk are scattered throughout the county, and after adjustment for social class, none are adjacent to one another. No census tract stands out on the basis of a particularly high number of excess cases.

Thumbnail Interpretation

We do not know the reasons for the gender and racial disparities that are seen in the pattern of occurrence of malignancies in pre-school children, which is driven by the pattern of acute lymphoblastic leukemia. No evidence is found that risk differs according to place of residence.

Figure 1: Age-adjusted incidence rate by place.

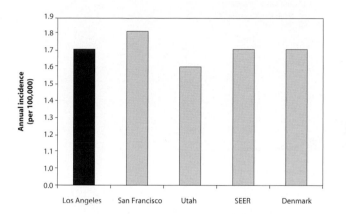

Figure 2: Age-adjusted incidence rate over the period.

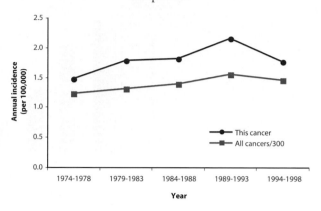

Figure 3: Age-adjusted incidence rate by age and race/ethnicity.

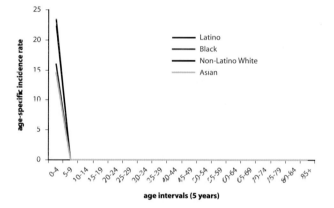

Figure 4: Age-adjusted incidence rate by social class.

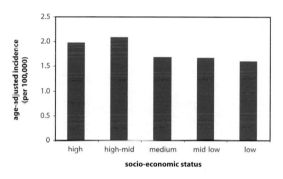

Figure 5: Distribution of the relative risk values for all census tracts.

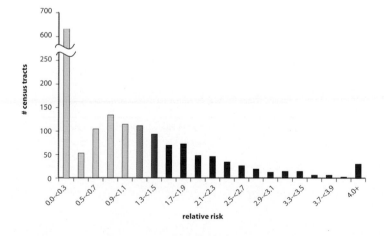

Figure 6: Census tracts by the number of cases per tract.

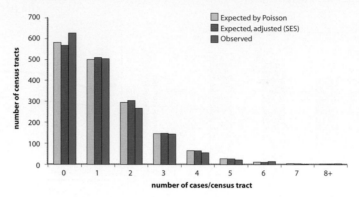

Figure 7a and b: Census tracts at high risk by the number of cases. (a) Unadjusted and (b) adjusted for social class.

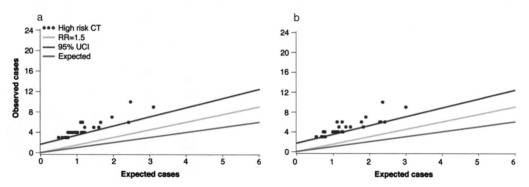

Figure 8: Risk over the period for high-risk census tracts relative to all census tracts.

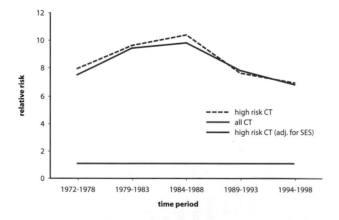

Figure 1: Age-adjusted incidence rate by place.

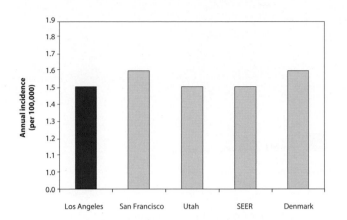

Figure 2: Age-adjusted incidence rate over the period.

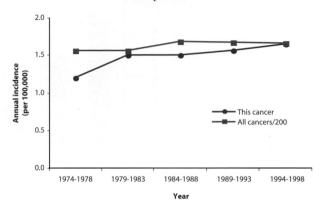

Figure 3: Age-adjusted incidence rate by age and race/ethnicity.

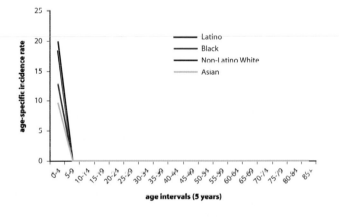

Figure 4: Age-adjusted incidence rate by social class.

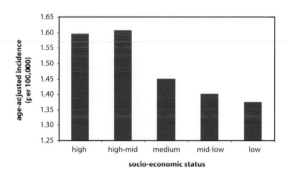

Figure 5: Distribution of the relative risk values for all census tracts.

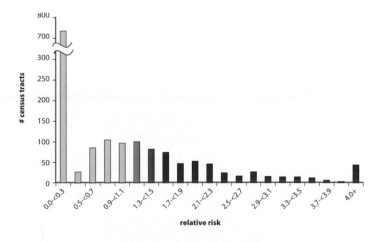

Figure 6: Census tracts by the number of cases per tract.

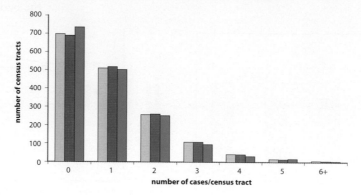

Figure 7a and b: Census tracts at high risk by the number of cases. (a) Unadjusted and (b) adjusted for social class.

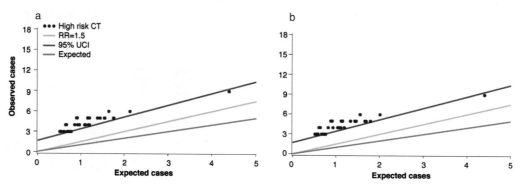

Figure 8: Risk over the period for high-risk census tracts relative to all census tracts.

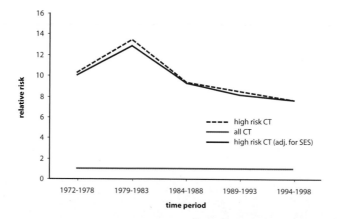

Figure 9: Map of census tracts at high risk.

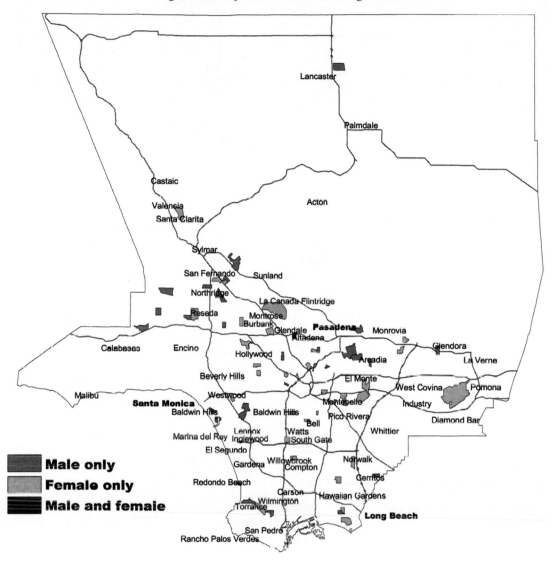

Figure 10: Male-female correlation between the relative risks for high-risk census tracts.

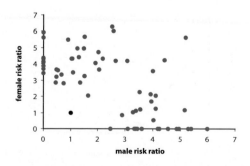

Figure 11: Map of census tracts at high risk, adjusted for social class.

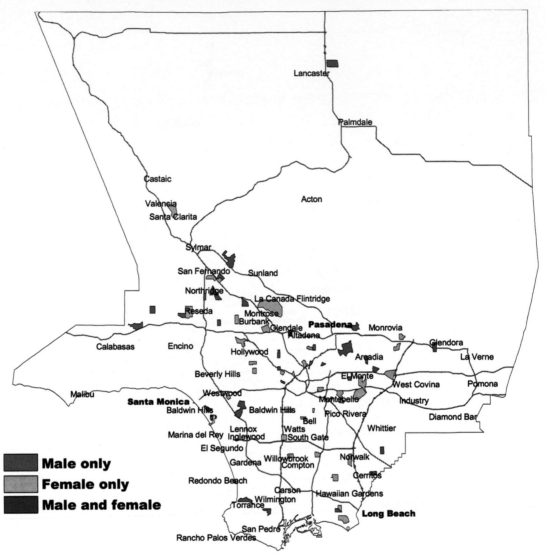

Figure 12: Male-female correlation
between the relative risks for high-risk
census tracts, adjusted for social class.

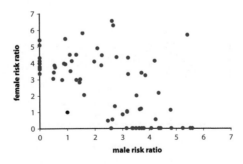

All Malignancies of Older Children

ICDO-2 Code Anatomic Site: C 0–80
ICDO-2 Code Histology: All
Age: 5–14
Male Cases: 1921
Female Cases: 1663

Background

Cancers in children aged 5–14 are diverse in histology, in pattern of occurrence, and undoubtedly in etiology as well. There is only a limited overlap in cell type between these malignancies and those of younger children. In addition to leukemia, Hodgkin lymphoma, non Hodgkin lymphoma, brain malignancy, and sarcoma are all prominent among the cancers in this age group.

Local Pattern

Malignancies of older children are somewhat more common in boys, and are less common in Los Angeles County than in San Francisco or the country generally. Whites and Latinos are at higher risk and Asian-Americans at lower risk than African-Americans, and there is no consistent relation between risk and social class. Over time, incidence in boys has not changed consistently, but in girls there has been a slight increase, in the county as a whole as well as among the residents of high-risk census tracts. Figure 6 shows only a slight non-random excess of census tracts with unexpectedly few or unexpectedly many cases. Before and after adjustment for social class, the high-risk census tracts are scattered throughout the county with few contiguous census tracts occurring in no specific pattern. No census tract stands out on the basis of a particularly high number of excess cases.

Thumbnail Interpretation

The reasons for the gender and racial disparities in the occurrence of these malignancies are unknown, although one suspects that gender-specific risk is becoming more equal as girls and boys adopt a more uniform behavior and lifestyle. As with the malignancies of younger children, no evidence has emerged to suggest that risk of cancers among school-age children varies according to place of residence in Los Angeles County.

Figure 1: Age-adjusted incidence rate by place.

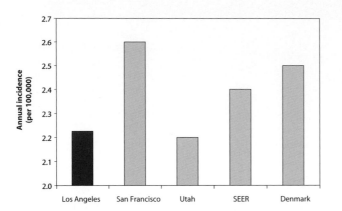

Figure 2: Age-adjusted incidence rate over the period.

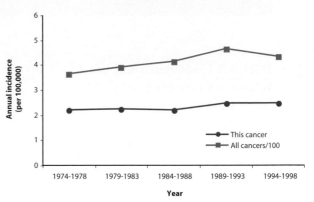

Figure 3: Age-adjusted incidence rate by age and race/ethnicity.

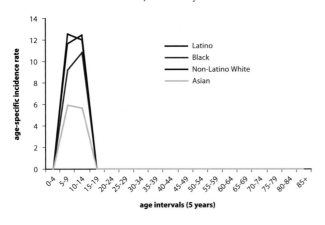

Figure 4: Age-adjusted incidence rate by social class.

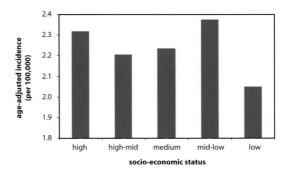

Figure 5: Distribution of the relative risk values for all census tracts.

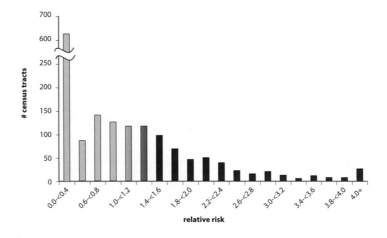

Figure 6: Census tracts by the number of cases per tract.

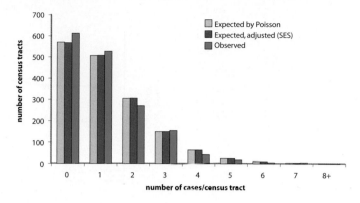

Figure 7a and b: Census tracts at high risk by the number of cases. (a) Unadjusted and (b) adjusted for social class.

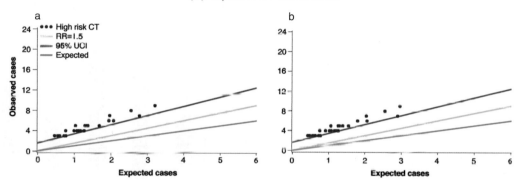

Figure 8: Risk over the period for high-risk census tracts relative to all census tracts.

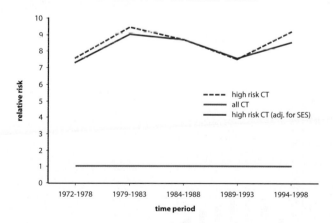

Figure 1: Age-adjusted incidence rate by place.

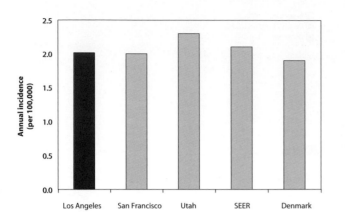

Figure 2: Age-adjusted incidence rate over the period.

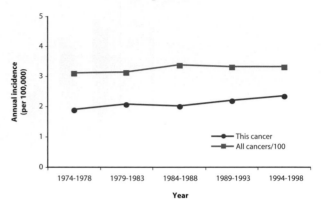

Figure 3: Age-adjusted incidence rate by age and race/ethnicity.

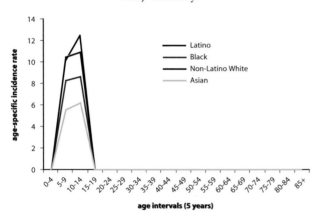

Figure 4: Age-adjusted incidence rate by social class.

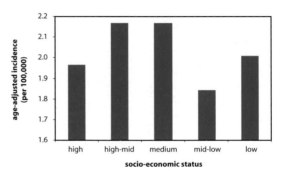

Figure 5: Distribution of the relative risk values for all census tracts.

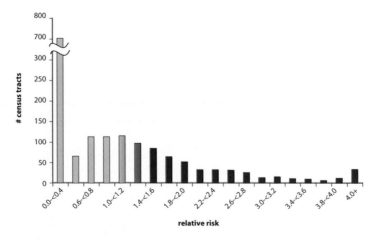

Figure 6: Census tracts by the number of cases per tract.

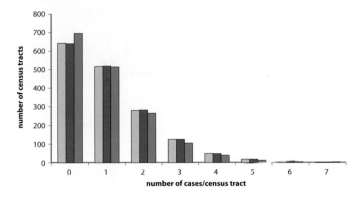

Figure 7a and b: Census tracts at high risk by the number of cases. (a) Unadjusted and (b) adjusted for social class.

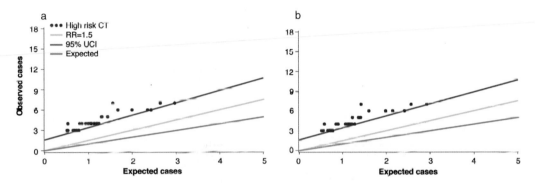

Figure 8: Risk over the period for high-risk census tracts relative to all census tracts.

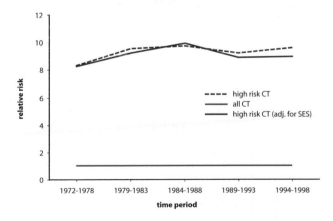

Figure 9: Map of census tracts at high risk.

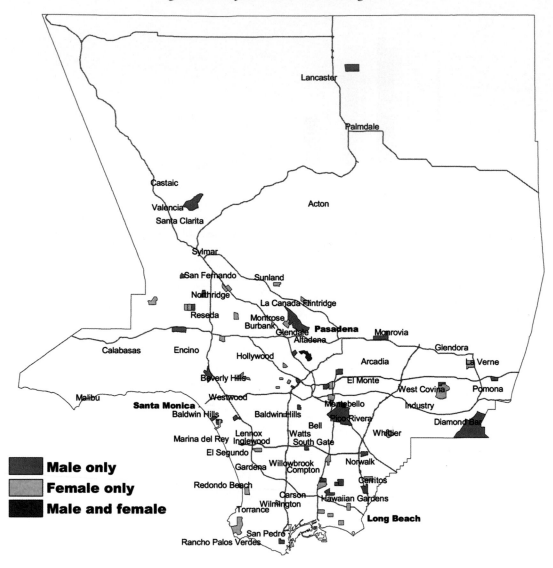

Male only
Female only
Male and female

Figure 10: Male-female correlation between the relative risks for high-risk census tracts.

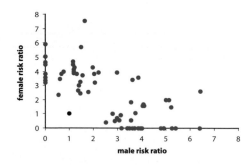

Figure 11: Map of census tracts at high risk, adjusted for social class.

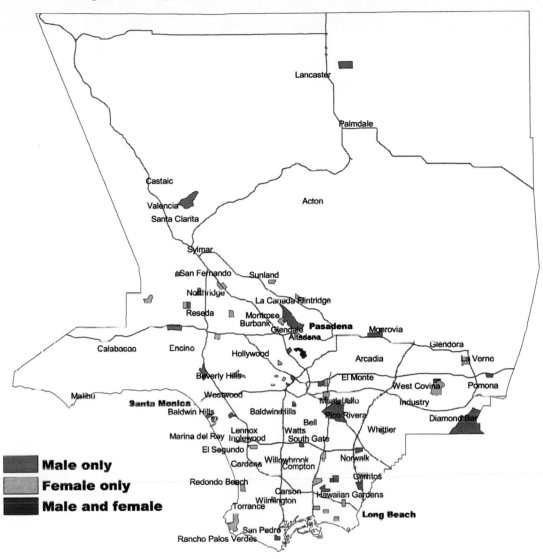

Figure 12: Male-female correlation between the relative risks for high-risk census tracts, adjusted for social class.

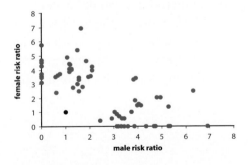

All Malignancies of Young Adults

ICDO-2 Code Anatomic Site: C 0–80
ICDO-2 Code Histology: All
Age: 15–49
Male Cases: 49011
Female Cases: 73649

Background

Among adults aged 15–49, breast cancer and thyroid cancer in women, testis cancer in men, and, in both sexes, Hodgkin lymphoma, non-Hodgkin lymphoma, malignant melanoma, brain malignancy, and colon cancer occur commonly. Kaposi sarcoma and those non-Hodgkin lymphomas related to AIDS are concentrated in men of this age.

Local Pattern

Malignancies of young adults as a group are slightly more common among women than men, and those of men are less common in Los Angeles County than in San Francisco. Among men, whites are at higher risk until about age 40, after which incidence is higher among African-Americans. Among women, whites are at slightly higher risk. No clear social class pattern is apparent among males, but among women slightly higher risk occurs among those residing in higher social class neighborhoods. Among men in Los Angeles County, malignancies increased in frequency until the introduction of effective antiretroviral therapy for AIDS. Incidence among men then declined, and that is especially evident among young men residing in high-risk census tracts. Among young women, a higher level of incidence (based partly on breast cancer and melanoma) has remained constant over time, both in the county as a whole and among residents of the few census tracts at high risk. Figure 6 shows a moderate nonrandom excess of census tracts with unexpectedly few or unexpectedly many cases, especially among men. On the map, before and after adjustment for social class, a large number of census tracts at high-risk for men are aggregated in a large complex between Beverly Hills and Silver Lake, with a smaller cluster in Long Beach. The few census tracts at high risk for women bear no relation to each other.

Thumbnail Interpretation

The pattern of all malignancies in young adult males reflects the disparities among the relatively frequent Kaposi sarcomas and non-Hodgkin lymphomas precipitated by the appearance of AIDS. No meaningful pattern of risk among young adult women is apparent.

All Malignancies of Young Adults: Male

Figure 1: Age-adjusted incidence rate by place.

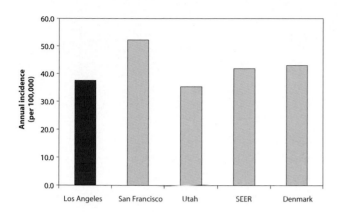

Figure 2: Age-adjusted incidence rate over the period.

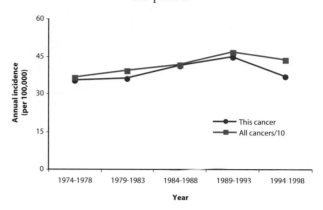

Figure 3: Age-adjusted incidence rate by age and race/ethnicity.

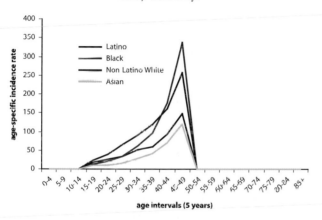

Figure 4: Age-adjusted incidence rate by social class.

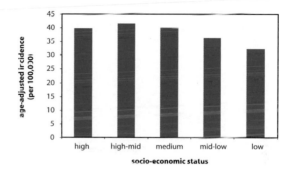

Figure 5: Distribution of the relative risk values for all census tracts.

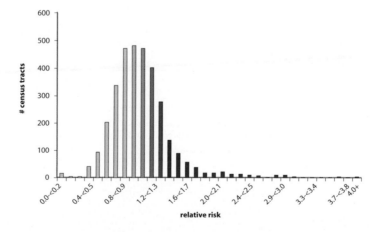

Figure 6: Census tracts by the number of cases per tract.

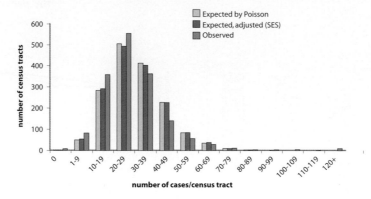

Figure 7a and b: Census tracts at high risk by the number of cases. (a) Unadjusted and (b) adjusted for social class.

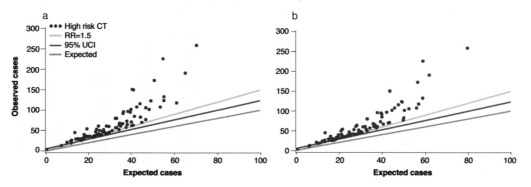

Figure 8: Risk over the period for high-risk census tracts relative to all census tracts.

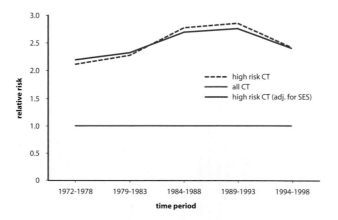

Figure 1: Age-adjusted incidence rate by place.

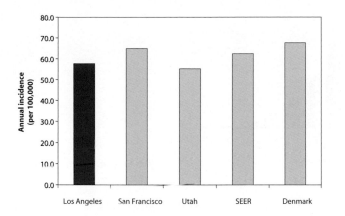

Figure 2: Age-adjusted incidence rate over the period.

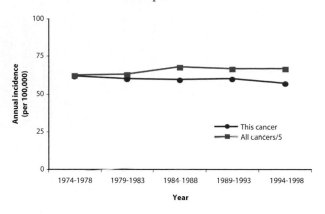

Figure 3: Age-adjusted incidence rate by age and race/ethnicity.

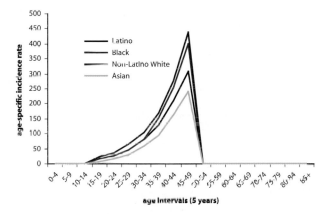

Figure 4: Age-adjusted incidence rate by social class.

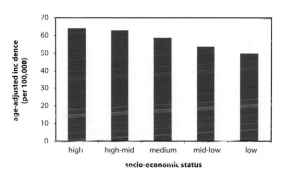

Figure 5: Distribution of the relative risk values for all census tracts.

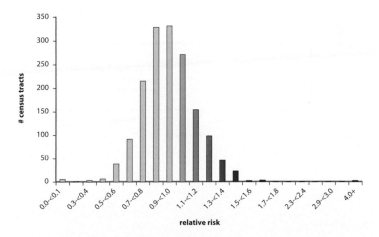

Figure 6: Census tracts by the number of cases per tract.

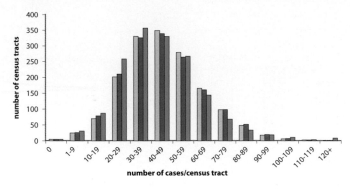

Figure 7a and b: Census tracts at high risk by the number of cases. (a) Unadjusted and (b) adjusted for social class.

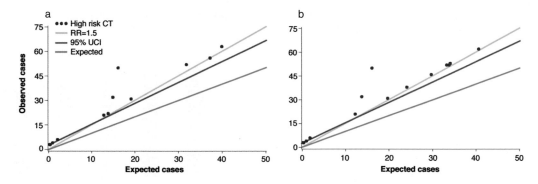

Figure 8: Risk over the period for high-risk census tracts relative to all census tracts.

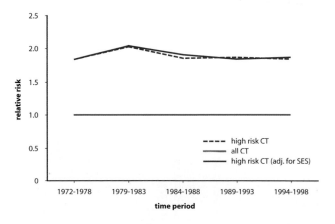

Figure 9: Map of census tracts at high risk.

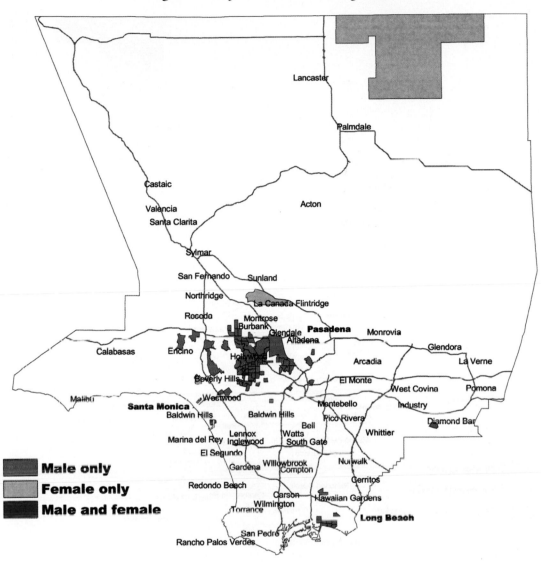

Figure 10: Male-female correlation between the relative risks for high-risk census tracts.

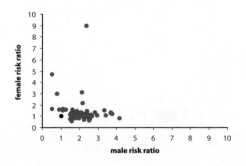

Figure 11: Map of census tracts at high risk, adjusted for social class.

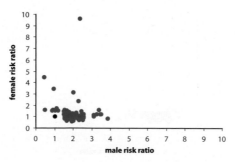

Lancaster

Palmdale

Castaic

Valencia
Santa Clarita

Acton

Sylmar

San Fernando   Sunland

Northridge
                La Canada Flintridge
Reseda   Montrose
            Burbank
            Glendale   **Pasadena**   Monrovia
                Altadena
Calabasas   Encino   Glendora
            Hollywood   La Verne
                        Arcadia
Beverly Hills   El Monte   Pomona
                                    West Covina
Malibu   Industry
            Westwood
Santa Monica   Montebello
Baldwin Hills   Baldwin Hills   Pico Rivera   Diamond Bar
            Bell
        Lennox   Watts   Whittier
Marina del Rey Inglewood South Gate
El Segundo
            Willowbrook   Norwalk
Gardena   Compton
Redondo Beach   Cerritos
            Carson   Hawaiian Gardens
Wilmington
Torrance   **Long Beach**
San Pedro
Rancho Palos Verdes

■ **Male only**
■ **Female only**
■ **Male and female**

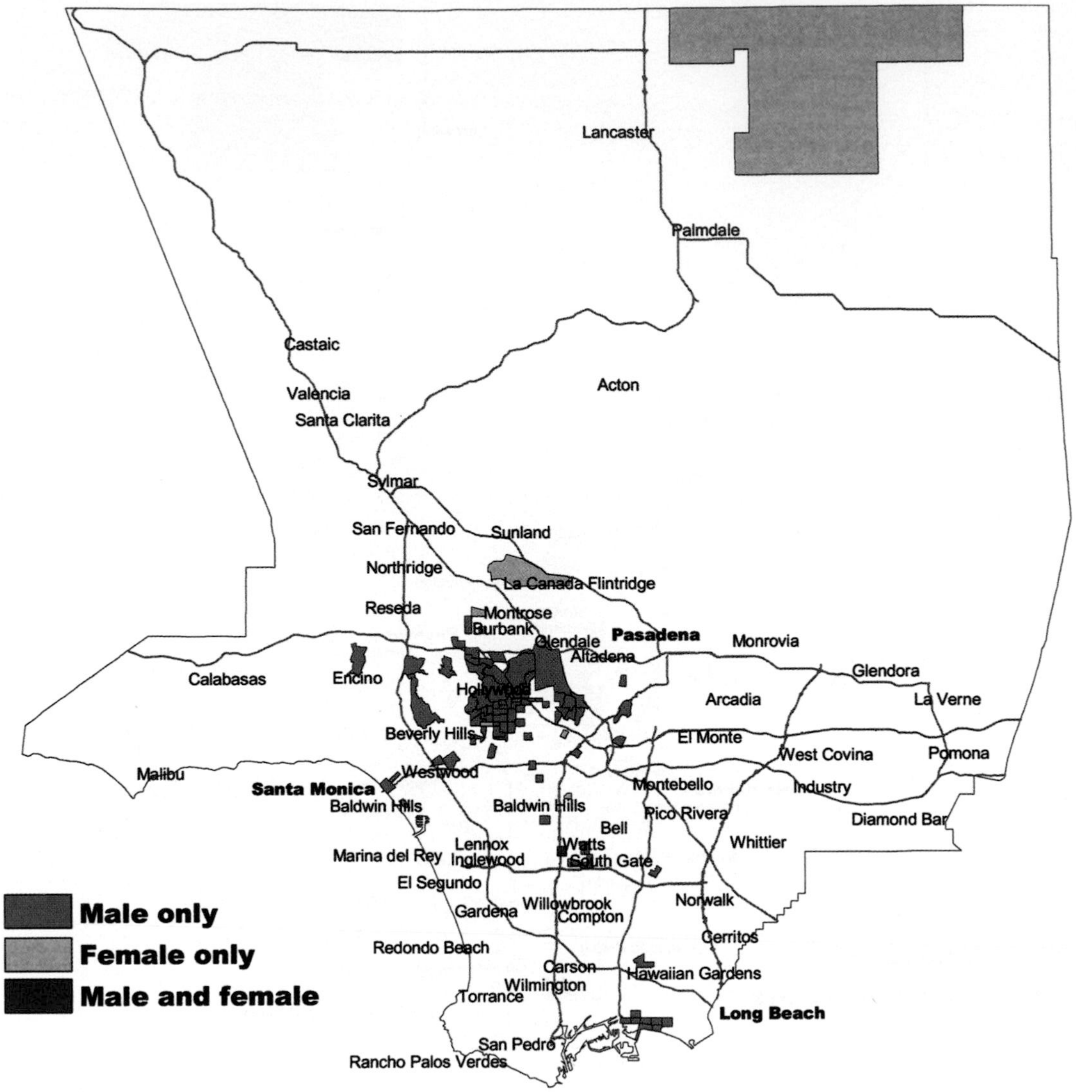

Figure 12: Male-female correlation
between the relative risks for high-risk
census tracts, adjusted for social class.

All Malignancies of Older Adults

ICDO-2 Code Anatomic Site: C 0–80
ICDO-2 Code Histology: All
Age: 50+
Male Cases: 323392
Female Cases: 312891

Background

The most important cancers in persons 50 and over are the most common cancers overall, namely breast cancer in women, prostate cancer in men, and lung and colon cancer in both genders.

Local Pattern

As a group, cancers among older men are less common among the polyglot population of Los Angeles County than among other populations, even including Utah. In contrast to malignancies among younger adults, those among older adults are more common among men than among women. African-American men and white women have higher risk than other racial/ethnic groups, but in both sexes risk is higher in those of higher social class. Incidence was constant over time, both in the county as a whole and among those residing in high-risk census tracts. Figure 6 shows almost no nonrandom excess of census tracts with unexpectedly few or unexpectedly many cases. Census tracts at apparent high risk were very few and widely scattered. No census tract stands out on the basis of a particularly high number of excess cases.

Thumbnail Interpretation

The common cancers of advanced age represented here have different patterns of occurrence and showed no tendency to concentrate within the same population groups. No systematic pattern of geographical occurrence is apparent among the few high-risk census tracts, and therefore older persons seem to experience roughly the same risk in all neighborhoods.

Figure 1: Age-adjusted incidence rate by place.

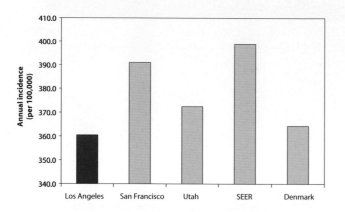

Figure 2: Age-adjusted incidence rate over the period.

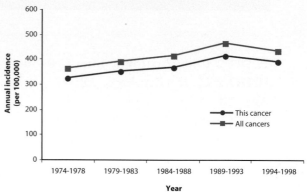

Figure 3: Age-adjusted incidence rate by age and race/ethnicity.

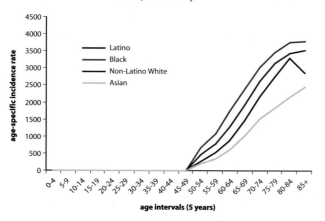

Figure 4: Age-adjusted incidence rate by social class.

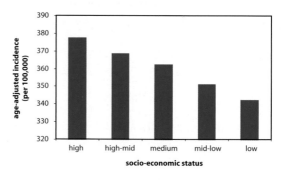

Figure 5: Distribution of the relative risk values for all census tracts.

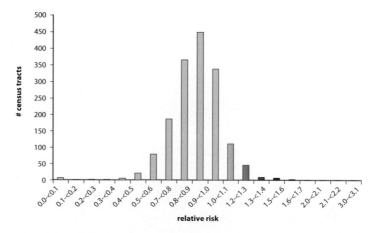

Figure 6: Census tracts by the number of cases per tract.

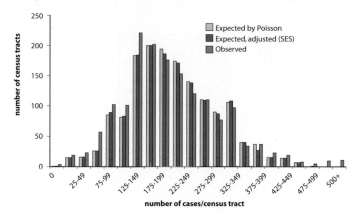

Figure 7a and b: Census tracts at high risk by the number of cases. (a) Unadjusted and (b) adjusted for social class.

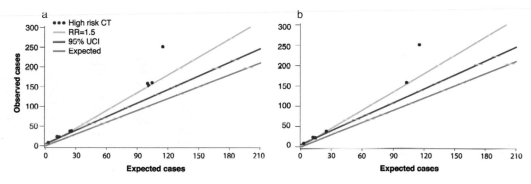

Figure 8: Risk over the period for high-risk census tracts relative to all census tracts

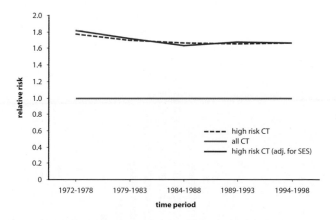

Figure 1: Age-adjusted incidence rate by place.

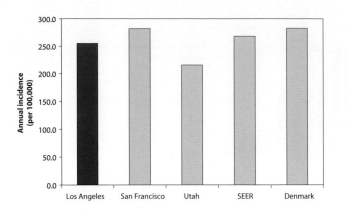

Figure 2: Age-adjusted incidence rate over the period.

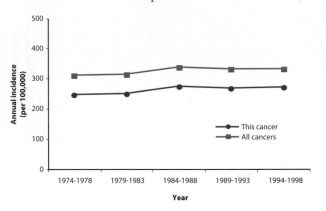

Figure 3: Age-adjusted incidence rate by age and race/ethnicity.

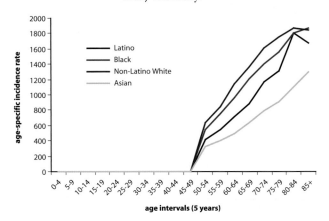

Figure 4: Age-adjusted incidence rate by social class.

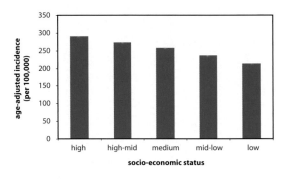

Figure 5: Distribution of the relative risk values for all census tracts.

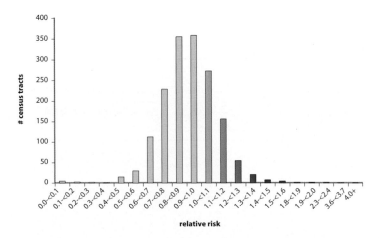

Figure 6: Census tracts by the number of cases per tract.

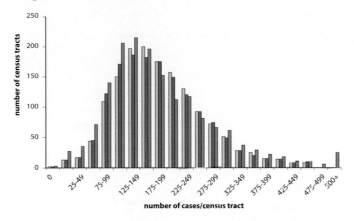

Figure 7a and b: Census tracts at high risk by the number of cases. (a) Unadjusted and (b) adjusted for social class.

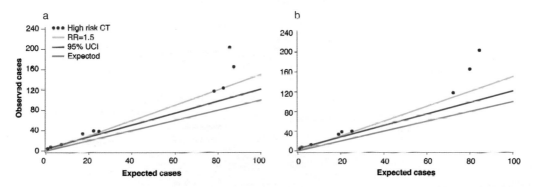

Figure 8: Risk over the period for high-risk census tracts relative to all census tracts

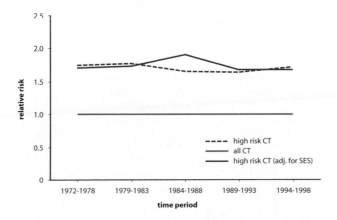

Figure 9: Map of census tracts at high risk.

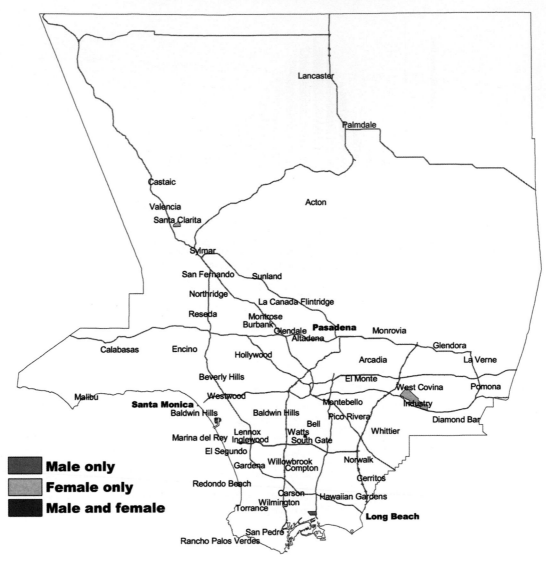

Male only
Female only
Male and female

Figure 10: Male-female correlation between the relative risks for high-risk census tracts.

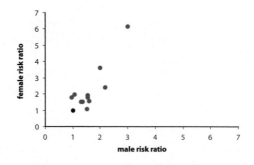

Figure 11: Map of census tracts at high risk, adjusted for social class.

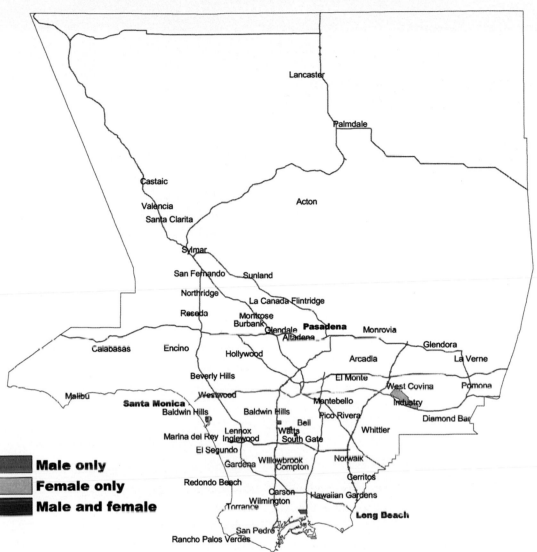

Figure 12: Male-female correlation between the relative risks for high-risk census tracts, adjusted for social class.

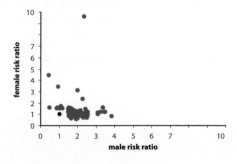

All Malignancies

ICDO-2 Code Anatomic Site: C 0–80
ICDO-2 Code Histology: All
Age: All
Male Cases: 376292
Female Cases: 389767

Background

The pattern of all malignancies combined is the sum of patterns of all the individual malignancies in all the individual age groups. Because the malignancies in the older age groups are more numerous, they tend to dominate this pattern, and as among the members of that group, the different cancers predominating in different localities tend to cancel each other out.

Local Pattern

When all forms of malignancies at all ages are combined, the rate in Los Angeles is almost identical to rates in other regions of the developed world. Although rates begin to increase earlier in women than they do in men, overall they are about a third higher in men than women. Rates have been relatively stable over time, both in the county as a whole and among residents of the few high-risk census tracts. Among men, overall incidence is highest among African-Americans, then, in order, among whites, Latinos, and Asian-Americans.

Among women, the relative positions of whites and African-Americans are reversed. In both sexes, higher risk falls upon persons of higher social class. Figure 6 shows only a slight non-random excess of census tracts with unexpectedly few or unexpectedly many cases. A few seemingly high-risk census tracts are widely separated and exceed the criteria for males. No census tract stands out on the basis of a particularly high number of excess cases.

Thumbnail Interpretation

The disparities by gender, race/ethnicity, and social class are as described under each of the common malignancies, and the combined malignancies among those of advanced age. No systematic pattern of geographical occurrence of malignancies overall is apparent. It is of interest that one of the few census tracts at high risk is the relatively unpopulated census tract in which the Los Angeles County General Hospital is situated. It happens that when cancer is diagnosed in a homeless person, the address of the hospital is recorded as the patient's address of record.

Figure 1: Age-adjusted incidence rate by place.

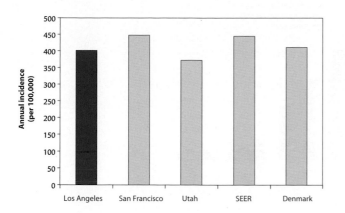

Figure 2: Age-adjusted incidence rate over the period.

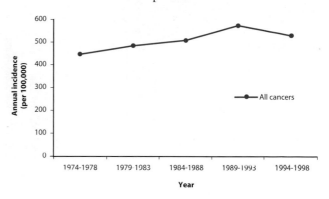

Figure 3: Age-adjusted incidence rate by age and race/ethnicity.

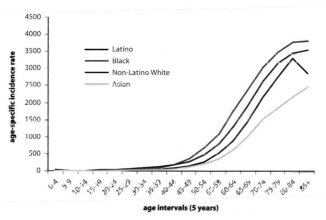

Figure 4: Age-adjusted incidence rate by social class.

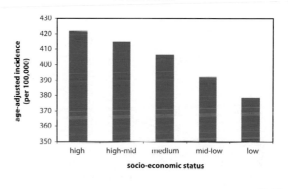

Figure 5: Distribution of the relative risk values for all census tracts.

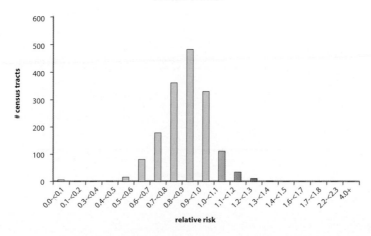

All Malignancies: Male

Figure 6: Census tracts by the number of cases per tract.

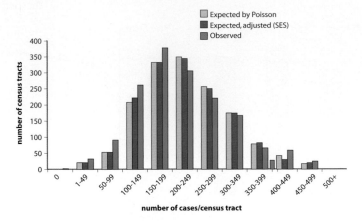

Figure 7a and b: Census tracts at high risk by the number of cases. (a) Unadjusted and (b) adjusted for social class.

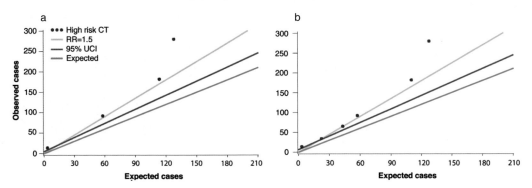

Figure 8: Risk over the period for high-risk census tracts relative to all census tracts.

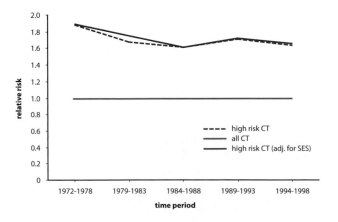

All Malignancies: Female

Figure 1: Age-adjusted incidence rate by place.

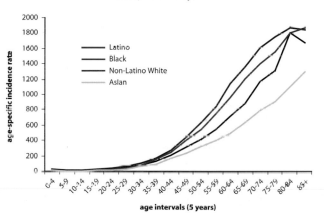

Figure 2: Age-adjusted incidence rate over the period.

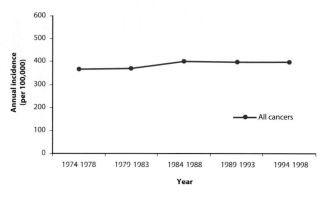

Figure 3: Age-adjusted incidence rate by age and race/ethnicity.

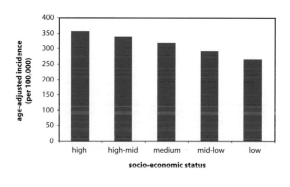

Figure 4: Age-adjusted incidence rate by social class.

Figure 5: Distribution of the relative risk values for all census tracts.

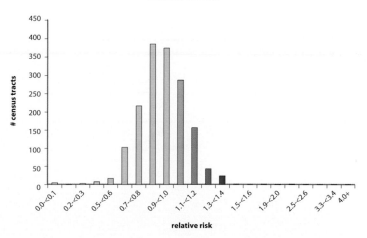

Figure 6: Census tracts by the number of cases per tract.

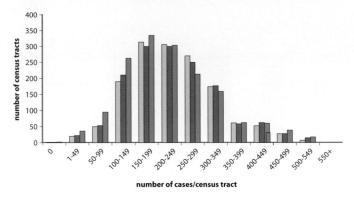

Figure 7a and b: Census tracts at high risk by the number of cases. (a) Unadjusted and (b) adjusted for social class.

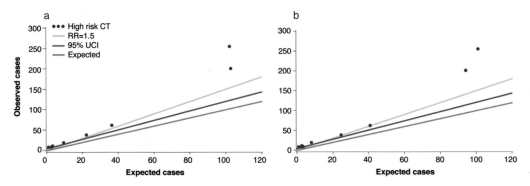

Figure 8: Risk over the period for high-risk census tracts relative to all census tracts.

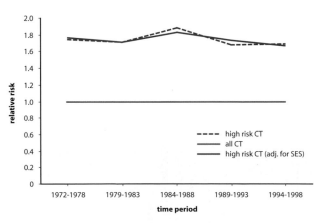

Figure 9: Map of census tracts at high risk.

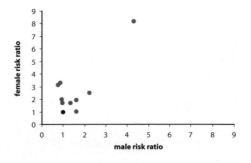

Figure 10: Male-female correlation between the relative risks for high-risk census tracts.

Figure 11: Map of census tracts at high risk, adjusted for social class.

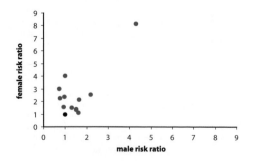

Male only

Female only

Male and female

Figure 12: Male-female correlation between the relative risks for high-risk census tracts, adjusted for social class.

Overall Summary

By describing the patterns of occurrence of malignancies in Los Angeles County, we have demonstrated that "cancer" is not a single disease, but a collection of many different diseases, each occurring because a different type of cell has grown out of control. We have tried to acquaint readers with the factors, notably chance and bias, which make it difficult to verify a local increase in incidence. We have explained that dramatic nonrandom patterns of occurrence sometimes are produced by exposures that are very personal and have nothing to do with pollution. Malignancies with different patterns of occurrence can be safely assumed to have different causes, whether or not the latter are all known.

A total of 72 different malignancies plus 12 combinations have been examined in the various ways permitted by available information. Excluding geographical considerations, every malignancy in some way or other gives evidence of occurring in a systematic, that is a nonrandom, pattern, although in the case of a few very rare malignancies the number of cases is too small to be completely certain. In general, the degree of variability in risk goes far beyond age differences. In most cases, there are differences between persons according to sex, race/ethnicity, or social class, all indicating some form of personal or environmental factor.

In many cases there are characteristic trends over time, and about half the individual malignancies also provide evidence of some degree of systematic, i.e., nonrandom, geographic variation, thus indicating that factors other than chance determine the pattern of community incidence. Among the factors known to be responsible are personal experiences, such as occupational exposures, habits, recreational preferences, past reproductive and medical events, and genetic inheritance.

In at least six instances in this book the geographic distribution of high risk of disease was clearly nonrandom, but did not conform to the pattern that would have been predicted by available knowledge. The malignancies in question include oropharyngeal carcinoma, small cell carcinoma and adenocarcinoma of the lung, papillary carcinoma of the thyroid, squamous carcinoma of the bladder, and diffuse mixed B-cell non-Hodgkin lymphoma. The true explanation for none of these patterns is currently known, although educated guesses provide tentative hypotheses that are currently under evaluation. As of this writing, no evidence of a malignancy caused by a strictly environmental carcinogen has yet been confirmed.